Christopher Brown Jr.

May 1999

GW00703033

NAFEMS World Congress '99

EFFECTIVE ENGINEERING ANALYSIS

Conference organised by NAFEMS
at
The Newport Marriott Hotel
Rhode Island, USA

25-28 April 1999

Volume 2

ISBN
1 874376 25 5

Conference organised by NAFEMS at
The Newport Marriott Hotel,
Rhode Island, USA

25-28 April 1999

NAFEMS was launched in 1983 by the National Engineering Laboratory with support from the UK's Department of Trade and Industry. In December 1991 NAFEMS was transferred from the DTI, to its new owners, the NAFEMS Members. The main aim of NAFEMS was, and remains, to promote the safe and reliable use of Finite Element Analysis and Related Technology.

Today NAFEMS is the leading International Association for the Engineering Analysis Community, with corporate members in forty countries world-wide.

NAFEMS
Whitworth Building
Scottish Enterprise Technology Park
East Kilbride
Glasgow G75 0QD
United Kingdom

Tel:- 44 (0)1355 225688

Fax:- 44 (0)1355 272749

Web site:-   http://www.nafems.org

KEYNOTE ADDRESS: Principal Invited Speaker
Professor E Onate,
University of Barcelona, Centre for Computational Mechanics, Spain

ENGINEERING INTEGRATION CHALLENGES

INDUSTRIAL APPLICATIONS

INDUSTRIAL APPLICATIONS (Cont'd)

COMPUTATIONAL DEVELOPMENTS

PROCESS SIMULATION

BIO-MEDICAL APPLICATIONS

PREDICTING DYNAMIC RESPONSE

MULTI-PHYSICS ANALYSIS & SIMULATION

DESIGN ANALYSIS METHODOLOGY

STRUCTURAL OPTIMISATION TECHNIQUES

QUALITY ASSURANCE & EDUCATION

LIFE PREDICTION TECHNIQUES

BIOLOGICAL STRUCTURES, MATERIALS & BIOMIMETICS

COMPOSITE STRUCTURES, MATERIALS & NON LINEAR PROBLEMS

CIVIL ENGINEERING APPLICATIONS

ANALYSIS INSPIRATIONS - PAST, PRESENT & FUTURE

SPEND A LITTLE NOW OR PAY A LOT LATER...
THE EFFECTIVE USE OF ENGINEERING ANALYSIS

Howard C. Schwartz, Senior Design Engineer/Analyst

Abbott Diagnostics Division, Abbott Laboratories
P.O. Box 152020 M.S. 2-21
Irving, Texas 75015-2020
Email: schwarh@ema.abbott.com

ABSTRACT

Just as product designers have integrated Solid Modeling and other Computer Aided Design tools over the past few years to decrease the product design cycle, there is now an increasing interest to integrate Computer Aided Engineering (CAE) tools as well. Although such tools have always been used by aerospace and other industries normally associated with high risk products, the typical product manufacturer has only until recently ignored its use. At Abbott Laboratories, we have spent the past number of years developing our CAE resources. Today, huge time and cost savings result from using our structural Finite Element analysis, injection mold analysis, and material selection software. By examining one case study, it will be shown how such software reduces design risk and product development costs.

1. BACKGROUND

It was not that many years ago that the term Computer Aided Engineering (CAE) or Finite Element Analysis (FEA) meant little to businesses except possibly to aerospace or other high-tech manufacturers. Such products were likely high risk, performance driven, and extremely reliable. Anyone involved in CAE back then probably recalls the weeks it normally took to construct Finite Element models which today take only a day or less. That was acceptable because it was reliability and not time-to-market that characterized the companies using these CAE products[1]. Computer simulation allowed the analyst to explore many "what if" design load conditions compared to the few allowed by hand calculation.

Elsewhere, you would not likely find much CAE being used. Because such CAE products were typically hard to master and lacked user-

---

friendliness, only a handful of company experts used them. Widespread use was limited especially for smaller companies not seeing the benefit of employing FEA experts or paying for expensive computer hardware and software. Also, the make-it-break-it-make-it-again principle ruled supreme.

Today, however, there is an ever increasing number of companies selling CAE products and trends[2] that suggest its use is filtering down to more mainstream types of products, such as consumer goods or medical device manufacturers. These trends include:

- Increased competition that necessitates shortening the product design cycle
- Widespread use of 3-D CAD modeling and parametric design
- Increased part design complexity due to design feature consolidation and metal substitution using high performance plastics
- Cheaper and faster computer hardware
- Cheaper, more user-friendly, and better integrated CAE software
- Development of analysis quality standards[3] through ISO9001 and NAFEMS
- Greater awareness and exposure of CAE in trade journals[4,5,6], seminars and conferences such as NAFEMS WORLD CONGRESS '99

One FEA company executive projected that within a few years such CAE tools would be found throughout every industry. Since 1991, it has already become a reality at Abbott Laboratories. Its Engineering Analysis Group supports a multi-divisional, diversified corporation that is a global manufacturer of diagnostic, hospital and other healthcare related products.

2. A LONG EXPERIMENT...

How a company uses CAE is as important as which ones it uses. In 1996, an internal survey was sent out to all Abbott Engineering Analysis customers. Its purpose was to reflect upon Abbott's experience using its mechanical CAE tools and measure a number of key elements, all of which are essential to successfully implementing CAE. This survey would:

- Measure the satisfaction level of our past customers.
- Measure CAE usage, i.e. Which types of CAE are requested the most?
- Measure the benefits derived from CAE usage, i.e. How much time or money has been saved or design risk reduced?
- From such metrics, determine how to improve our Engineering Analysis Group services and promote customer satisfaction.

Although such a survey could have been distributed much earlier to identify possible needs, the years of CAE use provided some valuable insights. Some conclusions drawn from the survey are:

- Of the many mechanical CAE tools used at Abbott Labs such as computational fluid dynamics and kinematic-dynamic analysis, our **structural FEA, injection mold-flow, and material selection software are used the most**. Since our products typically have:
  - fluid mass transport and heat transfer systems
  - actuators and pick-and-place mechanisms
  - structural panels, frames and brackets
  - components utilizing many types of molded plastics and elastomers

it is no surprise that the three CAE tools indicated above are used the most.

- Although a majority of engineers feel that Engineering Analysis should be used early in the design process, they also feel that its use should not be required but left to the engineer's discretion. This attitude is exemplified by the inconsistent use of our CAE tools. Some engineers have used them numerous times, whereas others very little. Most new customers often are a result of a past customer's "word of mouth" recommendations rather than by company policy to use CAE.

- Most analyses typically last about two weeks possibly saving projects a minimum of $10,000 or a month of design time. Given that designers may have as little as a week of "window of opportunity" to make design changes before manufacturing commits itself to tooling, **analysis must be performed quickly if it is to have any impact at all**. This is especially true if optimization or design sensitivity studies are conducted since these involve a lengthy iterative process. This important claim is also supported elsewhere in the survey results where the lowest customer satisfaction rating regarding the Engineering Analysis Group was in the group's commitment to meet schedules which suggests that customers wish they had results sooner.

If there is a lesson to be learned, it is: used wisely, CAE adds value to product design by increasing a company's profit margin. For a company the size of Abbott, it is a logical choice to internally develop CAE resources where 100's of parts, some of which are very complex, are designed each year. However, depending upon the number of products, the product complexity, and other factors[7,8], outsourcing to consultants may be a more logical choice for others thinking of using CAE.

3. DESIGN CASE STUDY

The following design case study shows how CAE can be used successfully in the design process. Cost or time savings, important analysis tips or design "rules of thumb" will be presented whenever possible and **highlighted in bold**. What is unique about this case study is that all three types of analysis, i.e. structural, material selection, and injection mold-flow were used concurrently to arrive at an optimum design.

Shown in Figure 1 is a Tube Carrier that is used in one of Abbott's newly developed diagnostic products. Tubes are typically hand-loaded into opposing pairs of cantilevered spring fingers. Originally, the Carrier consisted of a 12 metal spring finger assembly shown in Figure 2 whose unit cost was significant. The goal was to see if this high volume assembly could be injection molded into a single part thus simplifying the design and **reducing the total Carrier cost by at least 50%** according to one estimate. To begin, it was necessary to **understand the design requirements**, some of which were:

- The carrier must accept a variety of commonly available tube heights and diameters.
- The carrier must function reliably for many repeated uses over a long time period.
- Tubes must be held upright for long periods of time and visible from opposing sides.
- Human factors dictated tube insertion/removal loads not to exceed 1/2 pound force

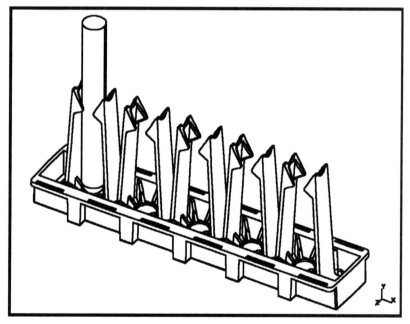

Figure 1   Tube Carrier with plastic springs

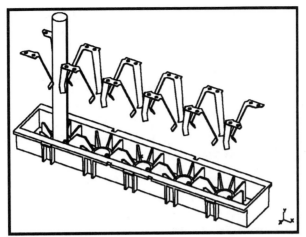

Figure 2   Original Carrier Spring design with 12 metal springs

From these requirements developed the following engineering principles:

- Long term reliability meant that **the design of any plastic spring was limited by its ability to resist fatigue and creep relaxation.** Spring tip permanent set caused by creep could not exceed .04".
- The tube diameter defined the maximum bending stress and tip deflection.
- According to simple beam theory[9], for any given tip deflection the **longest possible spring finger length and minimum thickness results in the least amount of bending stress.** This was the guiding factor in the design since creep and fatigue are greatly influenced by stress.

The first step was to determine if the use of any plastic was feasible. Assuming initial dimensions and using simple beam theory, the peak stress was calculated at about 3000 psi assuming a plastic such as Polycarbonate. Since many engineering grades of plastic, such as Polycarbonate, are in the 10000 psi yield strength range, a stress of 3000 psi corresponds to about 30% of yield strength. One rule of thumb found in some design guides states that **stresses in plastics should not exceed 25% of the plastics yield strength** when designing for creep or fatigue. Therefore, using any plastic was risky unless known creep and fatigue strength data existed, thus permitting analysis.

Using various plastics databases, the process of material selection began regardless of which vendors Abbott Labs used. **By using databases, the best plastics for a particular application can be identified thoroughly and quickly.** Based upon the structural design requirements, the available data suggested there were a number of possible candidates to choose from. All were 30% or higher glass filled. In general, **any fiber filled plastic is superior to its unfilled counterpart with respect to creep or fatigue resistance**[10].

Some Apparent Moduli (at 1000 hr) Comparisons					
Material and Strength (psi)	Applied Stress (psi)	Tensile Modulus (psi)	Apparent Modulus (psi)	Apparent Strain (in/in)	Ratio of both Moduli
Polycarbonate (PC) 30% Glass					
18500	3000	1.375e+06	1.163e+06	0.00258	1.1825
	5000	1.375e+06	1.053e+06	0.00475	1.3063
Polyphenylene Oxide (PPO) 30% Glass					
18500	5000	1.15e+06	8.000e+05	0.00625	1.4375
Polyetherimide (PEI) 30% Glass					
35000	5000	1.3e+06	1.116e+06	0.00448	1.1648
Polyester (PBT) 30% Glass					
20000	5000	1.2e+06	9.960e+05	0.00502	1.2048
Polyphenylene Sulfide (PPS) 30% Glass					
29000	5000	1.6e+06	1.429e+06	0.00350	1.1200
Polyacetal (POM) 30% Glass					
19500	5000	1.4e+06	7.463e+05	0.00670	1.8760

Figure 3   Tensile/Apparent Moduli ratios calculated from vendor data

To compare each plastic's apparent moduli[11], which is one measure of creep, a spreadsheet was constructed. This is shown in Figure 3. Although lacking the glamour of pretty pictures, **spreadsheets and other similar products can be terrific CAE tools**. The last column in Figure 3 lists a ratio that compares a plastic's apparent modulus at 1000 hours to it's tensile modulus. Simply stated, the closer this ratio is to 1.0, the more creep resistant is the plastic. This comparison was the basis of the initial material selection.

Although this paper highlights the importance of CAE, testing has its place too. Narrowing down choices further, a simple creep test was devised using standard 5.0" X .50" X .125" Flex Bars readily obtainable from the vendors whose plastics were being considered. Shown in Figure 4 is the setup for this test. By advancing the center screw a known amount, a known stress is applied to the flex bar simulating the actual spring finger response to a given tube size. Shown in Figure 5 is a chart which compares the amount of permanent set due to creep each bar experienced after 5 weeks of loading. Additional plastics recommended by the customer were also introduced into the test matrix. Because the Carrier could not have a permanent set greater than .04" tip deflection, clearly only a few plastics met that critical requirement. The customer agreed with the recommendation that 30% glass filled Polyetherimide (PEI) be used.

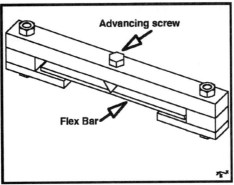

Figure 4   Creep test setup using Flex Bars

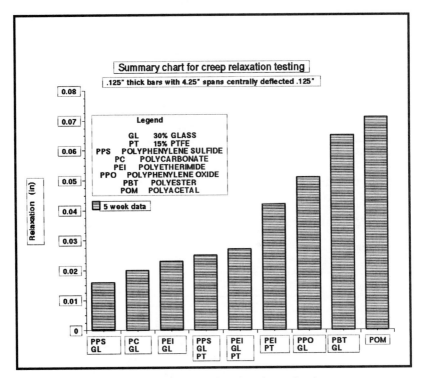

Figure 5   Creep test results of various plastics

Finally, to validate this design, a detailed stress analysis was performed using FEA. Shown in Figure 6 is a Von Mises stress plot that simulates a loaded spring. The peak stress in the fillet region is about 8,000 psi. Although a good stress analyst can approximate this stress value by hand calculation, **using FEA produces more accurate results**. Comparing this stress to the PEI fatigue life curve shown in Figure 7, one concludes that fatigue is not likely a problem. **Without FEA, the design risk would be greater** and the customer less likely to pursue an unproven, potentially faulty design.

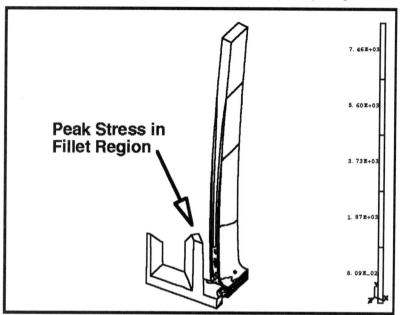

Figure 6   Von Mises stress hidden line plot of Carrier Spring section

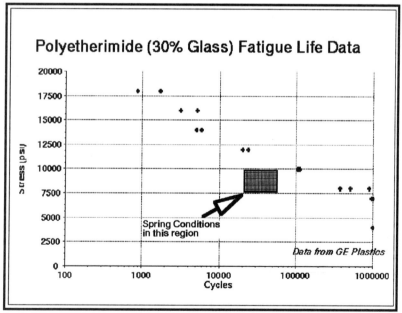

Figure 7   Polyetherimide (30% Glass) Fatigue Life

Likewise a mold-flow analysis was performed to determine if there were any molding concerns. Shown in Figure 8 is a fill time plot depicting the location of the weld lines (not apparent in the black/white vector plot shown here). Because these weld lines were in regions of minimum stress indicated in the prior FEA analysis, no problem was likely. Process conditions such as fill and hold time and cooling line size and location were modified until the optimum part resulted from the analyses.

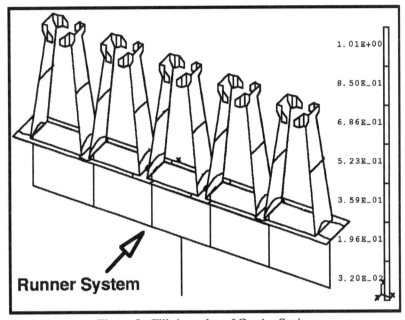

Figure 8   Fill time plot of Carrier Spring

Although adhering to sound mold design principles[12] is important, mold-flow analysis allows the designer to optimize the design. This is especially useful for high volume production. For example, in another mold-flow analysis performed at Abbott Labs, a part's thickness was reduced about 15% while maintaining moldability. **About 1 second was eliminated off a 10 second mold cycle translating into about a $100,000 savings per year in production costs**.

4. CONCLUSIONS:

Using CAE benefits everyone: from the designer who optimizes a product to the consumer who uses it. In this engineering age of "What if...?" and "How can...?", such tools are essential. If, however, everyone understood its capabilities, surely its acceptance as a design tool would equal that of CAD.

REFERENCES:

1 HALPERN M - Simulation Payoffs, *Computer-Aided Engineering*, pp 56, Aug. 1995

2 MILLS R. - Answering Design Questions Earlier, *Computer-Aided Engineering*, pp. 46-52, Oct. 1996

3 KENNY T - Keeping Analysis Users in Top Form: NAFEMS Registered Analyst Scheme, *ANSYS News*, pp. 9, Fourth Issue, 1996

4 CHRUMA JL, CRONIN KM, and CUNNINGHAM RR - Using CAE to Shorten Product Development Time, *Medical Plastics and Biomaterials*, pp. 14-23, 2:3 Summer 1995

5 RUSSELL R - Don't Trust The Pretty Pictures, *Machine Design*, pp. 68-84, May 23, 1996

6 EISENBERG B - Is FEA Headed For Prime Time?, *Product Design And Development*, pp. 48-49, April 1996

7 MANISCALCO M - Outsourcing Design Analysis, *Injection Molding*, pp. 32, Feb. 1997

8 CARLSON SE and TER-MINASSIAN N - Planning for Concurrent Engineering, *Medical Device & Diagnostic Industry*, pp. 202-208, May 1996

9 YOUNG WC - Roark's Formulas For Stress & Strain, 6th ed., McGraw-Hill Book Company, 1989, chapter 7, p. 100

10 HOWARTH C - Long Term Behavior of Reinforced Thermoplastics, *LNP Engineering Plastics Bulletin No.253-394*, 1994

11 GE Plastics Engineering Design Database User's Manual, section 3.1.2, Mar 1992

12 BEALL GL - Plastic Part Design For Economical Injection Molding, seminar course presented at the 1996 Plastics Fair, Dallas, Texas

Are You Building Better Products with Advanced Technologies?

Vince Adams
Director of Engineering; WyzeTek, Inc.

Abstract

Magazine articles and product literature have long extolled the potential benefits of simulation as a design tool. Few will question the potential for finite element analysis (FEA) to reduce product cost, shorten development time, and improve quality. However, when one probes deeper into the state of simulation at most manufacturing companies, it is apparent that few are even scratching the surface of the technology's potential and many engage, for various reasons, in counter-productive techniques that doom FEA to remain a minor player in their product development process. The fact of the matter is that acquisition of technology does not provide a competitive advantage or guarantee an improvement in bottom-line profitability. Wise implementation and use of the available tools separates successful integrators from the less successful users.

Many popular theories on product development, going by names such as concurrent engineering, integrated product development, and rapid product development, provide clues towards a more focused, productive employment of simulation. A common theme to many of these is a goal-centered implementation plan that focuses on the economic benefits of all product development decisions. In this paper, the product development process, with an emphasis on the use of simulation will be examined in the framework of the Theory of Constraints (TOC) as outlined by Eli Goldratt in his books *The Goal*, [1] and *Critical Chain* [2]. In this light, some current trends, including CAD integration of FEA and analysis-enabled designers, must be questioned. Metrics for evaluating the gains provided by simulation will be proposed.

1. What Is Success?

At several workshops for analysis in the design environment sponsored by WyzeTek, Incorporated, participants from major Midwest manufacturing companies, such as Harley Davidson, Kohler, General Motors, Caterpillar, and J.I. Case, as well as many mid-sized companies compiled this list of components that reflect a successful integration of simulation in the design process:

> User Controlled:
> - A valid analysis in a timely manner
> - Results predict actual part performance
> - Better insight into the design process
>
> Design Process Controlled:
> - Compression of the design cycle
> - A more cost-effective design
> - Improved overall product quality

After an in-depth discussion of the impact of these components, nearly all participants agreed that analysis was not successfully integrated into their design processes.

While most felt that the User Controlled components, which are in the control of the individual analyst, were commonly observed, it was the Design Process Controlled components that reflected the true measure of success of any tool or process: corporate profitability. The participants felt that the impact of these components were not readily apparent or that their companies did not have metrics in place to allow tracking of any improvement.

The consensus of the workshop attendees has been confirmed repeatedly by the author's interaction with manufacturing companies that use a variety of tools, in a variety of industries of all sizes. Independent surveys by Grant Thornton, LLP [3] and The Management Roundtable [4] determined that CAE tools had little or no impact on time-to-market or in establishment of market-share in manufacturing companies across the country.

How can this be when software developers are proclaiming unprecedented ease-of-use and accessibility? In an interview with *Design News* magazine, Victor Weingarten of SRAC announced that, using COSMOS/M, analysis is "…easier than developing a CAD model for basic analysis…a user only needs to know loads and boundary conditions…and the program does the meshing…" [5]. 'Only needs to know loads and boundary conditions'…that sure sounds simple. The simplicity of CAD-embedded analysis tools has appealed to many entry-level users based on their growing population. However, it is doubtful that these tools will produce any significant change in the level at which analysis improves the product design process.

2. Theory Of Constraints

The Theory of Constraints (TOC) provides a framework for improving throughput of a process by focusing on the constraints on that process. Constraints might be thought of as weak links or bottlenecks that, by definition, limit throughput and hand cuff both upstream and downstream processes. This theory, as introduced in *The Goal* [1] for manufacturing processes and expanded in *Critical Chain* [2] for project management, consolidates other popular, common-sense based, product development concepts such as Integrated Product Development [3], Rapid Product Development [6] [7], and Queuing Theory [8].

A common theme in all these concepts is that the success of any product development related process can only be measured by bottom-line profit or by metrics directly related to profitability. To maximize profitability, local efficiencies must often be sacrificed in pursuit of global cost reductions. A simple example would be an injection molding machine that can pump out an impressive 500 parts per hours (PPH). If this machine is run at maximum efficiency but the parts must then pass through a secondary operation that can only process 100 PPH, the total cost of the process will increase due to the stockpiling of inventory. Inventory ties up material and space, both of which have an associated investment cost. This cost does not even reflect the cost of handling and tracking as well as the penalty incurred if a change order is required that affects the parts in inventory. This is reminiscent of Just-In-Time (JIT) manufacturing. Where TOC differs is in the resolution of the efficiencies in a process.

By focusing on a system's constraints, TOC suggests these 4 steps to improve profitability:
1. *Identify* the constraints in the process
2. *Subordinate* all upstream processes to the needs of the constraint
3. *Exploit* the constraint by maximizing its throughput
4. *Elevate* the constraint, if further improvement is required, above its initial implementation to reduce its limitations

To understand these guidelines with respect to the role of FEA in product development, the process of developing products must be better understood.

3. The Process

Companies have spent millions on systems and tools to streamline the manufacturing process while agonizing over a $10,000 investment in FEA. A key reason for this is that manufacturing is a serial process with well-understood metrics. If an NC machining center can reduce part cost by 20% and improve cycle time by 10%, justification of the investment is straightforward. However, how do you quantify the return on investment from a seat of FEA. The difficulty rests in the perception or lack thereof, of product development as a process. Again, the lack of metrics becomes apparent. Designers themselves are partly to blame for the confusion by refusing to believe that creativity can be scheduled. Therefore, time-lines inevitably creep. Additionally, the absence of repetition in product design makes measuring improvement difficult. In manufacturing, changes in cycle-time can be readily seen. However, no two product development cycles are alike. Design is a one-time process [8]. Therefore, a metric that focuses on a portion of the process, CAD, CAE, FEA, etc., can not describe the profitability of the whole.

So, if profitability of a project is the desired goal, how can the product development process have the greatest effect? The answer lies in the three components of a successful FEA integration determined by the workshop participants: compressed cycle time, improved quality and reduced product cost. Quality improvement and cost reduction are commonly used metrics for justifying FEA. Reduced cycle time, however, warrants some further investigation. The virtues of up-front analysis or predictive engineering are casually tossed about. What is the impact of this on design cycle time...on product profitability? It is estimated that 95% of a product's cost is committed early in the design [9]. The majority of design decisions are made when the least is known about performance. Consequently, it is intuitive that a tool that can improve product knowledge early in the process will have a significant effect on the cost and flow of the process

Even *The Complete Idiot's Guide to New Product Development* by Edwin E. Bobrow [10] stresses the importance of up-front understanding. It states that the single most effective way to speed up the new product development process is...not to make mistakes! However, most companies still insist on completing a design, or taking it nearly to completion before using FEA to verify or improve its performance. The cost of delay due to identifying problems late in the cycle can be attributed to the fact that a significant investment has been made in the design by that point. The investment does not just reside in the part that is directly responsible for the problem but all parts that mate with it. This is called *Concurrent Commitment* [6]. Even if this commitment does not affect cycle time, it will, along with *Emotional Commitment*, impede the design team's ability to optimize the system. Emotional Commitment refers to a designer's sense of ownership in a part that took days or weeks to model in CAD.

Now that a case has been made for integrating analysis up-front in the design cycle, its role in the TOC framework can be explored.

4. IDENTIFY: Analysis Will Be A Bottleneck

The first step in TOC is to identify the constraints or bottlenecks in the system. It is important to note that a constraint can only appear on the critical path...by definition. If a company wishes to use simulation as a key technology for driving down cost, driving up quality, and compressing cycle times, it must be in the process for critical parts. In *Building Better Products with Finite Element Analysis*, we define critical parts as *"Product Musts"*. These are parts that have historically controlled cost or prototype iterations or have been identified as causing safety concerns. These parts may also be expensive or impractical to prototype and therefore can only be understood with FEA prior to design commitment.

A key assumption made in this paper is that, when utilized to the extent previously described, simulation will be a bottleneck. The justifications for this assumption come from several places. First of all, analysis has historically been a bottleneck. Rasna Corporation, the developers of MECHANICA, now Pro/MECHANICA from Parametric Technology Corporation, was formed by engineers because they could not get analysis results back fast enough. In Preston G. Smith & Donald G. Reinertsen's *Developing Product In Half the Time* [7], a case study at Senco Products noted that FEA slowed down a rapid development program and should only be used after careful consideration of its needs. Poor implementation was most likely the cause of this conclusion.

Secondly, despite the marketing hype about designer accessibility to FEA, for the foreseeable future there will be fewer design team members that are capable and competent with FEA than with CAD. Reasons for this are that the ratio of FEA seats per CAD seats at any given company will be low and users with the proper background and training will be harder to find. Just as an expert in FEA theory is not necessarily qualified to design parts if engineering judgement is lacking, a good engineer with a seat of FEA software does not an analyst make. The author recently completed a project for a company in which the engineering manager, a licensed professional engineer with a good understanding of mechanics, wanted verification of his analysis before committing his design. We found that he was limited to a CAD-embedded tetrahedral meshing tool and, due to resource constraints, was only able to mesh and solve one I-beam component in a frame structure. Our study, using a more appropriate beam idealization, showed 10 times greater displacement in the components of interest. It was not that he was a bad engineer. He was not trained to identify the appropriate model boundaries or that linear tetrahedrons would be inherently stiff. The beam model showed that most of the structure participated in the overall deformation. This intelligent engineer was limited by his choice of tools and training but, without outside intervention, did not even know there was a problem.

5.   SUBORDINATE - Adjust Your Design Processes To Provide Rapid Design Understanding

The above case study reflects a growing trend in the FEA industry. This is to force simulation to bend to the needs of CAD. TOC and common sense suggest the opposite approach. The greatest benefit comes from the earliest understanding of a product's behavior. Therefore, simulation must precede CAD or, at a minimum, detailed solid modeling. Today's tools have made it easy to mesh a complex solid. As the quote from Mr. Weingarten suggests, users are led to believe that the hard part is over when the mesh is automated. Instead of asking, 'Can I mesh?', users must begin asking, 'Should I mesh?' or better yet, 'What should I mesh?' Asking the latter question in the project-planning phase allows the design team to simulate the system at various points in the process. An initial, coarse model of shells, beams, or prismatic geometry might precede any detailed CAD. The appropriate gross features and sizes are fed to the designers for initial modeling where the 'fit' portion of *'fit, form, and function'* then comes into play. A second simulation phase can quickly identify errors or optimize features before any real CAD investment is made. Only when the simulation has confirmed the validity of the part, to the extent of the design team's knowledge of materials and boundary conditions, should detailed modeling proceed.

In a typical FEA implementation, analysts still spend a good deal of time in feature or dimensional reduction. This and the previously mentioned design commitments have turned analysis into more of roadblock than a bottleneck. Designs will bounce back to the CAD modelers instead of passing through with value added. To subordinate the CAD process to simulation, the needs of each particular simulation and idealization must be considered [11]. Planar idealizations or beam models will get little value from detailed solid models. In fact, they might be hampered by data in this form. However, properly configured wireframe geometry can allow the modeling to proceed quickly to the actual engineering.

6. EXPLOIT - Get Peak Performance Out Of FEA Tools

While a strong proponent of distributed simulation capabilities, it has been the author's experience that the best efficiency and results come from a design process that utilizes dedicated analysts. At the least, these individuals should have analysis as their primary role and all others subordinate to that. In *Critical Chain*, Dr. Goldratt discusses the importance of focus on reduction of cycle time [2]. When one's attention is divided among multiple tasks, the completion time, and maybe even the quality, of all will suffer. Additionally, the fact of the matter is that FEA can not be trivialized. There are too many assumptions that must be qualified to allow only a cursory understanding of the technology. There are four ways to exploit a constraint such as FEA [8]:

1.   *Increase Capacity* - This can be done by using faster hardware and software, adding users and systems, or by developing relationships with external resources. It is highly recommended that capacity be increased off the critical path. The worst time to learn new software or interview resources is when you need to. Many of the references clearly show that excess capacity is critical to maximizing the economics of product development.

2.   *Manage Demand* – In addition to "*Product Musts*" as defined earlier, most companies have "*Product Wants*". These are parts that can benefit from optimization but historically have worked when designed using just engineering judgement. The design process must clearly differentiate studies that are on one or more critical paths and those which can be made as part of an off-peak cost reduction program.

3.   *Reduce Variability* - Experience with the technology is the best way to improve the efficiency and consistency of the simulation process...another benefit of specialists. Well-documented projects and techniques will also help. Be careful when labeling a technique as a "Best Practice". When a product line or a family of parts are truly similar in use, a documented process for simulation can be invaluable. However, specifying a "Best Practice" precludes creativity and individuality, both of which differentiate a skilled problem solver from a mindless button pusher.

4.   *Use Control Systems* – Many companies with new or growing analysis capabilities let the FEA user determine methods and results acceptability. In the NAFEMS *Guidelines to Best Practice* [12] (the above discussion not withstanding) multiple references to third-party review of methods and results as key parts of an FEA quality control program are made. Nothing slows an analysis down more than aggressively pursuing the wrong approach. Design reviews and ongoing training will help control the tools' consistency.

7.   **ELEVATE – Make all members of the team understand the needs and limitations of the tool.**
Once the constraint, as initially implemented, has been made to run as efficiently as possible, it may further improve the cycle time to try to elevate it so that it is no longer a bottleneck. If the techniques for implementation described previously have been followed, an off-peak activity that will make simulation a mainstream task is to better educate all members of the design team, from management to draftsman, in the benefits, needs, and limitations of the technology. This improved understanding will minimize poorly formulated expectations and misconceptions. It will help companies choose the best tools; not the ones that are marketed to unrealistic desires. The now CEO of an FEA services firm in Kansas City, MO once responded to an objection about a

description in a brochure by saying, "It's not lying, it's marketing." However, when a company that makes thin walled plastic parts purchases a CAD-embedded tetrahedral meshing tool, one would have to wonder if a more educated consumer would have caught some salesperson in a big, fat "market".

8.   *Closing Remarks*

It is clear from the current body of work on product development that success with an FEA implementation must be measured in economic terms. More specifically, the technology must achieve one or more of the three objectives mentioned earlier, Compress Cycle Time, Reduce Product Costs, and Improve Product Quality. Cycle time reduction has received the least attention of these while companies have focused on the costs and efficiencies of isolated tasks. A study by McKinsey & Company showed that projects completed within budget but six months late to market reduce overall profit by 33%. On the other hand, projects that are 50% over budget but on-time are only responsible for a 4% profit reduction [3]. The staggering truth is that getting projects completed on-time or early may justify almost any expense. The contribution FEA can make to cycle time reduction has only begun to be explored. Its success has been hampered by education that lags far beyond the growth of technology accessibility.

As a tool, FEA is subject to the same rules of nature as all tools. If you give an accountant a power tool, he/she does not become a master carpenter. Similarly, if the only tool you have is a hammer, the whole world begins to look like a bed of nails. The right software must be given to the right people with access to the right training. People, tools and training...three legs of a platform that will improve product profitability. If one leg should fail, the individual strengths of the others can not begin to compensate.

While most of the responsibility does lie with the user, those with the advantage of better knowledge can not continue to mislead the growing and enthusiastic. Anything less than a complete commitment to the technology will fail to effect a change in the way products are developed. For their part, development companies need to educate their design team in all the tools and techniques required to build better and more profitable products. Without goals, an implementation plan driven by these goals, and an evaluation or measurement process for verifying success, terms such as 'Collaborative', 'Concurrent' and 'Integrated' are nothing more than buzzwords. The only way to properly achieve successes, beyond those task-limited successes touted in articles and vendors brochures, is through an evaluation process whereby you measure how well your "well-aligned" implementation plan drove you to the defined goal. Are you building better products with advance technologies? It all begins with the goal!

Acknowledgements

The author would like to thank Catherine Campbell and Abraham Askenazi for their input and constructive criticism for helping focus this discussion.

References

1   GOLDRATT, ELIYAHU M. – *The Goal*, The North River Press, Great Barrington, MA, 1992

2   GOLDRATT, ELIYAHU M. – *Critical Chain*, The North River Press, Great Barrington, MA, 1997

3   GRANT THORNTON, LLP – *The Enterprise-Wide Impact of Computer-Aided-Engineering on Manufacturers*, pp. 24, 1996

4   VERMETTE, DAVID – *Product Development in the 1990s: Communication and Integration*, Management Roundtable, Inc. pp. 7, 11-12, 1997

5   Engineering analysis will enter the future hand-in-hand with design, Design News; October 10, 1998, pp194

6   ADAMS, V. & ASKENAZI, A. – *Building Better Products with Finite Element Analysis*, OnWord Press, Santa Fe, NM, 1998

7   SMITH, PRESTON & REINERTSEN, DONALD – *Developing Products in Half the Time –Second Edition*, John Wiley & Sons, New York, NY, 1998

8   REINERTSEN, DONALD – *Managing the Design Factory: A Product Developer's Toolkit*, The Free Press, New York, NY, 1997

9   CHARNEY, CYRIL – *Time to Market – Reducing Product Lead Time*, Society of Manufacturing Engineers, pp. 6, 1991

10   BOBROW, EDWIN E – *The Complete Idiot's Guide to New Product Development*, Alpha Books, New York, NY, 1997

11   ADAMS, VINCE – *Preparation of CAD Geometry for Analysis and Optimization*, NAFEMS 7th International Conference Proceedings, 1998

12   BEATTIE, G.A. – *Management of Finite Element Analysis – Guidelines to Best Practice*, NAFEMS, Glasgow, Scotland, 1995

Using Features for Effective Engineering Analysis and Design[I]

Chandrasekhar Reddy[II], Natarajan Sridhar[III]
and Anil Chaudhary[IV]
Concurrent Technologies Corporation, Johnstown, PA
and Applied Optimization Inc., Dayton, OH

SUMMARY

Detailed design of a product is typically comprised of a preliminary design phase, followed by an iterative phase involving engineering analyses (to verify the design) and design modifications (to improve the design). Manufacturing process design and integrated product-process design involve a similar set of activities. There are several Computer-Aided Design and Computer-Aided Engineering (CAD/CAE) tools that can be used by the designer in each of these phases; however, the designer still has to overcome certain design challenges.

Traditionally, detailed design is strongly based on geometry. A designer constructs the geometry and subsequently associates non-geometric details (material, surface finish, etc.) with it. The first design challenge is the difficulty of specifying, visualizing, and modifying this geometry and non-geometric information. Design verification by engineering analyses requires specialized expertise (analysis model building, analysis execution, interpretation of analysis results, etc.) and large setup and execution times. The current trend in engineering design is to make analysis tools available to the designer so that he/she can identify potential downstream problems very early in the design cycle. Thus, the second design challenge is to minimize the expertise and labor costs associated with engineering analyses. Design modifications involve making changes to the geometry and the associated non-geometric details in such a way that the design is "improved" as per some

[I] This work was funded by the National Applied Software Engineering Center, which is operated by Concurrent Technologies Corporation (*CTC*) and sponsored by the Defense Advanced Research Projects Agency. The content of this information does not necessarily reflect the position or the policy of the government, and no official endorsement should be inferred.

[II] Senior Software Engineer, Concurrent Technologies Corporation, Johnstown, PA

[III] CAE Specialist, Concurrent Technologies Corporation, Johnstown, PA

[IV] Scientist, Applied Optimization Inc., Dayton, OH

criterion. The third design challenge is to enable meaningful modifications of the design to ensure improvement.

This paper proposes a feature-based approach to address the above challenges. A detailed design is represented as an assembly of features. In this context, a feature is defined as any unit of semantic design information. For example, a feature may have a specific parameterized shape, material, and other design information. A feature is also "aware" of the nature of its interactions with other features. All aspects of the feature are parameterized. Feature parameters can either be input (i.e., specified by the designer) or output (i.e., determined by engineering analysis).

This approach enables a designer to quickly construct a preliminary design by creating and assembling parametric features following certain design rules. These design rules, which describe causalities, dependencies, and mathematical relationships among the feature parameters, capture existing design experience and knowledge. Enforcing these design rules automates most of the preliminary design activity. The feature-based design representation: (1) encapsulates the semantic information necessary to setup any desired engineering analysis for verifying the design, (2) serves as a common basis to setup and perform multiple engineering analyses, and (3) facilitates the mapping of analysis results back to the semantic information contained in the design. Design modifications can be quickly achieved by modifying the values of the feature parameters. Design correlations (in the form of quantitative measures of sensitivities among the feature parameters), design intent, and design specifications can be associated with design modifications to ensure overall design improvement.

The success of the above approach demands that the features contain all the semantic information that a designer needs to consider during detailed design. This rules out the practicality of developing a single design tool for all product-process design scenarios. The solution is to develop an object-oriented framework that can be used to construct custom design tools to meet the specific needs of a particular designer or enterprise. This paper also describes DECaF, a custom feature-based casting design application built using SimBuilder™ [V], an object-oriented framework for building simulation applications.

1. INTRODUCTION

Global competition and market pressures are forcing high-quality products to move from concept to delivery at reduced time and costs. To stay competitive, companies are forced to respond to the evolving user needs with minimum additional investment by leveraging existing infrastructure. The

---

[V] SimBuilder is a trademark pending for Concurrent Technologies Corporation.

task of detailed design, which is typically supported by engineering analyses, is the most important task during the design phase of a product's life-cycle. However, successful implementation of strategies for improving the effectiveness and efficiency of the detailed design task is impeded by the way design teams function and communicate. Most design teams implement a design process that is built around a particular analysis tool/technique, which limits their design contributions to the capabilities of the chosen analysis approach. Moreover, design teams do not communicate effectively due to a variety of reasons including diverse and differing design viewpoints and incompatibilities among the detailed design representations that they use. The key to overcoming these difficulties is the standardization of the detailed design representation and establishment of a common protocol to inspect and consistently modify a common reference design.

2. BACKGROUND

Traditionally, detailed design is represented and modified in terms of its low-level attributes (such as the points, lines, arcs, surfaces, etc. that define its geometry). There is limited representation of high-level semantic attributes such as shape, material, performance specifications, constraints, knowledge, and correlations. This poses the following concerns:

- Semantic Design – The traditional detailed design representation is difficult to construct and modify since a designer is forced to operate with low-level attributes instead of semantic attributes.
- Design Modification – A design modification can only be represented as a set of changes made to low-level attributes. Design modifications should be represented as changes in a high-level semantic attribute such as shape or performance and must be related to a design intent and a set of design specifications.
- Design Evolution – Due to the lack of documented design intent and design history, there is no assurance that a new design modification will lead to an overall improvement in the design.
- Design Knowledge – There is no facility to incorporate design knowledge (best practices, collective design experience, rules-of-thumb, etc.) that guides design modifications based on design specifications and design intent.
- Analysis Goal – Because design intent and design specifications are not formally represented, it is up to an analyst (and not the designer) to subjectively determine the type and granularity of the engineering analyses that are required to verify the design.
- Analysis Setup – The setup and execution of an engineering analysis and interpretation of the analysis results requires a great deal of specialized knowledge, skill, and experience, which a typical designer does not possess.

- Analysis Results – The results of an engineering analysis depend significantly on the assumptions and simplifications that were made by the analyst when the analysis was set up. Consequently, the analysis might not always generate information that is directly relevant to the design verification goals.

3. FEATURES

The use of features in detailed design is not new [1, 2]. *Shape features* have been primarily used in CAD to represent parametric, domain-specific geometric shapes that enable easy specification and modification of the design geometry and localize geometry changes. Examples of shape features include: (1) *cylindrical hole*, defined in terms of two parameters— radius and depth; and (2) *rectangular slot*, defined in terms of its length, width, and depth. Features have also been used in CAD/CAM software related to design for manufacture, particularly machining. However, both shape features and CAD/CAM features address only a subset of the overall detailed design goals. To be useful from an overall detailed design perspective, features must possess the following characteristics:

- Unbiased – A feature must represent information that is relevant to the overall detailed design goals. However, it should not restrict the representation to a particular downstream consideration such as analysis, manufacture, deployment etc.
- Extensible – The design information that is represented by a feature must be extensible in order to address any additional detailed design goals. For instance, if cost needs to be addressed during detailed design, it should be possible to associate cost-related information with features.
- Composable – Features must be amenable to assembly that will result in a complete and accurate description of a detailed design. There are two different ways features can be assembled to represent a design. One is akin to building a model out of blocks, where each block has certain characteristics, and perfectly locks into place with its adjacent blocks. The other way to look at features is layering as in CAD, where each layer captures certain aspects, and the layers overlap. Each layer is capable of impacting the entire design, but may be restricted to some locale.
- Semantic – A feature must encapsulate semantic design information related to one or more of: shape, material, design knowledge, design specifications, design modifications, design intent, design constraints, and design correlations.
- Parametric – The design information that is represented by a feature must be parametric, (i.e., defined in terms of mathematical variables). This enables the quick and easy specification/modification of a design.

Consequently, a designer/analyst can derive a desired view of a feature-based detailed design by considering a certain set of features and/or certain feature parameters. This view can be specified/modified/verified on its own, while maintaining its relationship to the overall detailed design. This ability to extract and manage multiple detailed design views facilitates quick and easy overall design improvements in an incremental and systematic manner.

4. CLASSIFICATION OF FEATURES

The following is a description of features and their classification aimed at addressing various detailed design issues identified in the BACKGROUND section. These features possess all the characteristics identified in the FEATURES section. These features are based on the encapsulation and inheritance paradigms of object-oriented analysis and design.

1. Basic Features

Almost all features useful for detailed design need to represent design information related to shape and material. *Basic features* are features that can be used to describe shape and material of a particular locale in the design. All other types of features are derived from the basic features by incorporating additional information that is relevant to a particular detailed design goal. Figure 1 shows how various kinds of features may be derived from basic features. Each basic feature contains information related to the following:

- Shape, shape parameters, and related constraints
- Position (location and orientation) of the feature in reference to the design, and related constraints
- Design information and related constraints
- Material and related constraints
- Shape, position, and material dependencies/relationships with respect to other features

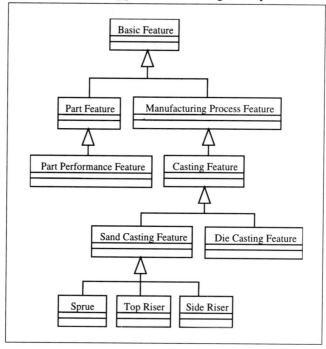

Figure 1. A Feature Classification Hierarchy (UML[VI] notation)

2. Part Performance Features

Part performance features are based on recurring patterns of information/knowledge that designers use during the detailed design of mechanical parts to satisfy in-service performance requirements. Part features can be assembled to produce a near-complete description of the part and its behavior/performance when subject to in-service conditions. Each part feature contains the following information in addition to the shape, position, and material information that it inherits from the basic feature.

- In-service conditions (e.g., loads)
- In-service behavior (e.g., vibration modes, deflection) and related constraints/specifications
- In-service performance (e.g., failure, safety factors) and related constraints/specifications
- Dependencies/relationships with respect to other part features

---

[VI] The Unified Modeling Language (UML) is a modeling language for specifying, visualizing, constructing, and documenting the artifacts of a system and/or process, approved by the Object Management Group (OMG) as a standard

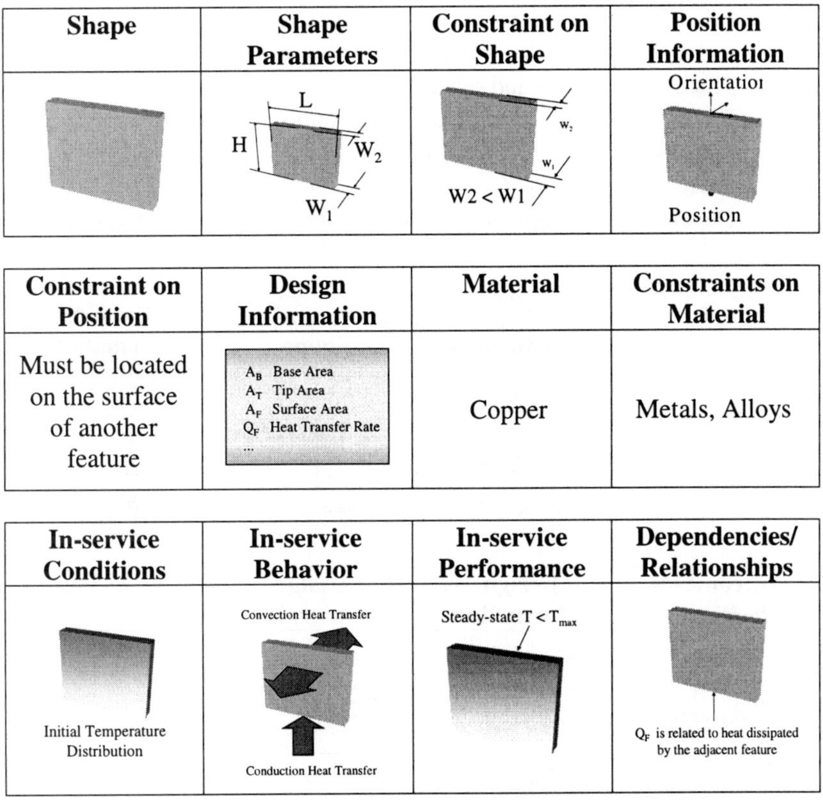

Shape	Shape Parameters	Constraint on Shape	Position Information
	H L W_2 W_1	W_2 W_1 $W2 < W1$	Orientation Position

Constraint on Position	Design Information	Material	Constraints on Material
Must be located on the surface of another feature	A_B Base Area A_T Tip Area A_F Surface Area Q_F Heat Transfer Rate ...	Copper	Metals, Alloys

In-service Conditions	In-service Behavior	In-service Performance	Dependencies/ Relationships
Initial Temperature Distribution	Convection Heat Transfer Conduction Heat Transfer	Steady-state $T < T_{max}$	Q_F is related to heat dissipated by the adjacent feature

Figure 2. Cooling Fin Part Performance Feature

Two examples of performance features are: (1) *stiffening rib*, which represents in-service conditions, behavior, and performance constraints/specifications related to stiffness; and (2) *cooling fin*, which represents in-service conditions, behavior, and performance constraints/specifications related to heat dissipation (Figure 2). Both of these features capture semantic information related to the in-service aspects of a part in addition to shape, position, and material aspects.

3. Manufacturing Process Features

Manufacturing process features [3] are based on recurring patterns of information/knowledge that designers use during the detailed design of manufacturing processes. As in the case of part features, process features are based on the basic features and can be assembled to produce a near-complete description of the operation and behavior of a manufacturing process. Each process feature contains the following information in addition to the shape, position, and material information that it inherits from the basic feature.

- Process conditions and related constraints
- Process phenomena and related constraints
- Process parameters and related constraints
- Process defects
- Dependencies/relationships with respect to other process features

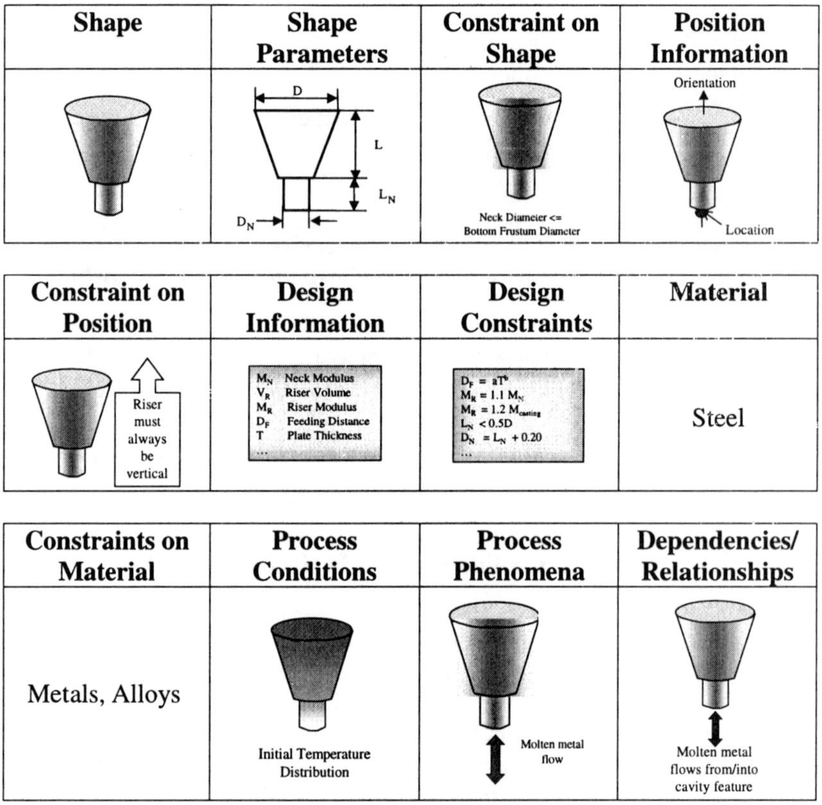

Shape	Shape Parameters	Constraint on Shape	Position Information
	D, L, L_N, D_N	Neck Diameter <= Bottom Frustum Diameter	Orientation, Location

Constraint on Position	Design Information	Design Constraints	Material
Riser must always be vertical	M_N Neck Modulus V_R Riser Volume M_R Riser Modulus D_F Feeding Distance T Plate Thickness	$D_F = aT^b$ $M_R = 1.1 M_N$ $M_R = 1.2 M_{casting}$ $L_N < 0.5D$ $D_N = L_N + 0.20$	Steel

Constraints on Material	Process Conditions	Process Phenomena	Dependencies/ Relationships
Metals, Alloys	Initial Temperature Distribution	Molten metal flow	Molten metal flows from/into cavity feature

Figure 3. Top Riser Manufacturing Process Feature

The *top riser* (Figure 3) is an example of a manufacturing process feature that represents a rigging element used in a sand casting process. The top riser is used to accumulate and hold molten metal in a casting, so that as the mold cools and the metal solidifies (and contracts), molten metal flows from the riser into the mold cavity.

5. WORKING WITH FEATURES

Features address the previously identified detailed design concerns in the following manner.

- Semantic Design – The use of features produces a detailed design representation that has significant semantic content. In addition to shape and material information, design knowledge, design specifications, design modifications, design intent, design constraints, and design correlations can be represented as well.
- Design Modification – The feature-based design representation is parametric and is amenable to easy specification/modification. Changes in the design are at a semantic level and constitute meaningful design modifications. These design modifications can be traced to design intent, design specifications, and engineering analysis results.
- Design Evolution – The representation of design history in terms of well-defined design modifications results in more effective and convergent design iterations towards an optimum.
- Design Knowledge – Features can be customized to encapsulate proprietary design knowledge in the form of relationships among feature parameters and acceptable ranges for parameter values.
- Analysis Goal – The information contained in features can be used to automatically generate the choice of analysis goals and types of analyses that are applicable for design verification. The designer can then choose the type and granularity of an engineering analysis based on a certain design verification goal.
- Analysis Setup – Based on the chosen analysis type and granularity, information contained in the features can be used to set up an engineering analysis. For example, the *top riser* (Figure 3) process feature encapsulates all the information necessary to drive a mold filling and solidification analysis of a casting process design using numerical analysis. The geometry of the feature can be used for discretization. Its associated fluid flow and heat transfer phenomena information can be used to setup the loads and boundary conditions to drive a fluid flow, heat transfer, or coupled analysis.
- Analysis Results – Since the designer chooses the analysis goal based on a certain design verification goal, the results of the engineering analysis can be directly used by the designer to verify the design and make design modifications (as necessary).

6. FEATURE-BASED SOFTWARE ARCHITECTURE-SimBuilder[TM]

SimBuilder[TM] [4] is an object-oriented software architecture developed

Practical Industrial Applications – Design Analysis Methodology at Concurrent Technologies Corporation (*CTC*). SimBuilder™ provides software design and code building blocks that can be used to build feature-based design and simulation applications (Figure 4). The centerpiece of the architecture is a layered framework of reusable software components. These reusable components are customized using the Application-Specific Class Libraries and are further supported by components that provide Knowledge Base access and GUI/Visualization Support. The framework interfaces with Legacy Simulations for analysis and uses Commercial Off-the-Shelf (COTS) Framework & Class Libraries for generic services.

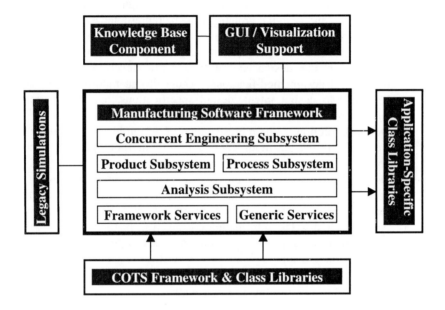

Figure 4. The SimBuilder™ Architecture

The overall theme for the SimBuilder™ architecture is software development by assembly and customization of reusable software components rather than developing a monolithic application [5]. The separation of the respective functions into separate object-oriented components provides suitable opportunity to customize the application based on the true needs of an enterprise.

7. FEATURE-BASED DESIGN APPLICATION – DECaF

DECaF (Design Environment for Casting using Features) [6] is a feature-based casting process design tool developed at *CTC*. DECaF was built using the SimBuilder™ architecture described above. DECaF provides an interactive environment for quick design and analysis of casting processes. DECaF enables designers to design a casting process by quickly creating and laying out a set of casting process features around the part to be cast. New process features can be quickly developed or modified to model specific processes or enforce enterprise-specific design rules. In addition, quick analysis options are provided to enable approximations to flow and solidification solutions. DECaF can also directly interface with detailed casting analysis software. DECaF is expected to reduce the number of analysis iterations, enable better communication between designers and analysts, and reduce the simulation setup time for detailed analysis from days to hours.

8. CONCLUSION

Feature-based detailed design enables the accurate, complete, and unique representation of a detailed design, which is independent of any specific downstream considerations such as analysis, manufacture, deployment etc. In fact, this representation serves as a common basis for deriving specific representations for downstream considerations. This ensures that the derivative representations are never in conflict with each other, and the verification/modification of any derivative representation can be directly mapped back to the basis representation. Moreover, a feature-based representation facilitates the association of design modifications to design intent, design specifications, and engineering analysis results. Consequently, the designer has a better understanding of the repercussions of new design modifications. This prevents the designer from making design modifications that might not significantly improve a design, thereby reducing the number of iterations required to achieve an optimal design. Detailed design is thus more natural, effective, and consistent.

Since the information contained in a feature-based design representation is independent of any specific analysis tool/technique, features can be used to develop an Application Programming Interface (API) for interfacing with external analysis tools that the designer desires to use, as demonstrated by DECaF. SimBuilder^{TM} can be used to quickly design and develop custom design tools, such as DECaF, that are productive even in the hands of novice designers.

REFERENCES

1. DIXON, J. R, LIBARDI, E. C. Jr., LUBY S. C, VAGHAL, M. V and SIMMONS, M. K – Expert Systems for Mechanical Design: Examples of Symbolic Representations of Design Geometries. <u>Applications of Knowledge-Based Systems to Engineering Analysis and Design</u>, ASME Annual Meeting, Miami Beach, Florida, pp. 29–46, November 17–22, 1985.

2. SHAH, J. J and WILSON, P. R, – Analysis of Knowledge Abstraction, Representation and Interaction Requirements for Computer Aided Engineering. ASME Computers in Engineering, Vol. 1, pp. 17–24, 1988.

3. ATHIRAJAN, P, SRIDHAR, N, GULEYUPOGLU, S, and REDDY, C – A Feature-Based Approach To Process Modeling For Simulation Applications. International Journal of Agile Manufacturing, Vol. 2, Issue 1, pp. 37–50, 1997.

4. CONCURRENT TECHNOLOGIES CORPORATION – Manufacturing Software Architecture–System Architecture Document. *CTC* Technical Report, August 1998.

5. APPLEY, G and CHAUDHARY, A – Making Engineering Analysis Software Easier to Use. Presented at Autofact, September 28–October 1, 1998, Detroit, MI.

6. SRIDHAR, N, REDDY, C, KRAULAND, M, and CHAUHARY, A – DECaF: Design Environment for Casting using Features. To be presented at the NAFEMS World Congress 1999, April 25–28, 1999.

DESIGN FOR ASSEMBLY USING FEM

Bojan Jerbić[1], Božo Vranješ[1], Miljenko Math[1]

SUMMARY

Product design for assembly usually follows the concept of minimizing the number of parts. This approach can result with fewer assembly operations and consequently with lower assembly costs. But, from the market and sale point of view the requirements sometimes demand different philosophy in product design. The packaging and transportation criteria can be expressed with the sentence "smaller is better". It means that large products should be designed applying the disassembling (modular) construction, the product has to be divided into smaller sections (more parts) joined by appropriate assembling techniques. Where are the acceptable dividing planes and what are the assembling techniques which would provide product functionality, quality and stiffness? These are the issues the paper deals with using the finite element method. This paper describes the designing of the handcart FE models for the simulation of assembling techniques, with the aim to find out the most appropriate solutions without the use of the conventional workshop prototyping. It addresses the problems of simulating the assembly FE models, bolt fits as well as friction and contact conditions.

1 INTRODUCTION

If a product needs to be built from more then one part, the designing of its assembly structure, the interfaces of the constituent parts and assembly techniques represent critical considerations in development process. They directly affect design functionality and manufacturing costs. With regard to assembly design engineers usually rely on experience, estimating assembly techniques and parameters. Such approach is no longer appropriate along

---

[1] The University of Zagreb, Faculty of Mechanical Engineering & Naval Architecture, I. Lučića 5, 10000 Zagreb, Croatia

with powerful engineering computer aided graphical and numerical tools based on finite element method which enable the virtual prototyping and analyzing of the geometrical, functional and physical characteristics of future product. Computer simulation provides the efficient optimization of product design contributing considerable development and manufacturing savings, and competitive market potential.

This paper is concerned with the application of computer simulation in the optimization of product assembly design. A handcart, used in construction and agriculture works, is taken as an example of large product whose homogeneous welded parts satisfy heavy duty characteristics. But if its packaging characteristics need to be optimized, enabling, for example, mail delivery on the basis of catalog or Internet ordering, the handcart must be designed as a modular (disassembling) product. It particularly relates to the handcart frame which obviously represents a heavy-handed piece for packaging. To retain the handcart original functionality, quality and stiffness, the redesign process have to research the acceptable dividing planes and consider the possible assembling techniques.

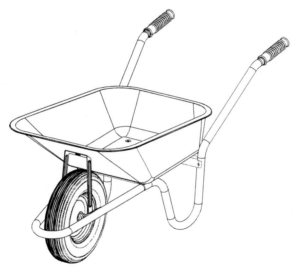

Figure 1. Handcart CAD model

The handcart FE models have been designed using the SDRC I-DEAS Master Modeler and Simulation software. They are based on triangular and quadrilateral thin shell elements mashed on the top surface of the handcart's solid model. The models were primarily investigated as a solid construction to find out the regions where the high stress was taking place, and accordingly to prevent the dividing of the cart construction in those critical sections. The boundary conditions were varied to conceive the various working situations. The resulting handcart frame segments have been designed and assembled by different assembling techniques. The paper presents the corresponding FE models and discusses the obtained simulation results and suggested approach with regard to the simulation of assembly models.

2 FE MODEL OF HANDCART ASSEMBLY

After the analyzing of the packaging requirements and stress distribution along the handcart elements, the frame is divided on the basis of two dividing planes. The resulting cart construction with five frame segments is given in Figure 2.

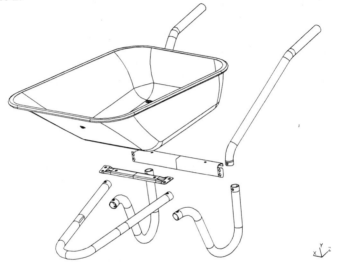

Figure 2. Modular handcart frame

The question is how to join the frame parts to keep the required rigidity. The simplest way is to tighten them by bolts and nuts which restrain the frame segments over two supports (Figure 3.).

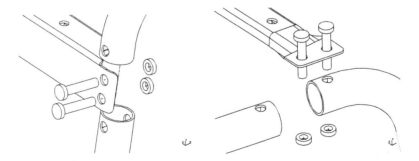

Figure 3. Bolted joints

The other way is to use an inner tube as an insert which strengthen the frame fittings (Figure 4.). There are also many other solutions which could be applied, but this paper is limited into the research presentation of these two mentioned assembly techniques.

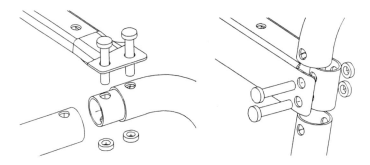

Figure 4. Bolted joints with the inserts

In order to investigate the stiffness of the proposed assemblies the corresponding FE model of the handcart assembly solid model has been designed. The handcart box transfers a load over the front and back supports to the frame. Therefore the box can be excluded from consideration, sparing the size of the FE model. By making use of symmetry, just the half of the handcart is taken into consideration.

2.1 Material properties

The handcart frame is made from Ø32/28 mm steel tube whose properties corresponds to DIN 1611 material number 1.020: modulus of elasticity $E = 210\,000$ Mpa, Poisson's ratio i.e. ratio between lateral contraction and axial elongation $v = 0.3$, yield stress $R_{p_{0,2}} = 260$ N/mm^2. The material of the front and back supports is a sheet metal with 2 mm thickness. The sheet metal properties are same as the properties of the frame material.

2.2 Mesh discretization

The FE mesh has been created by thin shell elements using the top surfaces of the symmetric half of the handcart assembly solid model. The surfaces were partitioned in the regions of contacts and high stress to allow finer mesh construction. The several types of thin shell elements were investigated, including triangle and quadrilateral elements with linear and parabolic definitions. The linear elements, both triangular and quadrilateral, showed some instability regarding stress dissipation and obtained values which were not consistent from case to case. Therefore, parabolic elements are accepted for analysis. Parabolic quadrilateral elements enable to create harmonic and smooth mesh if appropriate element size and the distribution of nodes were applied. However, they produce more nodes then parabolic triangular elements, requiring considerably longer calculation time. The level of accuracy for the given problem allows the use of relatively coarse parabolic triangular mesh, which can save computation time more then five times comparing to parabolic quadrilateral elements. The element size is critical only in the regions of the contacts where rotational and curved surfaces participate. Since the contact pairs produces very narrow touching areas, too large elements can cause convergence problems in the contact iteration algorithm.

In the case of parabolic quadrilateral mesh the model includes 5267 elements and 16246 nodes, including the mesh of inserts (Figure 5.). Parabolic triangular mesh contains 6097 elements and 12856 nodes. The elements have the tickness of 2 mm according to the material of the parts.

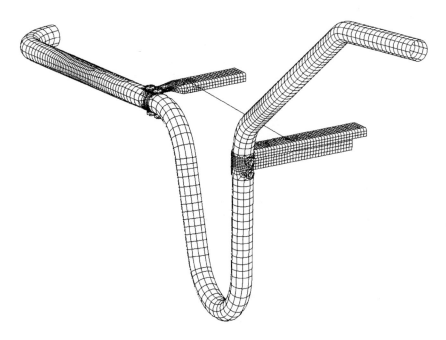

Figure 5. Parabolic quadrilateral mesh of the handcart model

The bolts and nuts are dicretized by solid mesh parabolic triangular elements to provide analysis of their deformations as well. The bolt includes 108 and the nut 33 solid mesh elements.

The front and back supports are joined on the handcart box constraining their relative movement. This condition is simulated by rigid elements which bind the holes of the supports.

2.3 Boundary conditions

Figure 6. illustrates the applied boundary conditions, including restraints, forces/loads and contacts.

Restraints
The restraints of symmetry are applied on the parts' faces which lie on the plane of symmetry, including the frame and both supports. Additional restraints are used to define a wheel support and handhold. The wheel support allows only z translation, as well as x and z rotation. The handhold constraints translation and rotation with the exception of x rotation. Such restraints simulate handcart in driving position.

Forces/Loads
Load of the handcart is distributed over the supports as surface force which represents a freight of 200 kg. Special forces are involved on the bolts' heads and nuts inducing the compressive forces in the joined parts and between contact surfaces.

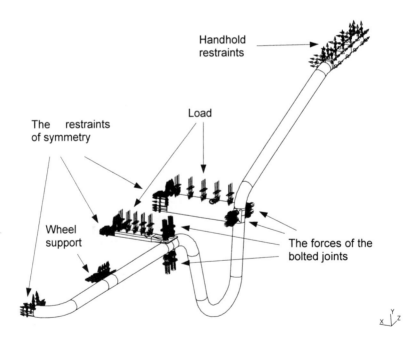

Figure 6. Boundary conditions

Contacts

The contact surfaces are defined between the frame, supports, bolts and nuts to simulate the transfer of forces/loads between the parts. The friction factor is assumed 1.0 to prevent the sliding between parts and consequently some inconsistency in the results of stress distribution.

2.4 Linear analysis

The FE model of handcart has been analyzed using I-DEAS linear structural finite element solver. Linear statics analysis simulates the behavior of a physical object or structure which has forces applied to it and solves for the displacements, stresses, etc., in the domain of elastic material properties. The solver assumes a small strain, small displacement formulation in which the finite element equilibrium equations are written with respect to the initial configuration of the structure. Linear statics analysis does not account for any nonlinear stress-strain relations, nonlinear strain displacement relations, or geometric nonlinear effects. these assumptions restrict the use of linear statics analysis to problems with locally small strains and infinitesimal rotations. If large stress, which exceeds yield stress for given material, was identified, the nonlinear plastic deformation analysis should be performed to research possible plastic deformation or potential crack zones. This work is limited to linear analysis with aim to explore problematic stresses and deformations in the assembly using different joining techniques.

The FE model of the handcart assembly has been analyzed for simple bolted joints and joints strengthened by the inserts. The applied solver parameters are as follows:

- Singularity removal algorithm is applied with a criterion value 10^{-20}. Singularity removal controls the singularity in the stiffness matrix during Cholesky factorization (a matrix singularity is detected when a zero or very small pivot is encountered), then a positive number is added to the diagonal of the matrix, and decomposition proceeds.
- The contact algorithm uses the penalty factors to control contact and sliding stiffness: the normal factor, controls the penetration stiffness of the surfaces that come into contact, is set to 10 and the tangential factor, controls the convergence of friction forces, is set to 1.
- Contact force convergence tolerance is 0.01.
- Minimum contact set percentage defines the percentage of contact elements considered active, and it is adjusted to 100%.
- The maximum number of iterations for a force loop (forcing a condition of zero penetration between parts) is defined 20 and a contact status loop (determining which contact elements are active) is set to 40.

3 RESULTS

3.1 Simple bolted joints

The first handcart assembly has the frame sections joined simply by the bolts and nuts on the front and back supports. The postprocesed result of linear statics analysis is given in Figure 7. Figure 7. presents the five times enlarged deformations of the assembly and the distribution of von Misses stress in the top layer of material. It can be noticed that high stresses are concentrated in the supports which keep the frame sections together. The stresses and deformations in the contact regions of the supports are responsible for the rigidity of the whole frame construction. Even the postprocessed result shows enlarged deformations, it can be perceived that deformations are relatively significant what indicates a possible problem of assembly rigidity. Furthermore, the back support and tubes suffer a high stress which exceeds elastic deformations for given material (more then 260 N/mm^2). Even they are localized, also they present dangerous zones (near the holes and corners, Figure 8.) where either crack or intensive wearing could be expected. Obviously, this assembly solution would not provide required frame stiffness for an expected load of 200 kg.

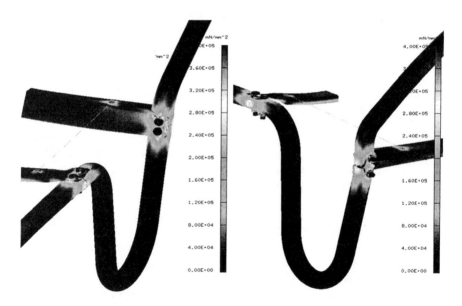

Figure 7. Von Misses stress distribution and deformations in the handcart
assembly

752

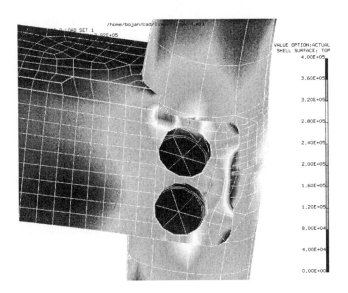

Figure 8. The high stress regions in the back support

3.2 Joints strengthened by inner tubes

In order to release the supports from bending stress, an inner tube can be involved in the joints. Figure 9. shows corresponding deformed model with von Misses stress distribution. Deformations and stresses are clearly lower then in the previous model.

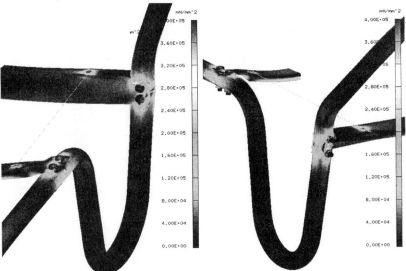

Figure 9. Von Misses stress distribution and deformations in the handcart assembly with the joints strengthened by the inserts

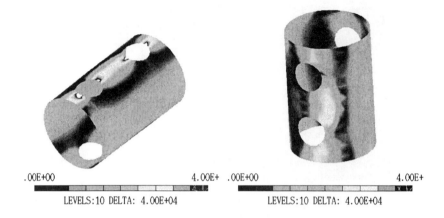

.00E+00   4.00E+ .00E+00   4.00E+

LEVELS:10 DELTA: 4.00E+04   LEVELS:10 DELTA: 4.00E+04

Figure 10. Von Misses stress in the front and back inserts

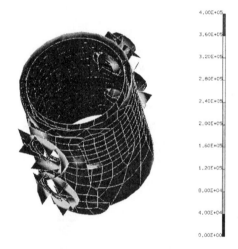

Figure 11. Contact stresses in the back joint

Figure 10. gives von Misses stress in the inner tubes (inserts) and Figure 11. shows contact stresses in the joint of the back support. There is no regions with high stress which exceeds elastic range, with the exception of the corners in the back support (Figure 12.). It is obviously inadequate design detail which has to be reviewed and changed by adding a fillet or chamfer, providing better stress dissipation.

Such assembly solution strengthened by the inserts clearly presents promising basis which could provide expecting handcart qualities.

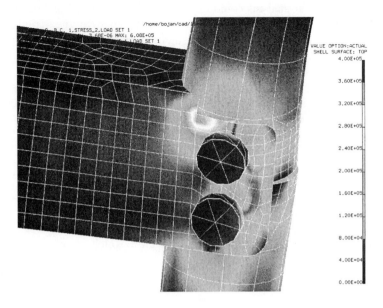

Figure 12. The high stress region in the corner of the back support

4 CONCLUSIONS

The designing of assembly structures is usually based on the engineer's experience and prototype analysis. The finite element method enables virtual prototyping of a future product. By using the contact boundary conditions which transfer loads between parts it is possible to simulate stresses and deformations in assembly CAD model.

This work presents the application of finite element method in the analysis of the handcart modular construction. Two different assembly techniques have been investigated: simple bolted joints on the front and back supports, which keep the frame segments together, and the bolted joints strengthened by the tube inserts.

The FE model has been designed on the basis of the handcart assembly solid model. The mesh has been constructed on the top surfaces of the solid model. The model has been analyzed for linear and parabolic elements, including triangular and quadrilateral forms. Linear elements have showed certain deviations with regard to the obtained results. Parabolic triangular and quadrilateral elements have given comparable results even relatively coarse element have been applied.

755

The results show expected model behavior concerning frame deformations and stresses. Finer discretization would give more accurate results and nonlinear analysis could more precisely indicate critical details in the construction. However, the obtained results approve the simulation approach and model reliability for investigation the designing of assembly techniques.

Further research would try to include solid mesh instead of thin shell elements and the simulation of stress in the bolts. It could enable the complex optimization of assembly techniques and design details.

ACKNOWLEDGEMENT

This work is a part of the research project "Machine Intelligence in Assembly", supported by the Ministry of Science and Technology of the Republic of Croatia and NOVOTEC, Croatia. The authors also wish to thank to the company LIMEX, handcart manufacturer, which inspired them to research the problem.

REFERENCES

1. Zienkiewicz, O C - The Finite Element Method in Engineering Science, McGraw-Hill, London, 1984
2. Seyerlind, L J - Applied Finite Element Analysis, John Wiley & Sons, New York, 1976
3. Butterworth, J W, Smyrell, A G - The Finite Element Analysis of a Structural Steelwork Extended End Plate Beam-to-Column Bolted Connection, Design, Simulation & Optimisation, NAFEMS world Congress '97, Stuttgart, pp. 410 421, 1997

FINITE ELEMENT ANALYSIS AND SYSTEMATIC BLADED-DISK DESIGN

Richard A. Layton[i] and John J. Marra[ii]

SUMMARY

A central issue in gas turbine engine design today is the demand for higher performance, greater reliability, shorter lead times and lower cost. The design of bladed disks (fans, compressors and turbines) is one area in which suitable design tools are sought to meet this demand. In this paper is presented a conceptual basis for a new, systematic, unified approach to bladed disk design. The approach is outlined in the context of a software based design tool, building on existing finite-element-analysis technology to perform system optimization in a novel fashion that promotes design integration among traditional functional disciplines.

1 INTRODUCTION

A basic obstacle to progress in the design of bladed-disk assemblies (fans, compressors, and turbines) is a lack of adequate design tools. First, current design methods incorporate knowledge-based rules (often contradictory) which could possibly be relaxed or eliminated if the physical aspects of engine technology were better understood. Second, current design methods require seemingly endless iterations among functional design groups, particularly between aerodynamic and structural groups.

The disciplines involved in designing a typical bladed disk, shown in Fig. 1, include gas dynamics, structural mechanics and heat transfer, with a substantial expenditure of effort in the areas of materials, airfoil design,

(i) Assistant Professor, Department of Mechanical Engineering, North Carolina A&T State University, Greensboro, North Carolina, USA.
(ii) Project Engineer, Design Systems Technology, Pratt & Whitney, West Palm Beach, Florida, USA.

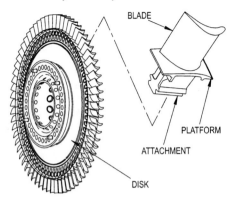

Figure 1: Typical bladed disk from the turbine section of a gas turbine engine. (Courtesy Pratt & Whitney.)

vibration, fatigue and manufacturability. Final designs are typically feasible rather than optimal and design procedures are not fully integrated among functional groups.

Current research efforts to address these issues tend to focus on specific problems within a discipline. (See, for example, [1-6].) However, important as such investigations are, they are unlikely (in our opinion) to significantly reduce design-cycle time because they approach design from a traditional, discipline-specific perspective. Such approaches do not systematically account for the strong coupling among mechanical, aerodynamic, and thermal processes in modern engines.

Developing approaches for systematic, multidisciplinary design requires a better understanding of the physical phenomena associated with the dynamics of bladed disks. Srinivasan [7] gives an in-depth study of blade vibration problems and concludes by emphasizing that design goals cannot be met unless "a dialogue is promoted and maintained among experts in analyses and testing in the fields of structures, aerodynamics, materials, fatigue and fracture, statistics, controls and diagnostics." New design ideas such as these, however, have not fully matured and integrated design tools have not been realized.

In this paper is presented a conceptual basis for a new, systematic, unified approach to bladed-disk design and an outline for future development of the approach as a software-based design tool. Technical tasks associated with this development are summarized. The basic contribution of this paper is to present the problem of bladed-disk design from a non-traditional perspective and to lay a conceptual foundation for true integration of design among traditional functional groups.

While some attempts have been made to approach the bladed-disk optimization problem in a systematic way, no all-inclusive technique has been previously proposed. Taking a system-dynamics approach to the problem leads to a design procedure which treats each component as an energy manipulator. This allows the behavior of the bladed disk to be understood from a perspective different than the prevailing discipline-specific practice. This energy-based perspective is intrinsically multidisciplinary, and holds the promise of improved design optimization.

2 CONCEPTUAL BASIS FOR A NEW APPROACH

In this section are presented the core concepts underlying the new approach to bladed-disk design. An energy-based perspective of physical systems is described and the design of bladed disks is posed as a numerical optimization problem. The relationship of this approach to prior research and current industry practice is discussed.

2.1 A physical-system perspective

A tenet of modern system dynamics is that the fundamental processes underlying a physical system's dynamic behavior are the storage, transmission and transformation of energy among the components of a system and between a system and its surroundings. Physical components are thought of as energy manipulators which, based on the manner in which they are interconnected, process energy injected into the system in a characteristic manner which is observed as the system's dynamic response [8].

For the purpose of illustrating how this approach is applied to the design of a bladed-disk assembly, consider a turbine rotor assembly of a gas turbine engine. As illustrated in Fig. 2, a turbine is an power-transformation system. A large fraction of the fluid power, where the fluid is air and power is the product of pressure and volumetric flow rate, is transformed by the action of the turbine into shaft power, where power is the product of torque and angular speed. The result of this energy-transfer process is that air at the turbine outlet has a lower energy state than the air at the turbine inlet and that the shaft power is sufficient to operate the compressor, fan and auxiliary machinery.

Power extracted from this energy-transfer process is both stored in and

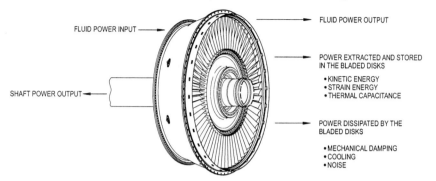

FLUID POWER INPUT

FLUID POWER OUTPUT

POWER EXTRACTED AND STORED IN THE BLADED DISKS

• KINETIC ENERGY
• STRAIN ENERGY
• THERMAL CAPACITANCE

SHAFT POWER OUTPUT

POWER DISSIPATED BY THE BLADED DISKS

• MECHANICAL DAMPING
• COOLING
• NOISE

Figure 2: Typical turbine rotor assembly, considered as a power-transformation system. (Courtesy Pratt & Whitney.)

dissipated by the bladed-disk assemblies. Energy storage takes the forms of kinetic energy due to rotation, kinetic energy due to vibration, potential energy due to strain, and internal energy associated with thermal capacitance. Energy dissipation through cooling, as well as mechanical damping, is an entropy production process where the product of entropy flow rate and temperature is thermal power. Dissipation via noise can be described in terms of acoustic power. Designing the bladed-disk assemblies to reduce the total amount of energy stored and dissipated via these mechanisms improves the efficiency of the turbine's essential function—the transmission and transformation of power. This perspective of turbine operation constitutes a basic framework for the systematic treatment of the entire physical system.

2.2 Proposed optimization problem

It is proposed that the design of a bladed-disk can be formulated as a finite-element based optimization problem where, for a given part or assembly, the objective function is a function of all energy stored in and dissipated by the part. This new energy-based objective replaces the traditional objective in the bladed disk problem which was to minimize weight consistent with constraints on stress, deflection, temperature, and so forth. The solution of the new optimization problem is the geometry or shape of the part. In essence, the design problem is to determine a geometry that minimizes the energy extracted from the energy flow through the rotating assembly while satisfying limits imposed by material properties, manufacturing, performance, and robustness.

The form of the objective function is being investigated. A possible formulation, neglecting dissipation, is total energy E, given by

$$E = T + U + V, \qquad (1)$$

where T is total kinetic energy due to rotation and vibration, U is internal energy associated with thermal capacitance, and V is potential energy due to strain. The optimization problem can be stated: given the load set P acting on the bladed disk, determine the set S of geometric parameters that minimizes the total energy E of the disk subject to the set of equality and inequality constraints C. Loads P can be the result of an approximate or detailed aerodynamic analysis. The set S can include such parameters as blade attachment geometry, rim width and depth, web thickness and depth, and spacer diameter. The constraint set C include limits imposed by material properties and manufacturing, performance limits such as those described on a frequency-speed diagram (Campbell diagram), and heuristic limits representing a manufacturer's knowledge base.

2.3 Relationship to existing design strategies

The proposed energy-based optimization problem unifies essential aspects of current design strategies. First, consider the problem of minimizing structural mass. In its simplest form, the kinetic energy T of a structure rotating about a fixed axis is given by

$$T = \tfrac{1}{2}\dot{Q}^T M \dot{Q}, \qquad (2)$$

where M is a consistent mass matrix and $\dot{Q}$ is a velocity vector.[iii] Minimizing kinetic energy inherently minimizes the mass of a structure at the rotational speeds of interest.

Second, consider deflection as a design criterion. In its simplest form, the strain energy (a form of potential energy V) of a structure is given by

$$V = \tfrac{1}{2}Q^T K Q, \qquad (3)$$

where K is a generalized stiffness matrix and Q is a global displacement vector. For a given value of K, minimizing strain energy V minimizes displacement Q. An upper limit Q_0 on deflection is imposed as an inequality constraint given by

$$Q - Q_0 \leq 0. \qquad (4)$$

A third design criterion is stress, particularly stress concentration at blade

---

(iii) Kinetic energy T is properly a function of momentum while its complement, kinetic coenergy T^*, is the function of velocity given in (2). The distinction is not significant in the present discussion. Further discussion is given in [9].

attachments and average tangential stress for burst margin. Stodola [10] shows that designing for constant strength, considering stress concentrations at the blade attachment only, produces designs with minimum mass moment of inertia, minimum strain energy and minimum deflection. The quantity in the proposed energy-based approach that subsumes such stress-based design criteria is again strain energy V, this time in the form given by

$$V = \int \sigma d\epsilon, \tag{5}$$

where σ is stress, ϵ is strain, and the integral is the area under a stress-strain curve. In the proposed optimization problem, stress concentration is limited by constraining energy density, and stress contours are limited by constraining energy contours.

Fourth, bladed-disk design is assessed according to thermal criteria involving temperature distribution and temperature gradients. A simple form of thermal energy (a form of internal energy U) for solids is given by

$$U = CT, \tag{6}$$

where C is a specific heat matrix and T is a temperature vector. By minimizing this component of an energy-based objective function, energy storage associated with temperature can be included in the general optimization problem.

Fifth, vibration, a critical aspect of bladed-disk design, can be expressed in terms of kinetic energy. The kinetic energy T, due to vibration, of a solid body comprising many discrete differential masses can be represented by

$$T = \tfrac{1}{2}\left(u^T M u\right) \dot{f}^2, \tag{7}$$

where u is a vector of coefficients, M is a mass matrix, and $f(t)$ is a harmonic function [11]. By including a function of this form in the general objective function of the proposed optimization problem, vibratory blade response at resonance can be addressed.

Last, a single energy-based objective function can incorporate both gas dynamics and structural mechanics, leading possibly to breakthrough advances in reducing the lengthy, iterative design cycle that characterizes current practice. This is a significant and difficult task, and is beyond the scope of this preliminary study.

As shown by these examples, significant aspects of bladed-disk design can be treated systematically and concurrently by expressing the basic physics of each in terms of energy and constraint. The main analytical challenges to be overcome in pursuing these goals are determining physics-based expressions for all important design considerations in terms of energy and constraint and developing an energy-based objective function that produces feasible part geometry.

3 AN OUTLINE FOR FUTURE DEVELOPMENT

3.1 Attributes of a new design procedure

A design procedure having the following attributes is proposed. In future work this procedure is to be implemented using commercial software.

a) Generate a parametric 3D solid model of the part as an initial estimate for the optimization problem, using a CAD package for basic geometry. Since part geometry changes with each iteration of the optimization routine, new models and meshes are generated each iteration.

b) Automatically generate a finite element (FE) mesh. Depending on the complexity of the design, this could be as simple as an axisymmetric model, or as complex as a periodic segment with cyclic symmetry and all relevant out-of-plane features.

c) Apply design loads to the solid model and compute resulting stress, strain, deflections, temperature gradients, and so forth for every element, using appropriate software for analysis.

d) Compute the terms of the energy-based objective function for every element and sum over the entire model.

e) Determine a set of geometric parameters that reduces the objective function, using either exploitive (hill climbing) or explorative (genetic-type) techniques, check convergence criteria, iterate.

The new design procedure does not, however, obviate the need for experienced designers and engineers. The procedure cannot add features or create new designs; it can only optimize a given design.

3.2 Technical tasks to be accomplished

Outlined below are the major technical tasks that would have to be accomplished to implement the new, energy-based design tool.

a) Physical systems theory: Determine physics-based expressions for important, discipline-specific, three-dimensional design considerations in terms of energy and constraint.

b) Objective function: Determine an energy-based objective function that produces feasible part geometry.

c) Numerical optimization: Develop a numerical optimization routine suited to a computational problem of this magnitude. Conjugate gradient and method of feasible directions as well as genetic or simulated annealing are candidate methods.

d) Computing environment: Select commercial software and develop the necessary architecture and interfaces to implement the prototype design tool. Write a supervisory program to control the flow of information among software packages and to and from the user.

Tasks (a) and (b) address the basic theory underlying the new approach and are both more challenging and more speculative than tasks (c) and (d). The second two tasks are areas of active research in turbine design and proposed future work can draw extensively on recent advances in these areas.

4 CONCLUSION

The new approach for bladed-disk design presented in this paper is consistent with the view that the dynamic response of a physical system is a consequence of the energy transactions occurring in the system. The proposed numerical optimization problem is of the same order of complexity as current structural optimization strategies and so should be tractable, although the efficacy of an energy-based objective function is unproved. Expressing important design criteria in terms of energy and constraint is conceptually straightforward but may prove to be a difficult task.

The new method promises to unify important aspects of current design methods from many disciplines, laying a foundation for an integrated design approach that could lead to breakthrough advances in bladed-disk design. Assessing the approach waits on future development of a prototype design tool and a comparison of the results of energy-based design to the results of conventional design. The outline of such a development, including the technical tasks to be accomplished, suggests that developing such a design tool is feasible.

Lastly, the authors hope that this study will facilitate interaction among practitioners of the various disciplines involved in bladed-disk design. The manipulation of energy provides a common, physics-based language and perspective that should, to echo Srinivasan [7], promote and maintain a dialogue among experts to meet modern design goals.

Acknowledgments

Our thanks to our many colleagues at Pratt & Whitney who offered critical comments and suggestions and to Bob Brockman at University of Dayton Research Institute for his interest and comments.

REFERENCES

1. ZBOINSKI, G., AND OSTACHOWICZ, W. - General FE computer program for 3D incremental analysis of frictional contact problems of elastoplasticity, Finite Elements in Analysis and Design, Vol. 27, No. 4, pp. 307-322, 1997.
2. CSABA, G., AND ANDERSSON, M. - Optimization of friction damper weight, simulation and experiments, proc. ASME Intl Gas Turbine and Aeroengine Congr. and Expo., 97GT115, ASME, NY, 1997.
3. FRISCHBIER, J., SCHULZE, G., ZIELINSKI, M., AND ZILLER, G. - Blade vibrations of a high speed compressor blisk-rotor, numerical resonance tuning and optical measurements, proc. ASME Intl Gas Turbine and Aeroengine Congr. and Expo., 96GT24, ASME, NY, 1996.
4. NATALINI, G., AND SCIUBBA, E. - Choice of the pseudo-optimal configuration of a cooled gas-turbine blade based on a constrained minimization of the global entropy production rate, proc. ASME Intl Gas Turbine and Aeroengine Congr. and Expo., 96GT509, ASME, NY, 1996.
5. GOEL, S., COFER, J. IV, AND SINGH, H. - Turbine airfoil design optimization, proc. ASME Intl Gas Turbine and Aeroengine Congr. and Expo., 96GT158, ASME, NY, 1996.
6. KODIYALAM, S., KUMAR, V., AND FINNIGAN, P. - Constructive solid geometry approach to three-dimensional structural shape optimization, AIAA Journal, Vol. 30, No. 5, pp. 1408-1415, 1992.
7. SRINIVASAN, A.V. - Flutter and resonant vibration characteristics of engine blades, Journal of Engineering for Gas Turbines and Power, Vol. 119, No. 4, pp. 742-775,. 1997.
8. WELLSTEAD, P. - Introduction to Physical System Modelling, Academic Press, London, 1979.
9. LAYTON, R.A. - Principles of Analytical System Dynamics, Springer-Verlag, New York, 1998.
10. STODOLA, A. - Steam and Gas Turbines, 6th ed., translation by L.C. Lowenstein, McGraw-Hill, 1927.
11. MEIROVITCH, L. - Analytical Methods in Vibrations, Macmillan, New York, 1967.

Parametric Programming for the Non-Specialist Engineers

A.N. Sainak [o]

Geotechnical Engineer

1- SYNOPSIS

Finite Element Analysis (FEA) is extensively sought after, at the present, in many engineering fields. The increase in its popularity is due to lack of closed form solutions or conventional techniques to complex engineering problem and advances in computing speed and cost. Irrespective of this increase in demand, the method remains as a specialist tool which can only be best handled by finite element experts. Engineers untrained in Finite Element Method (FEM) may find the technique difficult to apply. They would certainly not have the competence born of experience to manipulate models to check the validity of the assumptions made and their implications on the results. The method has also been advancing at a much faster pace than is perhaps appreciated by many engineers.

The aim of this study is to examine methods of narrowing the gap between finite element analysts and engineers, in particular civil engineers. These methods are described as Parametric Programming. The paper highlights the main steps taken in the pre-processing and post-processing stages, which can be coded to create interface programs which are easy to use by the engineers. It includes two case studies to show the applications of such programs and the ease and accuracy of the analyses carried out by non-specialist engineers. The paper also describes the steps taken in verifying the integrity and the accuracy of the programs.

---

[o] Geotechnical Engineer
Haswell Consulting Engineers

Civil engineering problems require safe and economical solutions based on the selection of the most satisfactory design for a particular structure. Once a trial design geometry has been produced, the performance of the structure in its likely environment during and after the construction stages must be predicted and assessed to check for adequacy. A suitable computational model should be created to provide some insight into the structural behaviour. A simple problem may permit "back–of-the-envelope" hand calculations, whereas a three dimensional problem may require highly sophisticated three-dimensional analysis by FEA. The choice of the model will depend on many factors, including time, money, expertise and software available and the degree of accuracy required. When a possible solution based on the initial model has been obtained, the implications of the results of the analysis are then examined. If the solution is inadequate the model must be modified and the verification process repeated until a satisfactory safe and economical design is achieved.

Figure 1 shows the stages taken to finalise an FEA solution to an engineering problem. The engineer and the FEA expert must work closely to arrive at a satisfactory solution or design. They may have different perceptions of the problem in hand. The FEA expert understandably does not have all the engineering skills required for complete resolution of structural design problems, whereas the engineer unqualified in FEM techniques finds the method difficult to understand and apply. Hence the need for bridging the gape between the FEA experts and ordinary engineers.

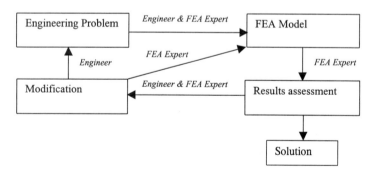

Figure 1 Route of engineering problem to solution and roles of individuals

The technique involves the development of simplified interface programs to generate the data files and handle the result files for a particular engineering problem. These programs require the input of the variables of the problem and the selection of the relevant results. They are developed by the FEM expert working in collaboration with the design engineer. Engineering terms rather than FEM terminology are employed in the programs. The accuracy of the programs should be verified by an appropriate independent institution. The steps taken in the developments are briefly described in the following sections.

An important feature of some FEA packages, such as LUSAS, is the ability to allow the users to automate analyses of structures having similar layout or arrangement. This is done by using the in-built scripting language. After defining an initial model, the scripting language can be used to develop a command file to regenerate data files for variable dimensions and attributes. Graphical menus and on-screen forms are added to allow engineers to enter data and build models of different structural dimensions and properties with the minimum of effort.

3.1 Pre-processing

The first stage in any FEA is the transformation of the physical problem into a computational model. A data file is generated to represent the problem to be analysed. This can be carried out with the aid of a simple and a robust parametric program in which the user is prompted to input dimensions of the structure, the material properties and choice of possible loading conditions. A general layout of the steps taken is shown below and, more details are given in the following case studies.

(1) Control Parameters; standard values to define the defaults used in the problem and type of analysis.

(2) Geometry; the dimensions of the problem are used in defining the nodal positions e.g.
DEFINE POINT PN=1 X=4*WIDTH Y=6*DEPTH Z=4*LENGTH

(3) Mesh; generation of mesh depending on the required accuracy and available computing facilities e.g.
DEFINE MESH BY_NAME IMSH=1 FEATYP=SURFACE LNAME=QTS4
MSHTYP=1 NDIVX=KH*P NDIVY=KV*1

(4) Boundary Conditions; definition of fixity of all parts of the model and their transition and rotation e.g.
DEFINE SUPPORTS ISUP=2 U=1 V=0 W=1 THX=0 THY=0 THZ=0 THL1=0
THL2=0 PHI=0

(5) Material Properties; choice of: initial properties and their variation in all directions, constitutive laws to govern the analysis and allocation of these properties to elements of the model e.g.
DEFINE MATERIAL 15 1 1 YoungF*(1+4*Ef/100) PossionF MassFs

(7) Loading Conditions; definition of type and size of loads applied, such as gravitational, dynamic, static, thermal …etc. e.g.
DEFINE LOADING ILDG=1 LTPF="Constant Body Force" AY=-9.81

(8) Data File; generation of data file from all above.

3.2 Post-processing

When the data file is analysed by an FEM package, the next stage is the interpretation of the voluminous amount of results. A comprehensive parametric program can handle the generated results files and extract the relevant output. This can be carried out by replying to a series of prompts giving the choice of the type, location and format of the required results. A general layout of possible routes to obtain and present the required results is shown in Figure 2 below.

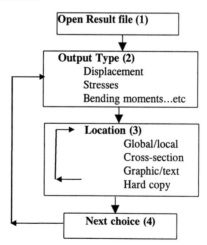

Figure 2 Flow chart to extract relevant results

770

To illustrate the application of the parametric programming technique, discussed in the previous section, two case studies are presented in this work. They are three-dimensional finite element analyses of buried rectangular and conical water treatment tanks. The case studies involve the modelling and analysis of soil/structure interaction problems. It should be noted that in these studies the soil had to be modelled elastically, with all the modelling limitations that this implies. However these limitations are considered acceptable in terms of their implications for the particular studies detailed here. It should be noted that the solutions to these problems are currently subject to the verification process referred to in Section 1 above.

The two example parametric programs included here were prepared using the LUSAS FEA package. Parametric programming is made easy in the LUSAS package where a fully featured parametric language is contained within MYSTRO, based on the syntax of the C programming language. The facility can be used in MYSTRO command files to prompt the user for a choice, an answer or to insert values. The command file can then build parts of the data file or locate the required results.

The design of underground water treatment tanks has traditionally been based on conventional hand calculation procedures which take into account the behaviour of the soil and of the structure independently. Hand calculation procedures also require three dimensional problems to be idealised in two dimensions. These two factors lead to inaccuracies in the soil/structure interaction and the effect on the structure. Parametric modelling of the soil and structure can combine the soil and the structure in one three-dimensional model, allowing engineers and analysts with only a basic knowledge of finite element methods to create tank models, perform analyses and produce results

With these parametric programs tank designs variables such as wall and slab dimensions, number and location of piles, concrete properties, backfill properties, and number of soil horizons are all entered via on-screen user-defined forms. Various loading conditions can be selected allowing for any number of the following: hydrostatic pressures from differing water levels in adjacent tank bays, surcharge, uplift due to pore water pressure and piled foundations with tension and/or compression capacity.

771

These are large reinforced buried concrete structures. They are designed to withstand high external forces, such as groundwater uplift and internal forces due to the water column in the tank. Tension and/or compression piles may be required depending on the ground conditions.

A pre-processing parametric program was developed to create the data file following the methodology described in Section 3. In step (1) the analysis control parameters and all variables are defined. Job title and other particular information relating to the job are also inserted. In step (2) the physical dimensions of the tank (length, width, depth, wall thickness and pile distribution and size) are input to generate the geometry of the numerical model, see Figure 3. The "features" of the model are defined by mathematical equations relating the actual physical dimensions of the tank to the position and size of these features. A visual check on the interpretation of these relationships into the FEA model is essential at this stage. Once a satisfactory model of the geometry is obtained, the next step (3) is to generate mesh with the appropriate type and distribution of elements. The choice of mesh size will be limited by the capacity of the computer and, together with the distribution of elements, will influence the accuracy of the solution. In step (4) the boundary conditions are decided to define the mechanism of the various parts of the structure. In step (5) the material properties of the soil and the concrete are input and assigned to the appropriate parts of the model and any variations with depth are allowed for. In step (6), which is the final stage in the computational modelling is the application of the loads to the structure. The loads are: self-weight, lateral pressures from backfill and pore water pressure on the walls and the base, water columns in the tanks and surcharge. These loads can be applied in all possible combinations. Figure 4 shows a cross-section of the representation of the problem by a numerical model.

Please input/modify dimensions of tank	
Length of tank (m)	72.6
Width of tank (m)	77.5
Depth of tank (m)	7.04
Thickness at top of outer wall (m)	0.3
Thickness at bottom of outer wall (m)	0.5
Thickness of slab (m)	0.8

OK   Cancel

Figure 3 A front page of a pre-processing parametric program for rectangular tank

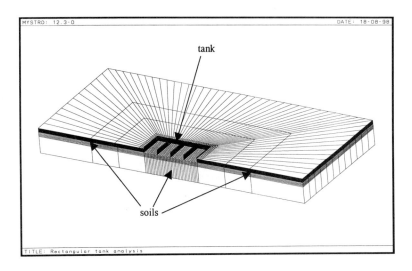

TITLE: Rectangular tank analysis

Figure 4 Cross-section of computational model

A second parametric program was developed to extract and present the results needed in the design process. The steps described in Section 3.2 were implemented. Bending moments, shear forces and deflections can be produced for the slab and each wall of the structure, see Figure 5.

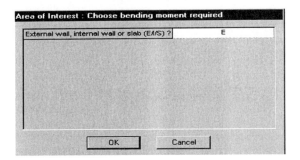

Area of Interest : Choose bending moment required

External wall, internal wall or slab (E/I/S) ?   E

OK   Cancel

Figure 5 A front page of post-processing parametric program for rectangular tank

4.1 CONICAL TANKS

A pre-processing parametric program was developed to generate the data files for a conical tank following the same steps described in the previous section. The only differences occur in step (2), where the physical

773

dimensions of the tank are: radius, depth and thickness of perimeter wall, sloping angle and thickness of slab, radius and depth of hopper. Figures 6a & 6b show the physical dimensions of the tank and a cross-section of the numerical model.

A post-processing parametric program was written to extract and present the results needed in the design of the various parts of the structure. Bending moments, shear forces and deflections can be produced for the perimeter wall, the slab, and the wall and the base of the hopper. Deformations and stresses in the model can also be extracted.

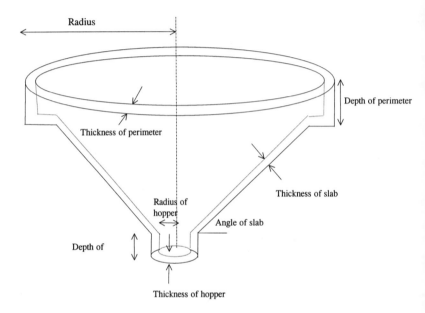

Figure 6a Geometry of tank

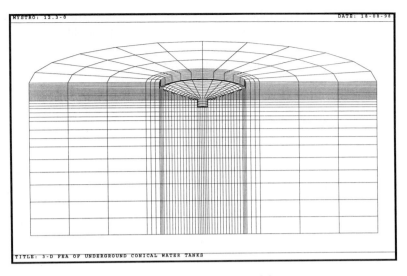

Figure 6b Cross-section of computational model

(5) Verification and testing

The parametric programs described in this paper are aimed at assisting Haswell civil engineers in the design of buried water treatment tanks. By their use, FEA is being introduced with a view to reducing the cost of these structures. Parametric programs like any other software can only serve their purpose when they are properly validated and verified.

The validation of the programs should be carried out in three stages:

1. The functionality of all built-in empirical relationships and/or calibration factors must be checked to ensure that the software is doing what is intended.
2. The data entered into the program must be validated; in particular the soil properties interpolated from site investigation and laboratory tests. The soils data must be validated by a geotechnical engineer.
3. The results must be compared with closed-form solutions, where possible, and all variations from hand calculations must be justified.

The programs must be accompanied by sufficient documentation stating all limitations, so that all users will be in no doubt as to how they work. Benchmark tests are also needed to establish confidence in them. The user should also be *knowledgeable* in soil mechanics, structural analysis and numerical analysis, as well as being conversant with the software (ref. 1*)*.

775

When the validation stages have been satisfactorily completed, the software must be fully tested for ease of use. It is also highly recommended that a specialist institution or individual experts carry out verification of the programs. The verification should cover stages 1 and 3 of the validation detailed above and all recommendations must be implemented.

(6) Conclusions

This work explores a technique in which the application of the finite element method can be simplified for practising engineers. It shows the steps taken to develop interface programs to generate the data files necessary for the FEA and to extract relevant results from the output files. It includes two case studies to illustrate the ease with which parametric programs can be written and executed. However, validation and verification stages are essential to establish reliability of this type of programming.

The use of parametric programming enables the FEM to be more readily introduced to problems within the Civil Engineering industry. It also helps to close the gap between specialised users of the method and practising engineers having real design problems requiring other than traditional calculation methods.

By using parametric programs for tank design many of the simplifications made in conventional analyses are unnecessary and more accurate results can be achieved. The soil/structure interaction can be investigated both globally and locally in 3D with stresses, strains, bending moments and displacements being obtained. The soil and structure can be analysed in one model with easy modification of geometry and material properties.

The use of parametric modelling techniques reduces the need for specialist finite element expertise and makes repetitive analyses easier to perform for less experienced users such as civil engineers.

(7) References

1. CRILLY, M.- Confirmation service, Report on the BGS meeting " Validation of geotechnical software for design" held at the ICE, Ground Engineering, November 1993.
2. SAINAK, AN - Parametric programming for three-dimensional analysis of underground fluid retaining tanks, The 13th LUSAS User Conference, September 1997, U.K., PP27-34.
3. SAINAK, AN and AJZENKOL, D - Programming for three-dimensional finite element analysis of underground fluid retaining structures, Ground Engineering, May 1998, PP30-32.
4. SAINAK, AN and JP Paterson – Bridging the gap between engineers and finite element analysts, The 14th LUSAS User Conference, September 1998, U.K., PP93-100.

Simulation Based Design: A Case Study in Automotive Application

T. Gielda

Visteon Climate Control System Division, Ford Motor Company, Plymouth MI

B. Webster

Simmetrix, Fenton MI

F. Shakib

AcuSim Software, Saratoga CA

Johnson, H. Fong

HAL Computer, San Jose CA

To be more competitive in the automotive marketplace it is imperative to reduce the time required to bring the vehicle to market. To accomplish this task, Visteon Climate Control Division is becoming more reliant on Computer Aided Engineering (CAE) methodologies to reduce the amount of physical testing performed on prototype vehicles. This effort paper will report on the development of an interdisciplinary design program to predict total vehicle thermal management, including passenger thermal comfort. The Unified Parametric Vehicle (UPV) system was developed to perform virtual testing of new vehicle platforms. In this paper the UPV system was utilized to perform vehicle thermal management analysis for a Taurus class vehicle. The UPV simulations will include engine cooling and climate control systems performance predictions.

Visteon Climate Control Division is committed to provide quality and value to our customers. To maintain customer satisfaction it is imperative that our vehicles provide an interior environment that maintains thermal comfort regardless of the external environment. To meet this requirement we have developed CAE engineering tools that can be used to analytically predict the thermal management of the vehicle, including passenger thermal comfort.

The UPV system is used to optimize climate control and engine cooling system performance months to years before the first prototype is built. UPVs not only predict the climate control system's objective performance but it also will predict how the passenger will "feel" in the vehicle. This is of tremendous value to Visteon because it allows us to put a customer satisfaction value on all improvements/degradation in objective performance. To perform this type of total vehicle analysis the following CAE methodologies were seamlessly integrated into a single operating environment.

1. Parametric Solid Model Representation of the Vehicle Interior/Exterior
2. Automatic Geometry Based Mesh Generation (Including Viscous Layer Resolution)
3. Solar Load Prediction Model (Required for A/C System Development)
4. Engine Warm-Up & Refrigeration Cycle Analysis Tools
5. Transient CFD Analysis of the Vehicle Interior and Exterior.
6. Objective-to-Subjective Human Comfort Correlation.

The key to he UPV system is the parametric solid model representation of the vehicle. Prior to UPV, geometric model preparation required anywhere from 3-6 months of engineering effort. With UPV, we can build a complete vehicle model in as little as 1 week. Figures (1-2) depicts the typical level of detail in a UPV geometry.

Figure (1) Side view of typical UPV model geometry.

Figure (2) View under typical UPV model geometry.

Once the vehicle geometry is constructed we will run a complete set of virtual wind tunnel tests for the vehicle. A summary of the type of vehicle evaluations we simulate with UPV is shown below:

1. Engine Cooling Evaluation
2. Heat Protection Evaluation
3. Heater System Testing
4. A/C Performance Testing.

The UPV system has achieved levels of solution accuracy that approach the repeatability of our testing procedures. An example of the accuracy of the UPV predictions for an A/C system evaluation is shown in figures 3-6.

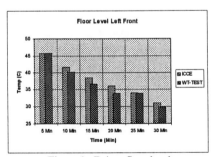

Figure 3. Driver floor level.

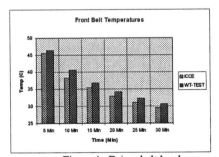

Figure 4. Driver belt level.

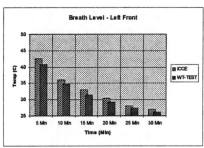

Figure 5. Driver breath level.

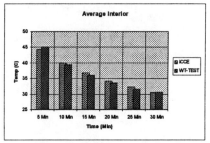

Figure 6. Vehicle average interior.

Figures 3-6 depict a comparison of the computed and measured floor, belt, breath and average interior temperatures during a standard A/C pulldown test. Figure 7 depicts temperature contours in the plane of the driver, at the 30 minute point of an A/C pulldown test.

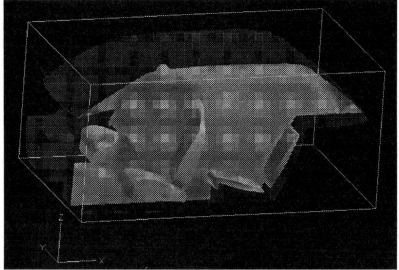

Figure (7) Temperature contours in driver's plane at 30 minute point of a/c pull-down test.

The A/C simulation results, shown above, were completed in approximately 1.5 hours on 8 of 16 processors of a HAL 375 cluster. The AcuSolve solver, from AcuSim, was utilized as the CFD engine.

In the final paper we will demonstrate a case study on how the UPV system can be utilized to improve vehicle thermal management performance. We will show results for engine cooling, heat protection and climate control system analyses. In addition we will provide comparisons of our computed solutions to the test results collected as part of our UPV calibration program. A full disclosure of the CFD methodologies, mesh generation and HAL cluster implementation we be discussed.

Using Mechanical Event Simulation in the Design Process

Ulises F. Gonzalez, Michael L. Bussler

ALGOR, Inc.

ABSTRACT

In this paper, we show how to incorporate Mechanical Event Simulation (MES) into the design process of mechanical components. MES relies heavily on nonlinear Finite Element Analysis (FEA). Our goal is to enhance the design process by considering the event that the part in question will undergo. Specifically, we aim to utilize virtual experiments on the computer to verify the validity of engineering designs. These experiments should simulate the interaction of the part with its surroundings.

Typical engineering designs rely on isolating the part from its surroundings by introducing boundary loads and conditions. These boundary constraints are relatively easy to determine for parts that experience only static conditions. Linear FEA has proven to be a valid tool in the design of such "static" components. For problems involving motion, engineers generally have also relied on linear FEA, but have had to both (i) utilize rule-of-thumb methods to estimate the magnitude of these loads, and (ii) artificially constrain the location of the part in order to ensure that the linear limit is not exceeded. The effects of a drop test represent a typical problem in which linear FEA has been utilized to analyze a dynamic event. By embracing MES, engineers are freed from having to make such compromises.

MES relies heavily on nonlinear FEA which has been regarded as notorious for its instabilities and computational inefficiencies. These drawbacks to nonlinear FEA have, in the past, greatly limited its use. Because of our ambitious goal to include nonlinear FEA in the design of a wide range of mechanical components, we must mitigate these drawbacks.

In this paper, we present results obtained with a nonlinear FEA processor whose state-of-the art adaptive solution method and fully automatic time- stepping scheme remove the stability concerns. These state-of-the-art techniques combined with the introduction of a new type of element, which we refer to as "kinematic," significantly reduce solution times. This robust and efficient processor is the first of our three-part approach to using MES to guide designs. Secondly, because we are particularly interested in dynamic events, the processor must be able to simulate large-scale motion and its consequences, from parts making contact to their fracture. This brings us to our third and final part: to closely duplicate material behavior. Besides fracture and the standard material models, we have added material-based damping. This damping allows us to consider real-world problems where such dissipation is always present.

The types of real-world problems that we can solve are extensive. As an example, we consider the design of a tank. We focus on the validity of the design if we demand that the tank's structural integrity withstand a high-speed collision. Our processor lets the engineer simulate the entire virtual event during which the intact tank collides and possibly even ruptures. From the results of this virtual experiment, the engineer can easily determine the adequacy of the design. Thus, the engineer can focus on the physics of the problem, rather than concentrating on modeling approximations such as the ad hoc estimation of critical loads.

1. INTRODUCTION

MES represents a paradigm shift in the design of mechanical components. It allows engineers and designers to simulate the actual conditions that a mechanical component will experience; that is, the event associated with its application. MES differs significantly from the general practice of using linear static FEA in the design process. Linear static FEA is the culmination of methods introduced by Galileo and da Vinci [1]. Even though it is well understood that linear static analysis has limited applicability, such methods are still an integral part of an engineering education. Linear methods are even used to model motion using pseudo-dynamic analyses that completely neglect inertia. MES is a true dynamic analysis tool because it considers inertia and is not based upon a linear static perspective.

The use of linear static FEA is generally rationalized by considering harsher than expected conditions. This is accomplished by using safety factors in conjunction with engineering estimates of the mechanical loads. When the component in question will experience only static conditions, linear static FEA may suffice as an analysis tool.

Of course, most static situations are the result of an event that included motion. MES is geared toward the design of components that experience motion, from the event that precedes a static situation to continuously moving operations. When motion exists, not only are the loads no longer constant, but their magnitude and direction are not easily estimated. MES completely bypasses the need to estimate these loads. That is because MES accounts for both (i) the interaction of the component with its surroundings, and (ii) the inertial forces generated by the motion of the component itself. Before describing how MES uses nonlinear FEA, it is important to show how even conservative estimates of loads in the design of moving components can result in severe under-prediction of mechanical stresses.

Dropping a nearly rigid object onto a flexible, simply supported beam clearly demonstrates how difficult it is to estimate loads when motion is involved. We are simply interested in obtaining a value for the maximum value of the stresses within the beam. The estimate that we need is the maximum impact force that the falling object has on the beam. Classical beam theory can then be employed to yield the maximum stress. Of course, classical beam theory is only appropriate under static conditions, but we expect the stresses within the beam to be much more dependent on displacement (deformation) than on the details of the motion. Specifically, we assume that the stresses are insignificantly affected by the inertia of the beam. This assumption will be revisited later in this section when we discuss the results of an MES of this problem.

The analysis is based on both energy conservation principles as well as on classical beam theory; a more detailed description of the analysis can be found in [2]. In order to obtain a closed-form analytical solution, we consider a problem with a simple geometry. The problem consists of dropping a small 4.0 lb. weight, W, from a height, H, of 1.0 in. onto the center of a simply supported steel (Young's modulus, E, of 30×10^6) beam with a length, L, of 23 in. and a circular cross-section with a diameter, d, of ½ in. The weight is assumed to make contact only along a plane at the middle of the beam and perpendicular to its primary axis. Finally, without sacrificing the accuracy of the results, the dropping weight is assumed to be rigid. In order to further verify our results, we also conduct a physical experiment of this problem.

We first pursue the classical physics solution for the maximum impact force. A simple energy balance between two end states can be used to obtain the maximum displacement of the beam. The initial state is defined by the weight, W, at rest at height H above the beam. The final state consists of the beam at its highest deflection and the weight having no velocity - the point at which the velocity of the weight changes

785

direction. This energy balance gives the following expression for the maximum displacement of the center-point of the beam

$$\Delta_{max} = \frac{W + \sqrt{W^2 + 2KHW}}{K},$$

(1)

where K is the classical bending stiffness of the beam. As long as we assume linear behavior (which is consistent with our approach), this bending stiffness is given by

$$K = 48\frac{EI}{L^3},$$

(2)

where I is the area moment of inertia. For a circular cross-section,

$$I = \frac{\pi d^4}{64}.$$

(3)

Equation (1) can be combined with Hooke's law ($F=K\Delta$) to yield an expression for the maximum force,

$$F_{max} = W + \sqrt{W^2 + 2KHW}.$$

(4)

Inserting the numerical values given above into Equations (2) and (3) yields: K = 363.1 lb/in and I =0.003068 in^4. Inserting these values into Equation (4) yields a value of 58.1 lb for the maximum force. This is a striking result; how can a 4.0 lb weight generate more than 58 lb of force upon such a short drop? It should be noted that the maximum deflection Δ_{max} was only 0.160 in (0.7% of the length of the beam), thus we were well within the linear range. Finally, beam theory also gives us the surprisingly high value of 27,200 lb/in^2 for the maximum stress within the beam.

A Mechanical Event Simulation (MES) of the same problem was conducted. This provides for a numerical verification of the analytical solution and vice versa. Under this type of analysis the assumptions inherent to classical beam theory are not necessary. Thus, the beam can possess mass and, hence, inertia. Furthermore, the weight is not modeled to be rigid; instead it is assumed to be made of steel (E of $30x10^6$ lb/in^2). Contact elements were utilized to model the impact between the mass and the beam. Details regarding contact elements are discussed in section 2.

The MES yields a maximum force of 51.7 lb and stress of 24,700 lb/in^2. Note how the analytical solution yields slightly higher values for both the maximum force and stress. We hypothesize that the primary reason for the discrepancy is that the analytical solution assumes that the beam was massless. In the MES, as in reality, the beam possesses mass. This mass combined with the motion of the beam generates inertia, which absorbs some of the impact energy. We verified our hypothesis by performing a second MES in which we assumed the beam to be massless – an unphysical, yet mathematically valid assumption. This latter MES yielded a maximum impact force of 57.6 lb, which is within 1% of the analytical result. The physical experiment of the problem yielded a maximum impact force greater than 50 lb. Even though all three methods used to obtain the maximum force yielded slightly different results, we are confident that the short drop did generate at least 50 lb of force.

The example just discussed certainly demonstrates how the design of a component that will experience motion requires special handling. This example has been configured so that analytical methods would be applicable. Real-world problems are rarely that simple, and thus require numerical analysis. One could opt for linear FEA combined with significant loading related estimates, or one could choose MES. The latter choice does require a larger computational effort , but does mitigate the number of engineering man-hours spent performing estimates and assumptions. In the following sections of this paper, we discuss how we have enhanced nonlinear FEA to the level that a typical designer can use MES. This enhancement has required that the nonlinear FEA processor "underneath" MES be efficient, robust and able to simulate a wide range of mechanical scenarios. We first discuss how the processor is robust. Then we tie this robustness to how MES is efficient from both a user-intervention as well as from a computational perspective. The use of MES not only demands the simulation of motion, but also of its aftereffects. We discuss two such effects: impacts and accurate descriptions of material behavior. Finally, we discuss how MES is used to design an actual mechanical component.

2. MES SOLUTION METHOD

MES is heavily based on nonlinear FEA, which has generally been regarded as both highly unstable as well as computationally intensive. Both of these drawbacks have been addressed in the development of MES. The instabilities inherent to nonlinear FEA can usually be eliminated using a smaller time-step size, Δt. Of course, the smaller Δt, the larger the computing time. This trade-off between stability and efficiency can be significantly diminished by incorporating an "intelligent" (automatic) algorithm to change Δt throughout an event.

MES utilizes such an algorithm; Δt is automatically reduced during critical times and again increased if/when conditions no longer warrant such temporal detail.

The heart of the algorithm is the method used to identify "critical times." We use the rate of convergence of the nonlinear iterations to gauge stability. Such iterations are required to achieve mechanical equilibrium at every time-step. The rate of convergence is determined by examining how the iterative residual changes from one iteration to the next. During "critical" periods this residual will typically not experience a monotonic decline. Conversely, when the residual experiences monotonic declines, critical events are rarely occurring. These two types of behavior by the residual can be used to formulate criteria to trigger Δt decreases and increases. It should be mentioned that the scheme does not just reduce Δt at critical times, but it backtracks in time in order to ensure that the entire critical period is captured. Finally, Δt is also decreased during highly critical points at which not even one iteration is obtainable. Such highly critical points (i.e., the onset of buckling) generally require successive Δt reductions until a stable solution is obtained.

The automatic time-stepping scheme does have the added consequence of making MES efficient from the perspective of total analysis time. Specifically, the scheme diminishes user intervention by removing the common practice in nonlinear FEA to manually and intelligently change Δt throughout an analysis. Nevertheless, we still have to contend with the large computational efforts usually required by nonlinear FEA. After all, in order to achieve mechanical equilibrium, each time-step requires the solution of at least one equivalent linear FEA problem. Nonlinear mechanical equilibrium is achieved using Newton-Raphson iterative methods that generally require the solution of several linearized forms of the nonlinear problem [3].

In order to diminish run-times, MES includes a new type of continuum finite element. This element type, which we refer to as "kinematic," does not experience strains, and thus does not report stresses. Otherwise, kinematic elements behave just like their equivalent 2- or 3-D flexible counterparts. That is, they can have mass, have loads applied on their nodes and/or faces, and, more importantly, experience motion. Their advantage over their flexible counterparts is that they barely contribute to the size of the global stiffness matrix; thus their use can greatly improve run-times. Kinematic elements are particularly applicable in models with an appreciable number of elements known to experience insignificant deformations, but large scale motion, during an event. Such prior knowledge of a model's behavior does require some engineering expertise or prior experience with the model. Designers generally possess this knowledge. Hence, kinematic elements can be an integral part of using MES in the design of moving parts.

Using MES to design moving parts requires that the underlying nonlinear FEA processor be able to handle the consequences of motion. In this section, we concentrate on contact/impact. In the next section we discuss other possible consequences related to material behavior. MES supports two types of contact/impact: (i) surface-to-surface, and (ii) surface-to-rigid plane. Surface-to-surface interactions are accomplished using contact elements. These elements allow for any surface of any solid (including 2-D representations) to interact with any other such surface (including itself) during an event. Actually, it is even possible to begin an event with surfaces already in contact. This surface-to-surface interaction is particularly useful when designing mechanical assemblies. The surface-to-rigid plane contact provides for a highly computationally efficient means to simulate objects impacting and/or sliding on flat surfaces. The MES of a standard drop test represents an ideal application of this form of contact. It should be mentioned that both forms of contact are formulated using energy conservation principles.

3. NEW MATERIAL MODELS FOR MES

A further consequence of requiring an accurate simulation of motion is proper material modeling. Besides the material models generally handled by nonlinear FEA, we have focused on three specific kinds of material behavior: (i) detailed stress-strain curve with yielding, (ii) fracture triggered at threshold stress level, and (iii) mechanical dissipation through stiffness-dependent damping.

Case (i) is a generalized, curve description, form of the standard (bilinear) von Mises material model [4]; the curve portion is for strain levels above yield and is represented using a sequence of line segments (see Figure 1). This added detail makes the curve description model better capture material behavior beyond the yield point.

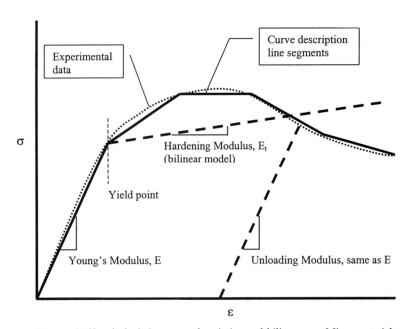

Figure 1: Sketch depicting curve description and bilinear von Mises material models. Note how both models behave identically, and in a linear manner, for strains below the yield point. Additionally, both models unload using the same technique. The curve description model can be made to closely follow the experimental data beyond the yield point. Even though in the sketch this portion of the curve is modeled using four segments, the model has no such limitation.

Case (ii) consists of allowing elements to lose their stiffness once a user-specified tensile and/or compressive breaking stress level is reached for a given material. This procedure is relatively simple to implement, but the sudden nature of such breaking/fracture does require the robust solver discussed in section 2. This ability to have elements "break" has allowed us to model the fracture of brittle materials. Figure 2 shows the results of an MES of shattering glass.

Case (iii) is an element-based form of Rayleigh damping [5] used to model mechanical dissipation within individual elements. Specifically, we aim to model the mechanical energy losses incurred solely because of material deformation rates without considering heat transfer details. Including such details is generally unnecessary and would significantly complicate as well as lengthen the duration of an analysis. Instead we chose to adapt standard damping theory to simulate the loss of mechanical energy into thermal. We refer to this form of energy loss as material-based damping. This damping is modeled as proportional to an element's strain rate and to its stiffness matrix, but not to its mass matrix.

Making the damping proportional to the latter matrix would have resulted in the damping of free motion – which is not how physical dissipation manifests itself. To utilize this material-based damping requires the specification of a multiplier. This multiplier is analogous to the stiffness-dependent parameter in Rayleigh damping. We take the multiplier to be a constant independent of operating conditions and to depend solely on the type of material. Thus, making this constant a material property.

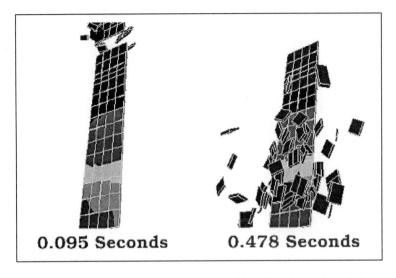

Figure 2: Results of an MES of a moving object shattering a plane of glass.

This material constant can easily be determined using a cantilever beam arrangement. Consider a beam mounted vertically on a floor; the primary axis of the beam should coincide with the direction of gravity. A mass should be affixed on the free end of the beam. The experiment consists of (i) manually displacing the free end of the beam a small distance perpendicular to the primary axis, and (ii) then letting the beam oscillate freely while measuring the displacement of the mass. To obtain the constant, several MESs of the experiment are conducted; each MES having a distinct value for the material constant for the beam's material. The MES whose value best matches the experimental data provides the value for the constant. The results of three such MESs are shown in Figure 3. Each MES uses an aluminum alloy 2024-T4 (with Young's modulus of 10.9×10^6 lb/in^2, poisson's ratio of 0.397, and specific weight of 0.101 lb/in^3) rectangular beam with dimension 20 by 1½ by ¼ in. The manual displacement was simulated using prescribed displacements in the direction of the minimum dimension of the beam. From Figure 3 one can observe the symmetric form of each time-trace. If the beam had been affixed horizontally, gravity would have caused an asymmetry – an unnecessary complication. It should be noted that because our material damping model does not include dependence upon operating conditions, the value obtained for the constant is most appropriate when simulating events under similar conditions (i.e., temperature) to that of the experiment.

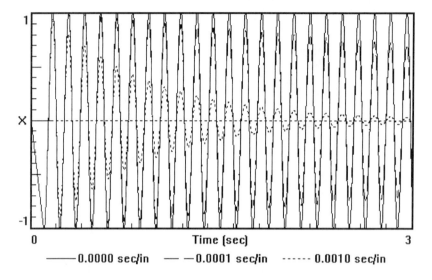

Figure 3: Results of MES of three cantilever beams. Each beam has a different value for the material-damping constant. The ordinate (X) corresponds to the displacement of the mass attached at the end of each beam.

4. USING MES IN THE DESIGN PROCESS

In the presentation corresponding to this paper, we discuss how MES can be used to design a tank. We focus on designing the tank such that it can withstand a high-speed collision when full with liquid. The collision is modeled using the surface-to-rigid plane contact discussed in section 2. The kinematic elements also discussed in section 2 are used only in portions of the tank known to experience insignificant deformations during the impact. These portions were identified during the design process by considering simpler versions of the model. The remaining portion of the tank is modeled using flexible elements with the curve description von Mises material model. As mentioned in section 3, this material model accurately describes post-yield behavior. The collision will certainly result in significant yielding, but the tank is designed to maintain its structural integrity. Specifically, care is taken so that ultimate (breaking) stresses are not reached. Thus, the MES does not need to account for the fracture of the material. Of course, we could consider an

MES where fracture does occur, and observe how the tank ruptures. This would be a possibly dangerous physical experiment, but certainly a very safe virtual experiment.

The liquid is represented using hydrodynamic elements. These elements allow for the simulation of the interaction of fluids with solids without considering the details of the flow. Such interaction is typically done using hydrostatic pressure loads; hydrodynamic elements provide for a much more accurate representation of fluid-solid interaction because they account for the inertia of the moving fluid.

This design clearly demonstrates how appropriate it is to incorporate MES into the design process. There is no need to estimate loads; all that is needed is a description of the operating conditions required by the design. So instead of spending valuable time making engineering approximations, the designer performs virtual experiments of the part "in action" using MES to gauge the validity of the design.

REFERENCES

1. Timoshenko, S P - History of Strength of Materials, Dover Publications, Inc., New York, NY, 1983.
2. A Finite Element Method for Problems Involving Motion, Algor Design World, Second Quarter 1998 (issue), 1998.
3. Spyrakos, C and Raftoyiannis, J – Linear and Nonlinear Finite Element Analysis, Algor, Inc., Publishing Division, Pittsburgh, PA, 1997.
4. Malvern, L E - Introduction to the Mechanics of a Continuous Medium, Prentice-Hall, Inc., Englewood Cliffs, NJ, 1969.
5. Bathe, K J – Finite Element Procedures in Engineering Analysis, Prentice-Hall, Inc., Englewood Cliffs, NJ, 1982, pp. 528.

ENHANCING THE CAPABILITIES OF VENDOR SUPPLIED POST PROCESSING SOFTWARE, AND ALTERNATIVE PRE AND POST PROCESSING METHODS - EXAMPLES AND BENEFITS.

A.Morris CEng, MIMechE.

Stress Engineer - Marine Power Division - Rolls-Royce Plc, Derby, UK.

Abstract

In this paper alternative methods of pre and post processing Finite Element models are described. Two examples are used to illustrate the benefits. The first example describes a method of automatically generating and analysing multi-variable Finite Element models. The second example describes a method of post processing user defined variables in impact analyses with the DYNA3D Finite Element code. Current vendor supplied software packages do not allow the post processing of user defined variables.

Example 1: Automatic Multi-variable Finite Element Analysis.

The analysis of pressurised water reactors is complicated by the influence of several key variables, which affect the inelastic stress response. The ABAQUS Finite Element code is used to model the inelastic stress response within the reactor. The analytical method involves automating the Finite Element model generation, sequential solution and post processing of results, within one controlling program. The controlling program is coded in UNIX C shell script. Automating the analytical route is required due to a combination of financial constraints, the necessity to ensure quality input data to a nuclear safety justification, and the considerable number of anticipated analyses. The UNIX C shell script program is supplemented by FORTRAN coding which provides links to the ABAQUS Finite Element software.

The pre processing phase involves the automatic generation of different Finite Element input decks from one nominal input deck. One set of analyses at one location within the reactor core requires the generation of

52 Finite Element models, with up to 15 variables in each model. The Finite Element models are solved sequentially and the post processing of stress data is achieved by direct access to the ABAQUS results files via FORTRAN coding.

Conventional methods of pre and post processing a set of 52 models take eight days elapsed time, the alternative method outlined reduces this to 16 hours. These methods can be adapted to suit other engineering problems and Finite Element codes.

Example 2: Visualising User Defined Solution Variables In DYNA3D Impact Analyses.

At the marine power division of Rolls-Royce the structural integrity under impact conditions is assessed against a conservative strain based failure criterion. An assessment of the integrity, against a more complex 'constraint' based failure criterion, could lead to more cost-effective designs. Currently DYNA3D is used for impact analyses, the standard post processing software packages do not have the facility to process user defined solution variables.

The ability to post process and contour such variables would greatly aid the interpretation of complex failure criterion and enhance the communication of the results to third parties. Initial methods development has been based on FORTRAN coding which accesses dummy ABAQUS results files. This enables the contouring of the variables with standard software packages. This method has been adapted from the methods discussed in example 1, and emphasises the generic nature of the methods discussed in this paper.

Keywords - Stress, Nuclear, ABAQUS, DYNA3D, UNIX, FORTRAN, Methods.

1.   **INTRODUCTION.**

This paper intends to show that cost-effective improvements in analytical performance can be achieved by more efficient use of existing finite element software. The methods described in the paper result in the following benefits:

- Reduction in total analysis time.
- Improvement in the accuracy of the analytical work.
- Extension of the capabilities of vendor supplied post processing software.

- Greater awareness of the potential for the cost-effective generation of improved analytical methods.

Finite element analysis usually involves the use of three types of software, pre processing, analysis and post processing. It is common practice to use software from different vendors. The choice may depend upon influences such as, corporate policy, software facilities, licence cost etc. The software is designed to operate within a variety of operating system environments, and with other software products.

For some types of analysis the extent of software diversification may result in a reduction in overall analysis performance. This is particularly relevant for analyses that require the assessment of the influence of several variables on the response. The overall analysis performance can be greatly improved by integrating these software packages into what is essentially a 'new product'. The first example in this paper describes an alternative solution method based upon the integration of different software packages.

Occasionally the best combination of standard software packages fails to enable the analyst to access the required results. Example 2 illustrates that cost-effective solutions can be obtained by enabling the interaction of different finite element codes. In both examples the coding has been validated by checking the output against that obtained from conventional procedures.

2. **EXAMPLE 1: AUTOMATIC MULTI-VARIABLE FINITE ELEMENT ANALYSIS.**

The method described in this example has been developed as part of the structural integrity safety justification of pressurised water reactors. The method can be modified to accommodate the requirements of other safety critical industries. Significant reductions in overall analysis time and improvements in accuracy are achieved. The main features of the method are as follows.

- Use of the operating system to provide complete control of the analysis process.
- Automating the generation of several finite element model variants based on one nominal model.
- Using a user defined post processing program to bypass the standard post processing route; this enables immediate access to analysis results.

Due to the confidential nature of the work the following description has been declassified to enable publication.

In these analyses the key response is the inelastic stress at critical locations within the reactor core. There are 15 key variables in the analytical finite element model. Statistical experimental design methods are used to define the distribution of these key variables within a set of 52 finite element models. Although outside the scope of this paper, this set of models is designed to enable a mathematical model of the response to be derived from the finite element results.

FEMGEN/FEMVIEW [Ref 1] software is used for pre and post-processing, and ABAQUS [Ref 2] for the analysis. A typical analysis for one location within the reactor core would take eight days. Alternative methods of analysis were required because of the following reasons.

- The complexity and scale of the problem required the analysis of several locations within the reactor.
- The desire to improve the accuracy, quality and reduce analysis cost.
- The desire to respond more promptly to questions from internal review panels, and external independent safety assessors.

2.1   Analysis method

It is not practical to model the complete reactor core with one finite element model. A representative section of the reactor core is modelled to ensure reasonable analysis times. The analytical method involves automating the generation of the finite element models, sequential solution and results post processing, within one controlling program. The models were analysed on UNIX/ULTRIX DEC 3000/800 workstations. Figure 1 illustrates the three main phases of the solution method.

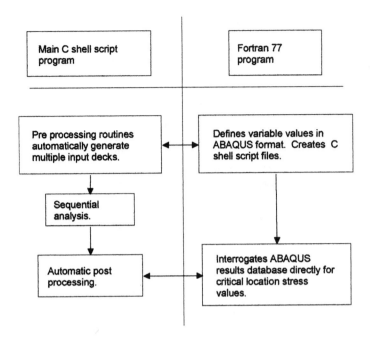

Fig 1. Main phases of the analytical method.

The controlling program is coded in UNIX C shell script. The UNIX C shell script program is supplemented by FORTRAN coding which provides links to the ABAQUS software. The UNIX C shell script provides a seamless integration between all phases of the analysis. The three phases of the analysis are as follows.

The first phase involves the automatic generation of different finite element input decks from one nominal input model. The generation of the input deck variants is achieved by using a FORTRAN program to generate 52 UNIX C shell pattern recognition/substitution script files, one for each proposed finite element model. The nominal model input deck is generated using FEMGEN. The nominal input deck is then edited to include file delimiters. These file delimiters enable automatic editing of the finite element model variables. The UNIX C shell program then cycles through each of the 52 pattern recognition/substitution script files. Each cycle results in the creation of a new finite element model input deck, which is automatically edited with the correct value of the variable.

The contents of a typical UNIX C shell pattern recognition and substitution script file, is as follows.

799

```
#1/bin/csh                            : file header
/**VAR1/,/**VAR1END/ s/x.xxx/y.yyy/   : line 1
..

..
/CVAR7/,/CVAR7END/ s/n/p/             : line 2
..
etc,
```

Line 1 instructs the UNIX C shell program to substitute variable 'x.xxx' in the nominal ABAQUS input deck with 'y.yyy'. This variable is located between comment lines in the ABAQUS input deck. These are represented by the **VAR1 and **VAR1END delimiters.

Here is an example of the relevant section of a nominal ABAQUS input deck.

```
*SHELL SECTION, ELSET=NN, MAT=MAT1
**VAR1              :ABAQUS comment line (start of
                     variable search)
x.xxx,
**VAR1END           :ABAQUS comment line (end of
                     search space)
etc
```

Within the ABAQUS FORTRAN creep subroutine the search space is bounded by the comment lines CVAR7 and CVAR7END. Line 2 instructs the UNIX C shell program to substitute variable 'n' with 'p' in the FORTRAN user creep subroutine.

The second phase is the analysis of the 52 automatically generated finite element models. This is again achieved by the controlling UNIX C shell script program. This section of the UNIX C shell script enables the current analysis log file to be monitored at specified short time intervals. The script file monitors the contents of the analysis ABAQUS .log file for a keyword that indicates that the analysis is completed. Once the keyword 'COMPLETED' is recognised the script file submits the next file in the sequence, and deletes any unwanted files from the previous analysis. The sequential solution optimises the performance on one workstation.

The final phase is the automatic extraction of the required inelastic stress history, through life, at the desired location within the model. The ABAQUS finite element software is structured so that communication with external programs is possible. The results database file (.fil) is accessible via standard FORTRAN commands, which enables users to create their own post processing program. At this stage of the analysis the UNIX C shell program performs the function of converting the .fil file extensions of the ABAQUS results files into appropriate filenames with FORTRAN unit number extensions. The UNIX C shell script program then calls the user defined post processing program. This FORTRAN program accesses the relevant stress data in the results database file. The stress data is identified by node number, through thickness section number, and a reference number related to the stress component of interest. This stress data is then compiled into a convenient form for further processing. This approach is significantly quicker than the conventional method of accessing results via the graphical user interface (GUI) of standard post processing packages.

2.2 Discussion

This analytical method has resulted in a reduction in elapsed time, for the analysis of one location, from eight days to 16 hours. There is also an improvement in accuracy, since the manual task of calculating the variable values, editing the input decks and post processing via the GUI has been removed. The analytical method is seamless and controlled by a single UNIX C shell script file.

This method is particularly useful for the analysis of complex engineering problems with many important variables. The benefits of this method of analysis are as follows.

- The methods can be applied to other safety critical analyses where the complexity and scale of the problem precludes the use of conventional analysis techniques.
- The methodology is not limited to the specific finite element analysis codes mentioned in this example.
- The methods are ideally suited for sensitivity analyses.
- The methodology is flexible enough to incorporate design optimisation logic in the FORTRAN and UNIX C shell script routines. This could enable convergence to optimum solutions for specific analysis types.

3. EXAMPLE 2: VISUALISING USER DEFINED SOLUTION VARIABLES IN DYNA3D IMPACT ANALYSES.

Post processing software should be regarded as a tool which is used to support engineers understanding of the analysis, and for conveying that understanding to interested third parties. It is important to note that these third parties might be quite remote from the engineer's work, so this accentuates the requirement for flexible and informative post processing tools. The engineer's task is made far more difficult if a parameter of significance is not immediately accessible for post processing.

At the marine power division of Rolls-Royce the post processing software used for DYNA3D [Ref 3] impact analyses is the D3PLOT and T/HIS programmes supplied by Oasys Ltd [Ref 4]. The current vendor supplied software is relatively inflexible. The available post processing options in the GUI do not allow the contouring of user defined variables.

3.1   Analysis method

A model has been created to explain the basic technique of visualising user defined parameters from a DYNA3D impact analysis.

Figure 2 illustrates the DYNA3D model geometry. This represents a quarter symmetry idealisation of the impact of a cover plate onto a section of main loop reactor compartment pipework. The pipe is constrained in all degrees of freedom, at the end plane perpendicular to the longitudinal axis of the pipework. Eight-noded brick elements are used throughout the model. The choice of a user defined variable for this demonstration is arbitrary. In this instance, a common measure of stress triaxiality is used. This is the ratio of two stress invariants, Hydrostatic stress and Von-Mises stress.

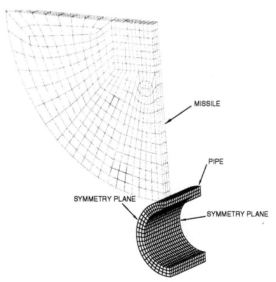

Fig. 2. DYNA3D model geometry

The method is illustrated by visualising the user defined variable for a subset of nodes at a specific time 't' in the impact event. The node subset consists of main loop pipework nodes located in the impact region. Figures 3 and 4 illustrate Hydrostatic and Von-Mises stress contour plots respectively, at time 't'.

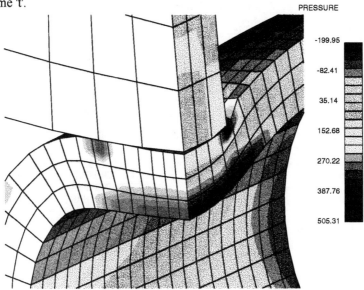

Fig. 3. Hydrostatic stress contour plot

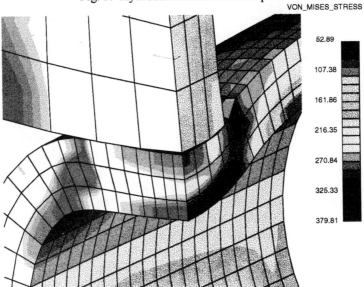

Fig. 4. Von-Mises stress contour plot

Figure 5 depicts a flowchart detailing the adopted methodology.

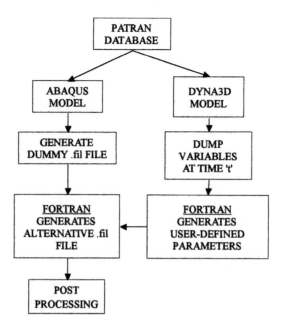

Fig. 5. Methodology for visualisation of user defined variables in DYNA3D.

The model geometry is created in the PATRAN [Ref 5] general pre-processor and a DYNA3D analysis is submitted in the normal manner. The Hydrostatic and Von-Mises stress data is dumped into a data file, at time 't'. FORTRAN coding is then used to calculate the user defined variables at each node in the subset. An ABAQUS formatted input deck is now created from the PATRAN geometry database. This ABAQUS model will eventually be 'seeded' with DYNA3D user defined variables and post processed. The ABAQUS model is defined as a heat transfer analysis with eight-noded brick elements. The elements have one nodal degree of freedom, which is temperature. A nominal nodal temperature of 1 is first imposed by analysing a steady state heat transfer model. At this juncture an additional piece of FORTRAN coding is used which accesses the nominal ABAQUS results database and seeds it with the DYNA3D user defined variables at impact transient time 't'. This FORTRAN program is in effect a user defined post processing program which converts an ABAQUS results file into a DYNA3D results file. This alternative results file can now be post processed in the conventional manner via ABAQUS/POST, or in this case FEMVIEW. In the post processing software, the nodal variable is labelled as nodal temperature, however it actually represents the user defined variable.

Figure 6 represents a contour plot of this user defined variable for the prescribed subset of nodes. The gradients at the edges of the node subset region necessarily occur since a nominal value of 1 has been preset for all the nodes in the model. The plot clearly shows the magnitude of the user defined parameter, being highest in the tensile region and progressively decreasing into the compressive region.

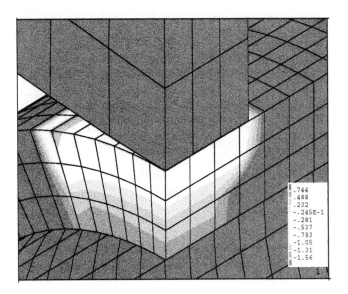

Fig. 6. Contour plot of user defined variable in region of interest.

3.2 **Discussion**

With this method it is possible to create and contour any user defined variable, as long as it depends upon the basic solution variables available within DYNA3D. This facility could be beneficial in several ways, such as:

- Use as a general post processing tool that enables further investigation of the response of impact models.
- To aid the investigation of alternative methods of assessing impact structures. These alternative methods would probably be based on the variation of some complex user defined function.

805

- It could be used in analyses of existing equipment if current assessment techniques failed to satisfy the regulatory authorities.

It is important to emphasise that being able to easily post process any function is very powerful and liberating for the engineer. This facility would enable the engineer to extract more from the analysis code, and in due course encourage the generation and application of new analytical tools.

4. CONCLUSIONS

A common message has emerged from the description of the two examples in this paper. This is that the integration of essentially different software codes at the operating system level has great benefit. These benefits explicitly manifest themselves as reductions in overall analysis cost and improvements in accuracy and quality. The methods make the analyses more affordable for the customer. This is important since it is envisaged that the customer will require additional analysis in the future. These are benefits that every industry strives for. The techniques described in this paper enhance the capabilities of standard finite element codes.

The use of these methods can also lead to further innovation, the techniques used in example 2 are a direct result of the methods developed in example 1.

Example 2 demonstrates that knowledge of the capabilities of more than one analysis code can lead to benefits, since solutions to seemingly intractable problems can be obtained at little cost.

In order to fully capitalise on the potential benefits of such analytical methods, companies must actively encourage their engineers to develop such tools. Also the engineer should always consider challenging and extending the capabilities of standard analytical and pre and post processing methods. The cost and performance benefits are real.

REFERENCES

1. FEMGEN/FEMVIEW V4.1 software : Femsys Ltd, UK
2. ABAQUS V5.6 software : Hibbitt, Karlsson & Sorensen Inc, USA.
3. LS DYNA V 940 software: Livermore Software Technology Corporation, USA.
4. D3PLOT & T/HIS V7.0 software : Oasys Ltd , UK
5. PATRAN V6.0 software : The MacNeal-Schwendler Corp, USA.

WHY YOUR ANSWER MIGHT BE WRONG

R P Johnson NRA BSc MSc MIMechE CEng
Tickford Consultancy Limited/DAMT Limited
11 Keynes Close, Newport Pagnell, Bucks, MK16 9AT, UK
Phone/Fax: +44 1908 217930
Email: bj@tickford-cons.demon.co.uk

SUMMARY

The answers from a single, isolated, FEA run are likely to be in error because of a simple mistake or misjudgement made at some point or another. The problem with a single run, in spite of the rigorous checks that can be carried out, is that it stands alone and cannot be compared with anything else. Understanding the problem comes from performing a number of runs and spotting trends as the input parameters are changed. One set of results cannot provide such sensitivity information. This paper describes an imaginary FE project where an otherwise well-executed analysis is compromised by one simple mistake.

The correct way to approach the simulation task is to forget the single/definitive run approach and to adopt a solution strategy based on the output from a number of sub-runs. The analyst should develop a series of runs with complexity increasing from run to run. Hand calculations should start the process, followed by perhaps a spreadsheet or calculation tool, then 2D FEA, followed by 3D FEA. Most experienced analysts already adopt this approach because they are being asked to develop complex non-linear simulations where the overall result will be affected by a wide variety of inputs and (hitherto) uncharted "funnies" in the program being employed.

1. INTRODUCTION

A number of papers and articles have been produced outlining consistent and quality-conscious ways of performing Finite Element Analysis. Most elaborate on a "rubbish in - rubbish out" principal which states that the quality of the output is in direct relation to the quality of the input.

NAFEMS has been responsible for the FE industry adopting a very structured, well-organised and professional approach to computer simulation and this is very much to the benefit of us all. NAFEMS have encouraged a questioning philosophy where the first assumption in any analysis is that the answers are wrong. WRONG UNTIL PROVED RIGHT might seem to be more bothersome in the short term but the total cost of reaching the best design will be much less if no mistakes are made along the way.

This paper examines the progress of an imaginary FEA project from initial assumptions and ideas all the way through to results assessment and reporting. Extensive checking procedures are followed throughout the analysis and these are reported herein. No problems are found and the run is reported as usual allowing the 'customer' to take the necessary action in order to re-direct the design process. In spite of all the careful checking, the final section uncovers an embarrassing error which has been allowed through: An otherwise excellent piece of work completely undermined by a single human oversight. This paper does not give further checks to be undertaken (because rigorous checking was done satisfactorily) but rather a different solution approach so that the inevitable human errors do not have such an impact.

2.   THE PROBLEM
The problem consisted of a locking pawl subject to cracking and, on occasion, complete failure. The locking pawl was used as part of the tipping method for cradles used to transport molten steel around a steel works. No safety concerns were relevant as fail-safe devices tripped in when the pawl failed. Nevertheless the workers operating the system were not impressed and valuable production time was lost as the men made heavy of their task to get around the problem. The stressing of the offending pawl was referred back to the design department where FEA could be carried out in order to modify the shape of the pawl.

3.   ANALYSIS PLANNING
Drawings of the forged pawl showed a catch-type structure basically consisting of a boss at one end and a hook at the other end. The pawl pivoted about a shaft running through the boss and the hook was made to engage a pin so that locking was secured. Not satisfied with just the drawings, the design office visited the works in order to see the pawl in action. The method of operation became clear and, armed with the weight capacity of the cradle, the FEA was started as a matter of urgency. The geometry was approximately symmetric about the vertical centreline and a half model in solid elements was decided upon. The cracking experienced in practice occurred in the hook very close to the contact with the pin. It was therefore decided that local contact conditions may have a bearing on the cracking failure experienced and that a non-linear contact analysis was required. This increased the complexity of the analysis but

it would also allow the analyst to investigate the "sloppiness" between the pivot shaft and the boss at the same time. Computer resources were checked and it was confirmed that such an analysis could be performed without any significant impact on other, equally important, computing tasks.

4.   ANALYSIS PREPARATION

A new version of the usual FEA code was to be used for the analysis. One of the many enhancements in the new version was the inclusion of contact surfaces and contact pairs instead of the old gap element approach. In spite of the code passing all the required installation checks the analyst working on the pawl decided to carry out a test run to satisfy himself that the contact pair approach was valid. A 3D model of a simple rectangular-section cantilever was built with a contact surface defined on the bottom face of the beam. Also modelled below the beam was a round abutment such that, when the beam was loaded by a downward end load, bending would occur until the clearance was used up and contact made with the abutment. This analysis was performed twice to compare the difference between the contact surface approach and the old gap element approach. The results of the two small-sliding analyses showed good agreement and the work was written up to be included as an appendix in the main report. The work was showing good progress.

5.   MESHING AND CONSTRAINTS

Linear 8-node brick elements were used for the main analysis. These elements were not ideal for stress analysis but recommended for contact and, later on, plasticity if required. A fine mesh was used especially in the areas around the hook/pin contact (where cracking was observed) and the blended transition into the boss. There was a temptation to model the locking pin and the pivot shaft with rigid analytical surfaces but, since close inspection of the stresses would be required near the hook/pin contact, explicit elastic components were meshed instead. The three separate parts of the model (the pawl itself, the locking pin and the pivot shaft) were subsequently meshed and contact surfaces were identified where contact was most likely. The contact surfaces were made larger than necessary in order to ensure that one contact surface (of a pair) did not drop off another. Symmetry constraints were imposed on the vertical centreline with minimal constraints applied elsewhere on the pawl in order to prevent rigid body motion. Minimal constraints were also applied to the pivot shaft and latch pin. (Constraints would not pick up any load as a balanced set of loading was applied – see below). Constraints and contact surfaces were checked graphically. The mesh was checked to ensure that individual components did not have splits (edges drawn) or duplicate elements. A plot of the free faces was also obtained to ensure that there no buried faces within the mesh. Element quality checks were carried out and no element had a score less than 0.32 (where

809

0.0=corrupt and 1.0=perfect). A geometry check was also completed and modelled dimensions showed good agreement with drawings.

6. MATERIAL PROPERTIES

Linear elastic material properties were applied to all constituent parts of the model. The properties assigned included the Young's Modulus (E), and the Poisson's ratio (v). The value of Young's modulus was not reduced in view of the elevated temperature on the grounds that any reduction would only affect the displacements. The density (ρ) was also provided in order to check on the mass of the model. The units used in the analysis were; mm for model dimensions and displacements; loads in N; pressure/stress in N/mm^2 (MPa); with a consistent unit of mass of 10^3 kg (tonnes). The supplier of the forged pawls also provided stress-strain data from a tensile test at 200 °C (thermocouple measurements in situ' gave 180 °C). A file note was issued to ensure that the nominal stress/strain values supplied were converted to real stress and real strain as required by the FE program. Material sets were plotted graphically to check that all three components had been isolated properly.

7. LOADING

The loading was a simple tensile load acting between the pivot shaft and the latch pin. The total load was calculated from the weight of the molten metal acting through an angle of 5 degrees. The same load was applied to the pivot shaft and the latch pin (one positive X load and one negative X load). The action of the contact pairs would transfer the load into the actual pawl and the correct load path would be predicted.

8. FE SOLUTION PHASE

All input parameters were checked for a second time and a data check carried out. The data check issued warnings about the shape of some thirty elements but these were passed off on the basis that none of the elements occurred in important areas of stress assessment. Checks were carried out at the end of the run to ensure that the summation of applied loads came to zero and that the summation of reactions also came to zero. No other warning messages were issued by the FE system. No error messages were issued.

9. RESULTS POST-PROCESSING

The first task was to ensure that the results were correct: Firstly displaced mesh plots were produced to see what the mesh looked like in the loaded state. The pin and the pivot shaft had moved apart from each other in the line of action of the applied load and the pawl had shown a tendency to straighten just as expected. Tensile stresses (P1) were high (order 1360 MPa) but this was to be expected as cracking was the mode of failure under investigation. The location of the high tensile stress gave good agreement with the crack location found. Detailed inspection of the P1 stress trajectories showed them flowing around the radius in the hook. The P3 stress was found to be the through-thickness stress and the P2 was the stress normal to the free surface. A localised area of high compressive stress was evident between the hook and the pin (-2560 MPa) although this would be reduced if plasticity had been modelled. All results seem to be reasonable and the analysis was reported in full.

10. WRONG UNTIL PROVED RIGHT

In this case an otherwise excellent piece of work was spoiled by one simple mistake. The analyst was drawn away from the simple tasks by the fact that a new contact algorithm had to be tested first. The contact method was tested as it should be but success here lulled the analyst into a false sense of security: Other parts of the project were rushed once the "tricky" bit had been resolved. In this case the full total load was applied to a half model and stress levels were twice as high as they should have been. Had preliminary work been completed, say a 2D plane stress model, the chances of the simple human error getting through would have been much reduced. Had simple calculations been carried out before the work started (rough estimate of the nominal stress) the error would have shown itself. The problem now is that the Engineering Director has signed off major modifications to the pawl as it is deemed to be much too small for the job. Are you going to tell him or shall I?

11. CONCLUSION

The analyst should develop a series of runs with increasing complexity from run to run. Hand calculations should start the process, followed by perhaps a spreadsheet or calculation tool, then 2D FEA, followed by 3D FEA. Most experienced analysts already adopt this approach because they are being asked to develop complex non-linear simulations where the overall result will be affected by a wide variety of inputs and hitherto uncharted "funnies" in the program being employed.

The safest approach is to adopt a series of self-calibrating runs which all contribute to the answer. One of these can be fully reported as the final "production" run with the supporting runs included, maybe, as appendices in the main report. The important thing is that the supporting runs have been completed and show solidarity with the reported "answer". An isolated result is a dangerous result. A collection of results which illustrate the importance and relevance of all the key assumptions will lead to overall confidence and reliability.

Fred J Bell
Honeywell Space Systems
USA

ABSTRACT

"Analysis Methodology of Military and Aerospace Navigation Equipment
Structural Design"

This paper deals with the analytical method, processes, and technical issues
encountered while verifying the structural integrity of navigation systems for
aerospace and military use.

Starting with an SDRC-generated 3D solid model for reference, a Finite
Element Model (FEM) is constructed. Next the FEM is analysed (using) either
SDRC or NASTRAN) for pressurisation, vibration, and shock load cases. By
analysing the initial design parameters, it can be determined where the structure
should be strengthened, or where material can safely be removed, to maintain
adequate margins of safety based on the constituent materials properties. This
is performed concurrently with the goal of optimising the design in terms of its
overall weight and spectral characteristics.

In addition to the individual load cases mentioned there are also combined
environments, which represent what the products are most likely to experience
during their service life in the real world. The combined environments
represent a challenge to the analyst because they are handled differently: as
either static or dynamic loads, and thus require multiple solutions. Following
the analyses, the results must then be combined, which is often a complex task
for today's large FEMs. Another challenge is that most static load cases can be
analysed using SDRC, while the dynamic loads require NASTRAN solutions.
Thus the results have different formats, which require special post-processing.
Whatever tool is used, the most common analytical results desired are: modes
of vibration, accelerations, displacements, and stresses.

A typical navigation product design consists of a cast chassis, cast or machined
cover (s), an inertial instrument cluster, its support electronics, plus various
fasteners and connectors. All of these major sub-components of the design are
modeled in the FEM, exerting most fasteners and connectors, which are
analysed more effectively by manual techniques. For better FEA accuracy, the
effects of these fasteners and connectors must be included as restraints, or their
inertial propertied somehow retained in the FEM. Other important aspects of

good modeling include correct boundary conditions and proper dynamic tuning using amplification ratios.

Once the modeling is complete and an initial performed, the results are passed back to the designer. Design Improvements can then be made, within the constraints of available space or mounting patterns, etc. Then the FEM is updated, and the new design re-evalued. Initial evaluations often include the dynamic response characteristics and resultant stresses. To avoid adverse modal interactions, we design using the octave rule. That is, to prevent separate structural modes from coupling together, their resonant frequencies should be an octave apart (twice the frequency). For example, the chassis should not resonate near a printed wiring board's first natural frequency.

As mentioned, analysis is highly dependent on boundary conditions. For example, the boundary between the navigation equipment and its mounting platform in the flight vehicle. If this joint is not properly modeled, it will make a difference in the predicted modal characteristics. At the component level within the structure, accurately simulating the joint between the cover(s) and the chassis will affect the frequency characteristics of both these components, and thus the overall structure.

On a more detailed level, the type of elements used also makes a difference in successful analysis. Obviously, no FEM can be infinitely detailed to represent the real structure. But modern computational platforms have the ability to handle models with large numbers of elements (>10,000). Advances in post-processing make it possible to visualise and rapidly evaluate the outputs from these large models. However, it is still important to understand what restrictions and/or advantages exist when using the different types of elements available to FEA.

a

Optimization Choices in Engineering

[I] Charles H. Roche, Ph.D., P.E.
[II] Peter P. Godston, MBA

Summary

Consumers enjoy more for less due to continuous engineering improvements. Mechanical engineering has evolved from its early days of pencil and paper boiler pressure design vessel to aerospace structural optimization - where CAE software significantly aids in the design process. Novel products and optimized systems are being developed by a subset of the CAE tools, referred to as optimization software.

A real-world company critical engineering problem is presented - a NASA funded gas turbine development program of a product that is considered by many to be part of an evolutionary process but clearly provided new challenges to the engineers. The common choices made in solving the proposed problem as well as their limitations are discussed.

In the sub-set of the CAE environment, optimization software is reduced to groups that highlight the evolution from rules based design to parametric optimization to multi-disciplinary optimization or to inventive software. The latest software tools to develop the next generation of technologically advanced mechanical structures are discussed. Obtaining information sources regarding optimization software for an engineering division is detailed with an aerospace emphasis.

The emphasis of the paper is to review of the software tools that aid in optimizing engineered systems and to challenge engineers and management to

[I] Dr. Roche is a senior analyst with FTS Inc. in East Hartford, CT.
[II] Mr. Godston is the European Managing Director of Software Emancipation Technology, Inc.

reconsider how they begin new product development especially since 90% of the cost is in the early design phase. Only optimization software solutions to the proposed problem are emphasized. In addition to parametric optimization, the next generation of structural products mandates consideration of topology optimizers, multi-disciplinary optimizers, and inventive software.

1. BACKGROUND

In his book relating structural failures with design improvements, Petroski [1] argues that mankind has engineered new products for millennia by enduring education from catastrophic failures. The concept of failure "is central to understanding engineering, for engineering design has as its first and foremost objective the obviation of failure." Indeed, earlier in this century one could argue *learning* ceased when the prototypes stopped breaking. Until recently, this widespread concept represented the development process of mechanical structures and justified the large experimental test facilities of many companies.

Design innovation has been dependent largely on the 'eureka' of one of a company's key engineers. Recall that Edison, Porsche, Eastman, and other engineering innovators founded entire major corporations on inventions. The trial and error approach that was once the foundation of engineering now is challenged by advances in software technology.

Because of unknowns or a matrix of unknown interactions, our desire to engineer products better, faster, easier, or stronger often would expend our experimental test resources. While time to market requirements and human resource constraints impose continually greater challenges upon today's engineering organizations, software tools available today enable thorough testing prior to prototyping - so when prototyping finally is conducted, causes for failure have largely been identified and avoided. Design, development, and testing all can occur within a computer environment. Concepts and innovations also can be critiqued within a computer.

The background of the two authors is very different but the occupations interact when engineering requires the latest software. One author brings MBA disciplines, coupled with significant systems product, marketing and sales experience; the other comes from a traditional mechanical engineering background - design, experimental test, and analysis. Thus, a background in software management and sales must unite with the engineer seeking to develop the next generation of products.

Product and service firms have developed sophisticated approaches for helping engineering organizations identify opportunities for improvement and understand the value of a process improving investment. The benefits of improving process and some powerful techniques for driving process improvement have been outlined by Smith and Reinertsen [2]. Beyond parametric design and finite element analysis (FEA), how well have companies put their efforts into utilizing software to advance the product in the design and concept stage? Recognizing 90% of the product cost is dictated in the first phase of the design, one must be committed to investments at the earliest stages of a development program.

2. PROBLEM EXERCISE

At this point in the paper, a problem similar to what many engineers and their managers have experienced is proposed by the authors to provoke thinking. The entire problem is somewhat fabricated and ad-hoc to facilitate understanding a decision making process:

Consider that NASA has given your gas turbine company $100M to design, develop, and test a low emission combustor for the next generation of commercial supersonic jet engines. You are in competition with two other jet engine developers who also have received $100M from NASA.

Recognizing the importance of the program, you list ideas to improve upon your company's fifty years of gas turbine combustor experience. A partial list may include:

- hire a top graduate from MIT.
- network to find a combustion engineer from General Motors
- network to find a catalytic converter engineer from General Motors
- seek out consultants in the gas turbine combustor industry
- convince your recently retired combustor engineer to come out of retirement
- network to find a commercial supersonic gas turbine combustor expert
- network to find a commercial gas turbine low emission combustor expert

These are reasonable ideas but there is clearly an underlying problem with this procedure. New and innovative technology requires new and innovative ideas. The development methods of relying on 'eureka' of one of your engineers, of relying on your competitors to fail, and of relying on simple design iteration put your company at a disadvantage. New technological challenges to an industry preclude there being any experts. Where are you to to find engineers experienced in new technologies like low emission gas turbines or commercial aircraft operating at supersonic speeds? The recently retired combustor engineer probably knows little about low NOx or supersonic operating regimes. What are the measurable gains from hiring top graduates for such a program? Today, there are substantial sources of commercial software to aid your development program.

The corporate decision making process may come under the leadership of the Operations Research and the Management Sciences (OR/MS) [4] department, the engineering management of the Internal Research & Development (IR&D), or even NASA because it is funding the program. OR/MS represent the professional disciplines that deal with the application of technology for informed decision-making and efficient management of engineering and manufacturing. They play a key role in how companies choose and implement software that should be integral in development programs.

3. INTRODUCTION TO SOFTWARE TYPES

The emphasis on this paper is that small underutilized subset of computer-aided engineering (CAE) software referred to as optimization software. Excluded from the groups would be any software utilized exclusively outside the preliminary product design phase. One could reduce the subset to the following groups:
- Rules Based Design (RBD)
- Finite Element Optimization (FEO)
- Multi-disciplinary Optimization (MDO)
- Inventive Software (or Computer Aided Innovation - CAI)

A curtailed review is provided for each group.

3.1 Rules Based Design

Rules based design methodology has existed for many years even before computers. It simply refers to routines containing a series of expressions which force one result from the input parameters. Design manuals can be considered rules based. RBD software would include user developed simple routines written on spreadsheet software and also would include the more advanced packages. For example, automobile manufacturers use a non-linear program for sizing a suspension system based on: vehicle weight, tire profile, tire width, tire material properties, rim sizes, anticipated terrain, desired response, wheel well clearance, type of shock, and type of spring.

Some general RBD commercial packages integrated with major CAD packages include Platinum's knowledge base system, AionDS, which can support many different CAD technologies but requires significant customization. StoneRULE from Prescient Technologies provides AionDS bundled with significant interface libraries. Rand Technologies also provides commercial-grade libraries that work with AionDS.

3.2 Finite Element Optimization

Many engineers are aware of the parametric finite element based optimization routines that have been in common use for over ten years. In most methods, *design variables* are varied in a series of finite element analyses until a desired *objective function* maximum or minimum is achieved without exceeding the *state variables*. For example, plate thickness (design variable) is varied in a structure until the maximum principal stress or maximum defection (state variables) is achieved while minimizing weight (objective function) [5]. Most Mechanical Computer Aided Engineering Analysis (MCAEA) vendors provide parametric optimization for common finite element analyses. Within a domain of topological definition, the codes can quickly converge to a solution. For background on developments in structural optimization, see Banichuk [6].

In addition to the parametric finite element optimization, note some FEA providers do other types of optimization. Topology optimization uses finite element methods to generate optimal designs. Given a finite element structure, all loads, boundary conditions and necessary points or volumes, the software intelligently attacks each finite element for its value to the structure until weight is minimized.

MacNeal-Schwendler Corporation offers topology optimization with their MSC/CONSTRUCT [7] package. Similarly, Altair Computing Inc. offers HyperShape/Pro [8]. With the ability to access an array of fatigue curves, some codes can optimize component life as the state variable instead of maximum principal stress. The authors are currently reviewing a larger set of finite element software packages and will make it available at the conference.

3.3 Multi-disciplinary Optimization

A more recent numerical or finite element optimization algorithm is referred to as multi-disciplinary optimization. MDO can be described as an optimization methodology for coupled or engineering systems. In order to solve complex engineering problems, the problem must be decomposed into smaller subsystems that can be concurrently solved while simultaneously accounting for the interactions among the different disciplines and subsystems

[10]. A common example is the aircraft wing aerodynamic profile; drag optimization competes with natural frequency goals, flutter, manufacturing limitations, geometric limitations, lift, fuel storage, and weight in the wing.

Engineous Software Inc. uses software to handle computational fluid dynamics (CFD) problems like an aircraft wing. DAA Solutions Inc. has software written in Intent! and refers to Rules Based Engineering as the name of the process to optimize coupled systems. The algorithms of the two companies are quite different but each may be considered MDO.

3.4 Inventive Software

Unanticipated and unique solutions may have dramatic impact on a design. Inventive software represents a very small part of the CAE environment and is quite new. In software form, inventive ideas are stored as objects in a data base ready to be utilized by an intelligence search tool [14]. More advanced techniques allow engineering problems to be solved by addressing the problem boundary conditions and problem type, and then seeking a large data base of solutions from many disciplines. For example, a biotechnology patent may be used to address an automobile tire development problem.

From Invention Machine Corporation, the term has been coined Computer Aided innovation (CAI). Their Effects Module provides the user with access to over 2,500 engineering and scientific effects coupled with associated design cases. Their Principles Module is based on the evaluation of over 2.5 million Patents. Their Prediction Module solves technical problems involving interactions between objects. The software tells the user how the system will behave when evaluating a new idea. Their Feature transfer module improves technological systems by allowing the user to transfer desirable features from one engineering system to another. Invention Machine may be unique to all others and mention is given to Rand Technology's ModelCheck [15].

4. SOURCES OF INFORMATION

We would be remiss not to discuss the size of the choices out there. With an aerospace background, the authors provide the following section highlighting information sources.

4.1 CAD Consultants

In the multibillion dollar industry of computer aided engineering, the emphasis has been on computer aided design and computer aided manufacturing - CAD/CAM. Companies like Daratech [16] and Dataquest [17] publish CAD/CAM vendor surveys, market research and technology information. Many engineering consultant firms exist that may have expertise in your field or your CAE needs. When a company begins the selection process for identifying the best software vendors for its needs, usually many of the major suppliers are invited to give a sales pitch to a select group of engineers, managers, and computer scientists. The process for selecting a CAD/CAM package deserves review in most major corporations.

4.2 NASA

For 30 years, NASA's Computer Software management and Information Center (COSMIC), operated by the University of Georgia at Athens has been a contractor to NASA, distributing software developed with NASA funding to industry, academia, and other government agencies. In 1999, COSMIC shall be a contractually independent, self-funding business partner helping NASA to commercialize its software. COSMIC has been responsible for many successful spinoffs, for example MacNeal-Schwendler. The NASA Commercial Technology Program is chartered to enhance the global competitiveness of U.S. companies. In fact, a congressional mandate ensures the widest possible dissemination of R & D results from NASA funded programs. Since 1976, the magazine of dissemination, *Spinoff*, has featured many down-to-earth applications of NASA technology. Visit the *Spinoff* web site at: http://www.sti.nasa.gov/tto/spinoff.html.

For a brief history of CAD/CAM, see the NASA Tech Brief, March 1998. Check out NASA's web page: http://www.nasa.gov.org.

4.2 The Internet

The World Wide Web has become a legitimate, fast, way to search for a wide variety of optimization information. The web is just in time as many of the Fortune 500 companies have downsized their libraries. The References section emphasizes this. The authors can provide more information obtained from web searches.

4.3 Engineering Societies

The American Society of Mechanical Engineers (ASME) offers 17 journals. Optimization topics can be found in all and the Journal of Mechanical Design may be most applicable for all conference attendees. The American Institute for Aeronautics and Astronautics (AIAA) offers six technical journals peer reviewed. The AIAA Journal may be a good reference for many conference attendees. Of course, NAFEMS is a growing organization in the finite element analysis community and directly offers practical publications, some of which are related to optimization software.

5   PROBLEM SOFTWARE SOLUTION

Without comment on the existing staff to solve the low emission combustor problem, the OR/MS team allows for optimization software investments on the order of $100K which represents 0.1% of the budget. Many CAD/CAM packages already have optimization program and were discussed in section 3.2. Surely, this software is ideal to parametrically design and analyze combustor concepts.

Invention Machine was contacted for our low emission combustor problem. The results of their effort are remarkable. Their Principles Module patent search on 'gas turbine' in any field of the patent abstract revealed 71 patents. Here are the titles of eleven deemed related to our problem:
1. Dry low NOx single stage dual mode combustor construction for a gas turbine.
2. Catalytic pollution control system for gas turbine exhaust.
3. Convectively cooled liner for a combustor.
4. Gas turbine combustor with means for removing swirl.
5. Method for preparing the working gas in a gas turbine installation.
6. Method and apparatus for keeping clean and/or cleaning a gas turbine using externally generated sound.
7. Gas-path leakage seal for a gas turbine.
8. Gas-path leakage seal for a gas turbine made from metallic mesh.
9. Gas-path leakage seal for a gas turbine.
10. Gas turbine arrangement.
11. Cooled turbine vane.

Some of these eleven were very insightful to the development program. In addition, other patents revealed concepts discussed in the effects module.

In the Effects Module, eight parameters were highlighted:
1. Increase chemical parameters
2. Decrease chemical parameters
3. Increase concentration parameters
4. Decrease concentration parameters
5. Increase temperature
6. Decrease temperature
7. Increase thermal parameters
8. Decrease thermal parameters

Many physical effects were generated in the categories and several were apart from the knowledge of gas turbine engineers. For those in the gas turbine industry, two effects are mentioned to demonstrate the creative thinking the software generates.

It may be advantageous to choke the flow of fuel to better atomize the fuel. Electrically charged combustor liners may effect the reaction nitrogen and oxygen. See Figures 1 and 2.

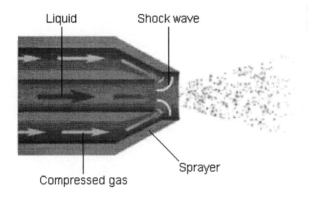

Figure 1
A shock wave atomizes a liquid

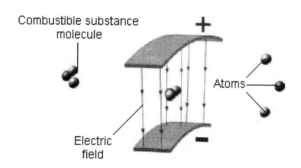

Figure 2
An electric field atomizes and vaporizes a liquid

823

The software also proposed several methods for measuring nitric oxide emissions and referenced appropriate patents. New products often require new experimental testing tools and new manufacturing tools. As a result, the CAI search reveals more uses for the company than originally thought. In Figure 3, a remote exhaust gas analyzer appears to offer a good way to measure concentration of NOx based on its radiation wavelength signature.

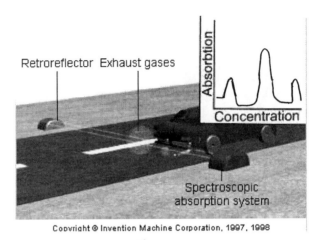

Figure 3
Remote exhaust gas analyzer

6. CONCLUSION

Invest as little as 0.1% of a program cost into optimization software and training at the outset of a major development program. The results of the CAI were presented during a brain storming session in the combustor group. This led to more ideas. Later, a few promising ideas were presented in an array for rig testing. The test facility incorporated a radiation wavelength measuring system for hydrocarbons and NOx.

In addition to parametric optimization, the next generation of structural products mandates consideration of topology optimizers, multi-disciplinary optimizers and inventive software. Considering the history of every new aspect of software technology to mechanical engineering, companies which take advantage of these types of software will be the leaders. CAD and CAM software went from aerospace and automotive to all engineering disciplines. Inventive software will allow expertise to traverse disciplines. Lastly, the Internet is a very efficient source of information.

7. REFERENCES

1. PETROSKI, H - To Engineer is Human: The Role of Failure in Successful Design, St. Martin's Press, NY, 1984.
2. SMITH, P G and REINERTSEN D G - Developing Products in Half the Time, Van Nostrand Reinhold, New York, 1995.
3. MARKS, P, and RILEY, K - Aligning Technology For Best Business Results, Design Insight, Los Gatos, California, 1995.
4. See the web page http://www.informs.org.
5. Design Optimization, ANSYS User's Manual, Vol. I, 1997.
6. BANICHUK, N V - Introduction to Optimization of Structures, Springer-Verlag, New York, 1990.
7. See the web page http://www.macsch.com.
8. See the web page http://www.altair.com.
9. GALLAGHER, R H and ZIENKIEWICZ, O C - Optimum Structural Design, Theory and Applications, John Wiley & Sons, New York, 1973.
10. SCHMIDT, JR., L A - Structural Optimization Symposium, ASME, Winter Annual Meeting, NY, NY, Nov. 17-21, 1974.
11. SOBIESKI, J. - Multidisciplinary Optimization for Engineering Systems: Achievements and Potential, DFVLR Symposium on Optimization, Bonn, West Germany, June 1989.
12. WANG, J, THOMSEN, N B and KARIHALOO, B L - Multicriterion Optimization: A Tool in Advanced Materials Technology, ASME Advances in Design Automation, Minneapolis, MN, Sep 11-14, 1994.
13. MISTREE, F, PATEL, B and VADDE, S - On Modeling Multiple Objectives and Multi-Level Decisions in Concurrent Design, ASME Advances in Design Automation, Minneapolis, MN, Sep 11-14, 1994.
14. See the web page http://www.invention-machine.com.
15. See the web page http://www.rand.com/prodserv/mda.
16. See the web page http://www.daratech.com.
17. See the web page http://www.dataquest.com.
18. MILLER, R B and HEIMAN, S E with TULEJA, T - Successful Large Account Management, Warner Books, New York, 1992.
19. FORESTER, T - The Materials Revolution, Superconductors, New Materials, and the Japanese Challenge, MIT Press, Cambridge, MA, 1988.
20. See also http://www.engineous.com.
21. See also http://www.sdrc.com.
22. See also http://www.DAASolutions.com.

Topology Optimization of Large Real World Structures

Authors: O. Müller, A. Albers, J. Sauter, P. Allinger

Presenting Author:

Dipl.-Ing. O. Müller
Institute of Machine Design
University of Karlsruhe
Kaiserstraße 12
76131 Karlsruhe/Germany
Phone +49-721-608-6495
Fax +49-721-608-6051
Email: ottmar.mueller@mach.uni-
karlsruhe.de

Co-Authors:

o. Prof. Dr.-Ing. A. Albers
Institute of Machine Design
University of Karlsruhe
Kaiserstraße 12
76131 Karlsruhe/Germany
Phone +49-721-608-2371
Fax +49-721-608-6051
Email: albert.albers@mach.uni-
karlsruhe.de

Dr.-Ing. Jürgen Sauter (VDI)
FE-DESIGN GmbH
Haid-und-Neu-Str.7
76131 Karlsruhe
Tel: 0721/96467- 0
Fax: 0721/96467-29
Email: juergen.sauter@fe-design.de

Dipl.-Ing. Peter Allinger
FE-DESIGN GmbH
Haid-und-Neu-Str.7
76131 Karlsruhe
Tel: 0721/96467- 0
Fax: 0721/96467-29
Email: peter.allinger@fe-design.de

Table of Contents

Abstract

The paper first introduces the understanding of the different structural optimization types and positions their usage in the product development process.

After that, the software and hardware opportunities given through High Performance Computing (HPC) are explained shortly. The following section shows how performance and efficiency improvements of the topology optimization process were achieved by using HPC techniques. This work was conducted within the HIPOP (HIgh Performance OPtimization) project.

Finally topology and shape optimizations of real world examples show the successfull implementation and their benefits.

1 Introduction

In the tough international competition, companies can only survive if, besides highly innovative power they can provide strongly cost optimized products. Therefore in new procedures like the Simultaneous Engineering, the calculation engineer is already integrated into the concept phase of the product development process. Efficient methods of working require powerful optimization algorithms to be provided in addition to the discrete methods (FEM/ BEM) proved worth while to support the calculation engineer in the draft and design phase. Almost all FEM codes have integrated sizing optimization capabilities to support the calculation engineer.

In this context new optimization criteria and control strategies for sizing, shape and topology optimization (Figure 1) were found at the Institute of Machine Design from the University of Karlsruhe, Germany, in 1991. Based on these new strategies the computer program CAOSS (Computer Aided Optimization System Sauter) was developed from the institute together with FE-DESIGN, Karlsruhe, Germany. The program was awarded in 1994 from the European Commission, DG XIII with the European Academic Software Award for the best program in the field of mechanics.

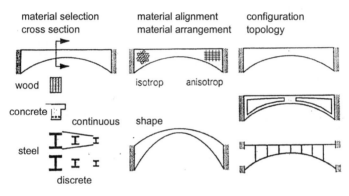

Figure 1   Types of Structural Optimization

For both sizing and shape optimization a first design proposal, which is used as the start design, exists. The objective of general structural optimization methods is to compute even this first design proposal (Figure 3). For topology optimization the designer creates only the design space, which includes the future component. Subsequently the functionally required boundary conditions are applied. The efforts for the modelling and preparation are extremely low. The optimum structural shape with the appropriate topology is calculated utilizing a FEM program and issued as a design proposal and might be refined by the designer.

Compared with the sizing and shape optimization the numerical efforts of this iterative process strongly increases. Therefore currently only components with 15.000 to maximum 30.000 elements can be calculated, even when powerful workstations are used. Due to the high number of iterations (typically between 20 to 30 iterations) and the fact that approx. 90 to 95% of the CPU time is used from the FEM program for the analysis, the performance and the resource requirements of the FEM program are of particular importance.

For the last two years certain FEM solvers (like MSC/NASTRAN) were ported to hardware platforms with distributed memory. With them a considerable reduction of the price/performance ratio was shown for the parallel code combined with impressive speedups. These facts make the utilization of these codes interesting for industrial users. Within the HIPOP (High Performance Optimization) project the coupling of the distributed parallel MSC/NA-STRAN and CAOSS was realized. The benchmarking of real world applications from companies like BMW, PININFARINA, AUSTRIAN AEROSPACE and FE-DESIGN showed the reliablity and efficiency of the approach. Even for models far beyond 200.000 degrees of freedom for a number of load cases speedups of 3 to 4 were achieved.

The current institute research focuses on the improvement of the efficiency of the whole optimization process. This includes e.g. the utilization of parallel iterative FE solvers as well as the improvement of the solver efficiency itself by an adaption of the solver to the particular optimization properties.

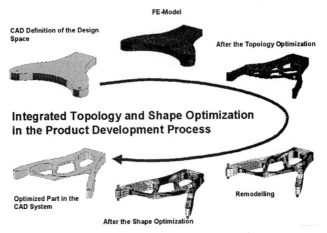

Figure 2   Topology and Shape Optimization in the Product Development Process

2   High Performance Computing and High Performance Optimization

2.1   High Performance Computing

PMN - Parallel MSC/NASTRAN

In the ESPRIT III project EUROPORT-1, Porting Work Area parallel MSC/NASTRAN (PMN), MSC started to port MSC/NASTRAN to distributed parallel hardware architectures. The distributed parallel modules developed in this project are available since September 1997 in the commercial MSC/NASTRAN release V69.2.

On serial computers and for large problems, the numerical solution modules typically require 85% to 95% of the total elapsed time. Therefore, parallelizing just these modules on their own gives a significant increase of performance. This strategy is well suited to gove a good return from moderate parallel systems. By taking this approach, it is possible to gain benefits from parallelization without the need for extensive modification to the code structure.

In the frame of this project a considerable reduction of the price/performance ratio was shown for the parallel code, which makes the utilization of PMN interesting to industrial users.

IBM RS/6000 SP2

The benchmarks for distributed memory computers were executed on an IBM RS/6000 SP2, located at the Scientific Supercomputing Center (SSC) of the University of Karlsruhe, Germany. This is the currently most powerful IBM parallel computer in Europe. The machine

830

was installed 1997 and comprises 256 nodes with a total peak performance of 107 Giga Flops, 130 Giga Byte Main Memory and 2.1 Tera Byte disk space.

Das HLRKA betreibt mit der IBM RS/6000 SP den derzeit leistungsfähigsten IBM-Parallelrechner in Europa und stellt Nutzern aus Wissenschaft und Forschung diese Kapazität zur Verfügung.

Figure 3   Topology and Shape Optimization in the Product Development Process

2.2   HIPOP - HIgh Performance OPtimization

Standard simulation techniques become increasingly popular while the field of available application software and problems' complexity are rapidly changing. Therefore the availability of an optimization tool which is fully integrated in the CAD and CAE environment is of great advantage.

In the past, due to computer resources and the approaches used, a bottleneck was that the optimization approaches forced the user to simplify the simulation model. The approach implemented in the MSC/CONSTRUCT optimization software is different and showed already in the past outstanding performance. This allowed even real world models to be optimized. For MSC/CONSTRUCT the FEM solver was and is the limiting factor

The HIPOP (HIgh Performance OPtimization) project was funded by the European Comission. The project's main objective was the sound performance improvement of the topology (factor 3-5) and shape (5-10) optimization software MSC/CONSTRUCT through faster algorithms and the excellent performance improvement through the use of distributed parallel MSC/NASTRAN (PMN). HIPOP directly addressed the heart of the optimization: the finite elemente solver and the optimization engine and their interaction.

As a result, large and therefore real world structures can be now optimized without the need for severe simplification. Therefore the pre-processing effort is very little.

Besides aspects of streamlining the optimization process through adaption techniques, the major task within the HIPOP project was to replace the serial MSC/NASTRAN with the parallel MSC/NASTRAN (Figure 5).

Within HIPOP, a fast and efficient coupling of traditional shape optimization tools was realized for the first time, which led to tremendous reductions in pre-processing effort as well as computing time savings.

MSC/CONSTRUCT addresses the needs of designers and analysts. The topology optimization capability allows new designs to be found either from scratch or based on an already existing design within a very short time frame. With the shape optimization capability, existing designs can be further optimized to reduce weight and component stresses.

831

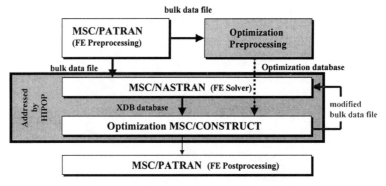

Figure 4   The HIPOP Approach

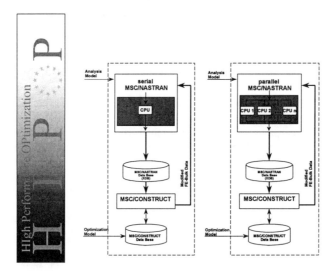

Figure 5   Embedding the Distributed Parallel MSC/NASTRAN

Summarizing the HIPOP project:
- within the same time frame real world models can be optimized without the time consuming need of further simplification
- much larger structures (>> 100.000 degrees of freedom) can be optimised in a reasonable time frame.

The efficiency and robustness was shown by benchmarking structures:
- from BMW where a half car model of a 3-series car, build of 80,000 solids and loaded with 6 different load cases, for running the topology optimization in one third of the time. Through the use of HIPOP software the solids could be meshed with the CAD derived voxel technique which reduces the preprocessing time from 1 1/2 weeks to a couple of hours and which allow the full integration in the CAE process (see below, See Section 3.1"CAE Embedded Topology Optimization" auf Seite 7.).
- from PININFARINA which reduced with topology optimization the weight of a car bonnet frame maintaining critical stiffness values and who did the optimization

within a forth of the time.

- from INA who performed a topology optimization on a chain tensioner containing more than 200,000 nodes and elements.
- from IABG where a front mount with more than 120,000 nodes and elements for 5 different loadcases was topology optimized.
- from BMW who used a coupled shape optimization approach for a crankshaft optimization which resulted in a time reduction from 2 weeks pre-processing effort to 5 hours and computing time reduction to 1/10.

3   Applications

3.1   CAE Embedded Topology Optimization

The target of the BMW benchmark was to evaluate the use of current topology optimization in the design of extremely large and complex structures. To show this the shell structure of a BMW 3-series car was selected.

In this approach the sheet metal from the wheel housing was completely removed and replaced by solid elements. The topology optimization software should then find the maximum stiffness for a given amount of material which satisfies all 6 load cases. The four different boundary conditions simulate various driving situations, for instance cornering.

To get an adequate resolution for the resulting structure an extremely fine mesh is required. For testing and evaluation purposes firstly the wheel housing was filled with a relatively coarse tetrahedral mesh. This resulted in a model containing 20,718 nodes and 14,821 elements, the shell elements of the remaining structure already counted.

To achieve the required resolution and to simplify the setup of the topology optimization model a second structure was created using the voxel technology. The voxel technology has its origin in the CAD/CAE-world where it is used e.g. to simulate engine maintenance processes. For our purposes it allows a very simple representation of the wheel housing and furthermore voxels are extremely easy to generate.

Like 2-dimensional pictures are printed with simple pixels, any CAD model can be described with the 3-d pendant of the pixels: the voxels. A voxel is simply a small cube with a certain edge length. The smaller the edge length the higher the resolution of the model. The data volume is reduced dramatically (often the reduction to 1 per cent of the original data volume is possible) by transforming the description of volumes, planes and points of the original CAD model to a common voxel model (Figure 6).

The difficulty with voxel-based FE models is that one comes easily to extremely large structures and therefore require high resources. The voxel model generated for the HIPOP project contains 91,162 nodes and 81,254 elements resulting in 287,965 degrees of freedom (Figure 7). This model needs to be solved for the above mentioned 6 load cases with 4 different boundary sets.

In Figure 8 the resulting elapsed computing times (left axis) and the corresponding speedups (right axis) versus the number of processors are shown.

The data recovery operations were not part of the parallelization and do not scale with the number of processors. This can clearly be seen from the diagram when looking at the elapsed times of the optimization loop. The curve of the times for one optimization loop with the HIPOP improvements is parallel to the one with the initial software (1 Opt.-Loop (old)).

From the diagram one can see that the elapsed time required for one optimization loop was reduced from 79,000 sec. to 25,000 on 8 processors. That means that the improvements lead to a speedup factor of 3.17.

833

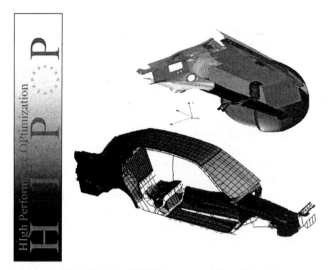

Figure 6   Half Model of the BMW 3-series and Voxel Model

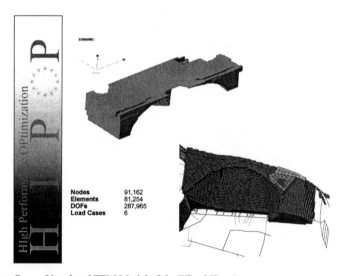

Nodes	91,162
Elements	81,254
DOFs	287,965
Load Cases	6

Figure 7   Voxel and FEM Model of the Wheel Housing

To fully integrate the process of topology optimization within the product development process it is necessary to get the resulting FEM structure (left plot in Figure 9) back to the CAD program. Within HIPOP this was done with an interim step using again a Voxeltool (see Figure 9). Today the final FEM model might be exported from within MSC/PATRAN as VRML file (Virtual Reality Markup Language) or as IGES.

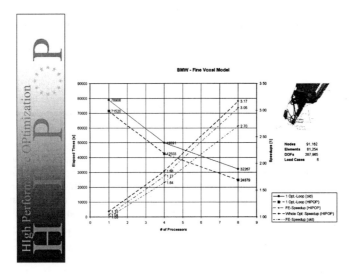

Figure 8   Performance Data

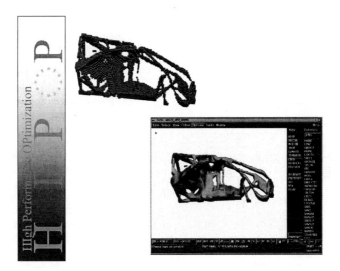

Figure 9   Exporting the final FEM Model to the CAD System

3.2 Coupled Shape Optimization with MSC/CONSTRUCT and SOL200

An optimization model is required for an optimization in the same way as an analysis model is required for a finite element analysis. For shape optimization this also covers the statement of allowable shape changes. In the MSC/NASTRAN SOL200 these shape changes are expressed by so called shape basis vectors ([MOO-94]). The numerical optimization algorithm then determines the best linear combination of these shape basis vectors.

To set up a SOL200 optimization model one major task is to derive shape basis vectors, which have sufficient influence on the optimization objective and constraints ([RAA-88], [CHA-90], [RAA-94]). Because the creation and definition of the shape basis vectors must be made manually, it is time consuming and costly especially for 3D structures.

Other optimization approaches are the optimality criteria procedures like the ones implemented in MSC/CONSTRUCT SHAPE. They often have the advantage to generate the new shape without the necessity of shape vectors. Their disadvantage is their lack of handling arbitrary object functions and constraints. Therefore the idea within HIPOP was to use an optimality criterion for the generation of shape basis vectors and to use the mathematical optimization approach of the SOL200 to fulfill arbitrary objective and constraint functions.

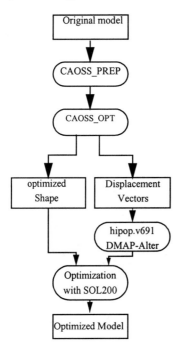

Figure 10   Data Flow for the Coupled
Shape Optimization

This idea was realized as shown in Figure 10. The optimization preprocessing was made within MSC/PATRAN. Using the MSC/CONSTRUCT GUI within MSC/PATRAN the modifiable design nodes were defined and the optimization control file was generated. The optimization control file includes e.g. the definition of the objective function, the constraints as well as the optional restrictions. Then the nonparametric shape optimization with MSC/CONSTRUCT SHAPE was started. This resulted in an optimum shape and a corresponding displacement vector, which describes the change from the original shape to the optimum shape of MSC/CONSTRUCT. With the hipop.v691 DMAP alter a modal analysis was run providing eigenforms which were then considered to be further shape basis vectors. The above mentioned displacement vector from the MSC/CONSTRUCT SHAPE run and these eigenform shape vectors were the parameters for the SOL200 optimization which was finally started.

This approach of coupling the easy and efficient modeling and optimization using MSC/ CONSTRUCT SHAPE with the robust and general shape optimization capabilities of the SOL200 was verified with the crank shaft model shown in Figure 11.

The left plot in Figure 12 shows the displacements through the shape optimization with MSC/CONSTRUCT SHAPE based on the start model. The objective of this optimization was the reduction of the stress levels of the design nodes and therefore the homogenization

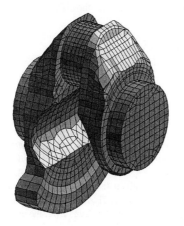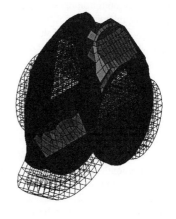

Figure 11   FEM Model and Design Surfaces

of these stresses. It led to a new shape of the crank shaft which formed the input model for a MSC/NASTRAN modal analysis for a couple of eigenfrequencies. The corresponding eigenmodes were then considered as further shape basis vectors for the following SOL200 shape optimization which included constraints on the eigenfrequencies.

Surface Changes through
MSC/CONSTRUCT SHAPE

Surface Changes through
MSC/NASTRAN Sol200

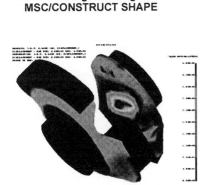

Figure 12   Shape Optimization Results

Comparing the plots of Figure 12 one may deduce that the shape optimization using MSC/ CONSTRUCT SHAPE led to a design which was already very close to the optimum design found by the following SOL200 shape optimization. SOL200 then found only minor shape changes.

Besides cutting the preprocessing time from 2 weeks to 5 hours the use of MSC/CON-STRUCT SHAPE led also to a reduction of the computing time for the SOL200 run by a factor of 10. This is due to the fact that running the SOL200 shape optimization based on a good start design reduces the number of required iterations.

The shape optimization using MSC/CONSTRUCT SHAPE and the SOL200 therefore shows the following benefits:

- The generation of shape basis vectors could be performed automatically.
- This process is fully embedded in MSC/PATRAN and easy-to-use.
- The combination of this software resulted in tremendous time savings without losing generality.

4 Conclusion

The work done in the HIPOP (HIgh Performance OPtimization) project resulted in sound improvements of the throughputs. With this software the topology and shape optimization of large real world models is possible and even much more important: efficient. With the performance improvements within MSC/CONSTRUCT the speedups of MSC/NASTRAN can directly be applied to the optimization and therefore result in the same speedups for the whole optimization. This was shown utilizing Parallel MSC/NASTRAN (PMN) on a distributed memory machine like the IBM-SP2 applying it to large test structures from the automotive industry.

Combining MSC/CONSTRUCT SHAPE with the SOL200 allows not only the automation of the shape optimization process but also results in tremendous time savings for the preprocessing and for the computing.

With the software modules presented in this article the design engineer as well as the analyst have tools to get clear design decisions throughout the product development process. The use of MSC/NASTRAN guarantees the robustness and reliability of the results.

The performance improvements and the functional extensions made in MSC/CONSTRUCT are available with release V2.5.

Acknowledgements

This work was supported in part by DG III of the European Commission under the ESPRIT Contract No. 24462. This support is gratefully acknowledged.

5   References

[ALL-94]   Allinger, P.; Brandel, B.; Müller, O.; Sauter, J.: Optimierung von Bauteilen mit CAOSS und VECFEM/S, Proceedings of the ODIN-Sympo-
sium in Karlsruhe, pp. 57-89, March 3-4 (1994)

[ALL-95]   Allinger, P.; Friedrich, M.; Müller, O.; Mulfinger, F.; Puchinger, M.;
Sauter, J.: Shape and Topology Optimization Using CAOSS and
MSC/NASTRAN
MSC World User's Conference; Universal City, California; May 8-12
(1995)

[ALL-96]   Allinger, P.; Bakhtiary, N.; Friedrich, M.; Müller, O.; Mulfinger, F.;
Puchinger, M.; Sauter, J.: A New Approach for Sizing, Shape and
Topology Optimization. SAE Congress, Detroit, Paper-No. 960814,
February 26-29 (1996)

[CHA-90]   Chargin, M.; Raasch, I.: Structural Optimization with MSC/NASTRAN
revisited in Version 66, MSC/NASTRAN European User's Conference, Paris (1990)

[KAS-94]   Kasper, K. ; Friedrich, M. ; Sauter, J. ; Albers, A.: Parameterfreie Formoptimierung von Bauteilen, Erfahrungen im industriellen Einsatz,
Infografik, 2/1994

[MOO-94]   Moore, G.J.: MSC/NASTRAN Design Sensitivity and Optimization,
User's Guide, Version 68 (1994)

[MUE-98]   Müller, O.: Final Report of the European funded project HIPOP (High
Performance Optimization - ESPRIT Contract No. 24462) (1998)

[RAA-88]   Raasch, I.; Irrgang, A.: Shape Optimization with MSC/NASTRAN;
MSC/NASTRAN European Users' Conference, Rome (1988)

[RAA-94]   Raasch, I.: Structural Optimization with Solution 2001 in the Design
Process; MSC/NASTRAN World Users' Conference, Lake Buena
Vista (1994)

[RAA-98]   Raasch, I.; Bella, D.-F.; Müller, O.: Weitere Fortschritte in der Topolo-
gie- und Formoptimierung unter Verwenduung von MSC/NASTRAN
als Analysepaket. 9th International Congress „Numerical Analysis and
Simulation in Vehicle Engineering; Würzburg; September 24-25
(1998)

[SAU-91]   Sauter, J.: CAOSS oder die Suche nach der optimalen Bauteilform
durch eine effiziente Gestaltoptimierungsstrategie. XX. International
Finite Element Congress, Proceedings, Baden-Baden, November
18-19 (1991)

[SCH-94]   Schreiner, A.: 5 Jahre ODIN, Ergebnisse einer Kooperation, Proceedings of the ODIN-Symposium in Karlsruhe, pp. 21-32, March 3-4
(1994)

Cost/Structural Optimization of a Complex and Composite Foamed Cabinet
Structure

Mr. Kenneth J. Rasche, P.E.

Lead Engineer, Whirlpool Corporation

(1)   SUMMARY

Historically, the appliance industry has not employed significant
engineering analysis to develop product cabinet structures. Instead, major
appliance manufacturers have extensively used the simple build and test
method. Components and small assemblies have been modeled and analyzed
for over a decade, but the expertise and effort to build a representative finite
element model of the refrigerator cabinet structure was simply not
economically feasible in the past.

The common household refrigerator is a complex and composite
structure. The basic structure consists of three different materials, steel,
polyurethane foam and high impact polystyrene (HIPS). The construction of
the cabinet does not lend itself to finite element modeling. The walls of the
cabinet are a composite/sandwich composed of steel and HIPS adhered
together with the foam. The steel components are joined together with
various fastening methods (lance locks, tox locks, toggle locks, clinching, self-
piercing rivets, screws, and welds). These construction methods induce much
uncertainty in the load path from the hinges to the floor.

This study shows that the finite element and solid modeling
technologies have advanced enough that building and using a representative
model is not only feasible but advantageous. The combined use of these tools
allows for much greater understanding of the products' structure and allows
engineers to better optimize the structure and thereby reduce cost without
diminishing quality. This is essential for continued cost reductions after all of
the "low hanging fruit" have been picked. The cost savings attainable with
these tools are simply not possible with only the build and test method.

(2)   THE REFRIGERATOR

The refrigerator is a common household appliance. The basic structure of the refrigerator cabinet is composed of steel, high impact polystyrene (HIPS) plastic and polyurethane foam composite walls. This basic structure is made of 5 components: wrapper, backpanel, freezer compartment food liner, refrigerator compartment food liner, and foam. The other major structural components are: front rail, glider rails (left and right), mounting plate, fill panels (left and right), compartment cover, center rail, and hinges.

The wrapper, backpanel, glider rails, fill panels and front rail are assembled in an automated manufacturing line. This subassembly then has the food liners, and center rail manually installed to finish the foam enclosure, and the assembly is then injected with the foam. The foam adheres to the wrapper, backpanel, both food liners, the center rail and the fill panels. See figure 2.1 for pictorial images of the components.

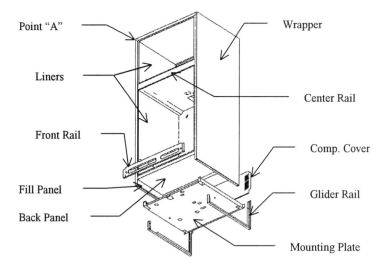

Figure 2.1: Refrigerator Cabinet Structural Components

The main purpose of the cabinet is to maintain the refrigerated compartments. In order to do this, the cabinet must be rigid enough to maintain the seal with the doors under adverse conditions. These conditions include, but are not limited to, large door loads and unlevel floors.

(3)   TEST METHODS

Whirlpool Corporation has found that the stiffness of the cabinet can be measured by the means of a "sway" test. The "sway" test was developed in the 1980's as a tool to evaluate and improve the stiffness of refrigerator cabinets that incorporated the recently introduced plastic food liners.

The test is performed with a fully assembled product. The cabinet is placed on metal blocks at each corner (this removes the rollers from the measurement). The cabinet of the product is loaded with a specific reference weight. The doors of the product are also loaded with a specific reference weight. The doors are opened to various angles and the total movement of the top front handle side corner of the cabinet is measured (this measurement point will be referred to as point "A"). In order for a new cabinet design to pass this test and proceed into production, the total deflection at point "A" for a given set of sample products (3 – 10, depending on the design) must stay below the test limit.

The "sway" test has been found to be very useful in the evaluation and comparison of different cabinet designs. However, there are two significant limitations. These are:

1.   Testing multiple fully assembled products is expensive (both time and money).
2.   The test can only discern fairly large changes in cabinet stiffnesses.

(4)   COST REDUCTIONS

Due to a highly competitive market, cost reductions are necessary for all consumer products. Refrigerator manufacturers have been more successful than most in reducing costs. The Producer Price Index for 17.5 – 19.4 ft^3 refrigerators has only increased 12.5%, (100.0 for 12/1981 to 112.5 for 12/1997) [1]. Inflation on the other hand, has increased 89.9% ($100 in 1981 was worth $189.87 in 1997) [2].

Historically, cost reductions of the structural components of refrigerator cabinets have been done by making small changes to prototype cabinets and then comparing the stiffness of these cabinets to "current production" cabinets by means of a physical test. If the test could not discern between the two different designs, the change was approved. This method has two significant limitations. They are:

1.   Multiple "Small" changes add up to significant stiffness changes.
2.   The changes that go into effect tend to be the ones that are easiest to prototype and implement, not the ones that lead to a more cost efficient structure.

(5)   FINITE ELEMENT MODELING TECHNIQUES

The refrigerator cabinet is a complex and composite structure. The cabinet is first designed to be mass produced, second to be aesthetic, and finally to meet its structural requirements. This fact drives some difficult finite element modeling challenges. The major challenges are listed in table 5.1.

1.   Foam geometry creation.
2.   Foam adhesion to structural parts.
3.   Wrapper front flange to food liner interface.
4.   Tox and Lance locks between sheet metal parts.
5.   Computation time restraints.

Table 5.1: Finite Element Modeling Challenges

These challenges were addressed by using a solid modeler and a finite element analysis package. These packages were used to complement each other in the effort to create the finite element model.

1   Mesh Creation

Several solid models were created in order to speed the preprocessing within the analysis package. The solid modeler was utilized due to its strengths of solid model creation and modification. The solid models of the various structural components of the cabinet were created as follows:

a)   Foam, Food Liners, Wrapper, Backpanel, Center rail and Fill Panels

The foam volume is defined by the food liners, wrapper, backpanel, center rail and the fill panels.
This volume was defined within the solid modeler and then tetrahedron meshed to define the foam elements. The solid modeler eased the geometry creation for the foam and this addressed challenges "1" and "2" listed above. The foam elements were output from

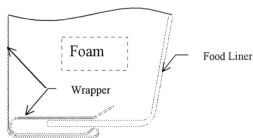

Figure 5.1: Wrapper and Food Liner Flange Interface

the solid modeler and read into the analysis package. These foam elements were then surfaced with food liner elements. A system of "dummy" constraints was used to aid/automate the conversion of the food liner elements to their appropriate properties (wrapper, backpanel center rail and fill panels).

844

The flange area of the foam was modeled in such a way as to allow the proper "slippage" between the wrapper flange and the food liners (see figure 5.1). This addressed challenge "3" listed above.

b)   Hinges and Doors

The hinges were modeled in the solid modeler and meshed as quad/tri shells and as beams. This mesh was output to the analysis package and substructured (created a super element). This allowed significant detail to be included in the hinge models without significantly penalizing computation time. This addressed challenge "5" listed above.

The doors were modeled with stiff beams and were created directly in the analysis package.

c)   Glider Rails, Mounting Plate, Front Rail, and Compartment Cover

The glider rails, mounting plate, front rail and compartment cover were modeled separately in the solid modeler and meshed with quad/tri shells and beams. The individual meshes for the components were input into the analysis package and assembled to the rest of the model. The tox and lance locks were modeled with degree of freedom coupling which allowed the proper relative translations and rotations for the different joints.

2   Boundary Conditions

The model had to be set up for two load cases. For both load cases the majority of the loading is applied through a beam model of the doors. Also for both load cases, the cabinet is constrained at the base as if setting on a level floor. Several other miscellaneous forces are added to both load cases to simulate the approximate weight/force of various components in a complete refrigerator (shelves, compressor, condenser, fan motors, etc.) and the acceleration due to gravity is also applied.

The first load case has the cabinet loaded with a reference load, the doors loaded with a maximum calculated load and the door opened 35 degrees. The cabinet for this load case is constrained in the vertical degree of freedom at three locations near the corners of the base of the cabinet. The front-to-back degree of freedom is constrained at the two front corners of the lower base, and the side-to-side degree of freedom is constrained at the corner of the base near the bottom hinge (see figure 5.2a). The rear corner of the base on the hinge side of the product is not constrained for this load case in any of the three directions. In both actual testing and in the finite element model, this location lifts off the floor and therefore does not carry any load into the floor.

The second load case is the same as the first load case except for three important differences.

1. Door is opened to 165 degrees.
2. The rear corner of the base on the <u>hinge</u> side of the product <u>is</u> constrained vertically.
3. The rear corner of the base on the <u>handle</u> side of the product <u>is</u> <u>not</u> constrained in any direction.

For this load case (2), it is the rear handle side corner of the base that lifts off and is therefore not constrained. Figure 5.2b shows the boundary conditions for load case 2.

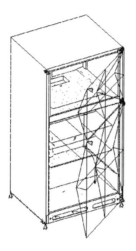   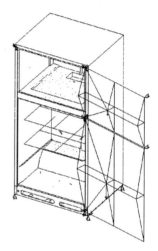

Figure 5.2a: Load Case 1   Figure 5.2b: Load Case 2

3   Solid Model to Analysis Database Automation

The analysis package used allowed for command input through script/input files. This capability was exploited to automate the translation of the solid model into an analysis database for solution. The task of writing the input files slowed the first "iteration" of the model into the analysis database, but the automation allowed components to be swapped in and out of the model relatively quickly after the input files were complete. The input files also allowed the analyst to consistently "clean-up" some of the "messy" results from the solid model mesher.

(6)   OPTIMIZATION TECHNIQUES

Two different optimization techniques were employed during the course of optimizing the refrigerator cabinet. The majority of the optimization work was performed with what the analysis package refers to as the subproblem approximation method. The analyst also used a partial derivative technique as the "optimum" design and manufacturing constraints were evolving.

1   Subproblem Approximation Method

The Subproblem approximation method (SAM) was used to optimize variables that behaved continuously across the design space. Variables that do not behave continuously across the design space (adding removing stiffening components) were handled by running a separate subproblem approximation for each state of the discontinuous variables. Due to space restrictions, details of the evaluation of the discontinuous variables are not discussed here. The continuous design variables (and their ranges) for this study are listed in table 6.1.

Design Variable	Range
Glider rail gage	0.036" – 0.043"
Front rail gage	0.056" – 0.063"
Mounting plate gage	0.044" – 0.050"
Wrapper gage	0.015" – 0.022"
Back panel gage	0.010" – 0.019"
Compartment cover gage	0.010" – 0.019"
Fill panel gage	0.010" – 0.019"
Center rail gage	0.017" – 0.022"
Liner extruded thickness	0.155" – 0.180"

Table 6.1: Design Variables and Their Design Range

The SAM defines the optimization constraints as the "state variables". In this case the one state variable was defined as the total cabinet deflection (total movement of measurement point "A" considering both load cases). The two load cases were analyzed with the design variables set at their current levels. The total cabinet deflection from this "current" evaluation was then used in the optimization as the maximum allowable for the state variable.

The SAM minimizes the objective variable. In this case cost, was defined to be the objective variable and the goal of the optimization was to minimize cost while satisfying the state variable and only varying the design variables within their respective ranges. The "cost" was defined as the

847

combined material cost of all the parts associated with the design variables. As each iteration of the optimization was run, the impact on cost of that iteration's settings were calculated and recorded.

The SAM operates by evaluating several random design iterations and then calculating approximate functions for each state variable and the objective variable. Each random design iteration incorporates changes in each design variable (v_{ds}). Each one of these iterations then becomes a data point for approximate functions for the state variables (V_{st}) and the objective variable (V_{obj}). For this project, the approximate functions took the form shown in equations 6.1.

$$V_{obj/st} = a_0 + \sum_{i}^{n} a_i v_{dsi} + \sum_{i}^{n} \sum_{j}^{n} b_{ij} v_{dsi} v_{dsj} \qquad \text{(Eq. 6.1) [3]}$$

The equation shows the inclusion of the cross terms. This allowed the optimization to include the effects of interactions between design variables.

The SAM continues collecting data points until it has enough to make an estimate on where a minimum of the objective variable may reside that also satisfies the state variable. The design variables are then set to these values for the next evaluation and data point. The solution of this design set is then compared to the estimate from the approximate function. If the results are within specified tolerances, the optimization is assumed to be at a local minimum and the subproblem is stopped. If the results are not within the specified tolerances, the results are added as another data point for the approximate functions and another minimum is estimated. This continues until the program converges to a local minimum or some predefined limits are exceeded (such as number of iterations, cpu time, etc.).

2   Partial Derivative Ratio Technique

The Partial Derivative Ratio (PDR) Technique was used to "hone" in on the best design and to "check" the subproblem approximation results. The first step of the PDR technique is that a "base" run be performed. This was done at the "optimized" design from the SAM. The second step is to then evaluate the effect on the state and objective variable for a small change in each individual design variable (interactions are not considered). The results of these two steps allow the analyst to perform step three that is to calculate the partial derivative of the state and objective function with respect to each design variable. This calculation is shown in equations 6.3 and 6.4.

$$\delta V_{obj}/\delta V_{dsi} = (V_{obj(1)} - V_{obj(0)})/(v_{dsi(1)} - v_{dsi(0)}) \qquad \text{(Eq. 6.3)}$$

$$\delta V_{st}/\delta V_{ds1} = (V_{st(1)} - V_{st(0)})/(v_{dsi(1)} - v_{dsi(0)}) \qquad \text{(Eq. 6.4)}$$

848

Equations 6.3 and 6.4 just show the calculation for one design variable and one state variable. These calculations were actually done for each design variable and state variable (only one state variable in this case).

For this optimization, these calculations (step three) resulted in numbers with units of $/"inches of thickness change" and "inches of sway"/"inches of thickness change" respectively for equations 6.3 and 6.4 for each design variable. Step four of the PDR technique then requires the results from equation 6.3 be divided (ratioed) by the result of equation 6.4 (for each design variable) as shown in equation 6.5.

$$(\delta V_{obj}/\delta v_{dsi})/(\delta V_{st}/\delta v_{dsi}) = \delta V_{obj}/\delta V_{st} \qquad \text{(Eq. 6.5)}$$

For this optimization the calculations from Eq. 6.5 result in numbers with units of $/"inches of sway". This calculation in effect results in an efficiency number for each design variable. In other words, the resulting number tells you that it will cost X dollars to improve the cabinet sway by Y inches for each design variable. This is very useful for checking out the optimization and making sure that the more expensive variables are being minimized and the less expensive variables are being used to satisfy the state variable(s). While the results from the two methods will not map exactly (due to nonlinearities and interactions that the PDR does not consider), the results should show some correlation. Table 6.2 shows that the most "expensive" variables (highest $/"inches of sway" – backpanel, compartment cover, fill panels) were generally forced to minimums, and that the least "expensive" variables (lowest $/"inches of sway" – glider rails, wrapper) were generally increased in size in order to maintain cabinet stiffness.

Design Variable	Efficiency ($/"inches of sway	Sub Prob. Results
Glider rail gage	11.21	Increase
Front rail gage	32.44	Decrease
Mounting plate gage	45.01	No change
Wrapper gage	57.16	Increase
Extruded liner thickness	92.03	Decrease
Fill Panel gage	99.61	Decrease
Compartment Cover gage	121.80	Decrease
Back Panel gage	200.01	Decrease
Center Rail gage	464.94	Decrease

Table 6.2: Design Variables, Their "Efficiency, and Sub Problem Optimization Results

(7)   MANUFACTURING INVOLVEMENT

Manufacturing involvement is essential in the optimization of any component, assembly or product. While no optimization project is complete or all encompassing, the project must include manufacturing considerations or fall risk to evolving/developing to a local minimum rather than the optimum. This is especially true for mass produced products such as refrigerators.

The limits used in the first passes of the subproblem approximation routine were obtained from manufacturing (see table 6.1). They were based mainly on engineering judgment on what was capable with current equipment and what might be obtainable with slight tooling modifications. The optimization process is by definition somewhat of an iterative process. For this reason, it was decided that it was best to first investigate the design portion of the project and then address the manufacturing considerations for the proposed changes.

The first optimization passes were performed, yielding the results listed in table 6.2. This provided some design direction and allowed manufacturing considerations to be addressed again, but this time to a greater level of detail since fewer changes were being considered. At this point in the project, all the proposed changes were still considered possible by manufacturing except for the thinner compartment liners. The responsible manufacturing engineers did not want to "sign up" for the proposed gage reduction due the relative timing of other projects. This resulted in a financial problem for the project. The loss of the savings from the liner reduction made the project financials less attractive. Fortunately for the project (and the company), the reduced density foam project (another unrelated cost reduction project) was having problems for different reasons. The reduced density foam project was having problems passing the cabinet sway specification (no compensation was included in this project other than a redistribution of the foam). It was decided to roll both projects into one.

The inclusion of the reduced density foam into the cabinet optimization was not a simple matter. The foam was not included in the initial optimization runs due to its highly nonlinear behavior (cracks, delamination, voids, etc.), variability and anisotropic properties. Assumptions were made to address the foam variable and the two optimization techniques were again utilized to develop a new optimum design. Due to the additional unknowns and assumptions, a larger number of prototypes were built and tested for correlation.

(8) CORRELATION TO TEST DATA

Correlation to physical test data is always important. In the process of performing any analysis or optimization, many assumptions need to be made. This is even more true on a complex and composite structure such as the refrigerator cabinet. However, modeling a complex, mass produced structure like the refrigerator cabinet also requires care when deciding what test or tests to correlate to. As test data show, variation occurs in test results due to different set ups, different lots of material, and even change of seasons. For the purpose of this discussion, three different examples of correlation will be presented.

1   Deformed Shape Correlation

The many assumptions necessary for this analysis and optimization leave some doubt as to whether the model is behaving (distorting) like an actual cabinet. In order to address this concern, a special test was set up. The exterior sides, back and top of a cabinet were marked off with a grid pattern. The cabinet was then tested with the same loads as used for a typical sway test and the displacements at each grid point were measured. The displacements from this test were then forced into a "dummy" shell model of a cabinet. The displacement contours of the special test were then compared to the displacement contours for the analytical model. Figure 8.1 shows the comparison.

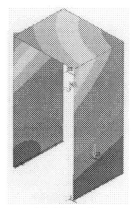

Test Results   Model Results

Figure 8.1: Results Comparison between the Special Deflection Test and the Analytical Model

As figure 8.1 shows, the deformed shape from the analytical model and the special deflection test do correlate. This indicates that the analytical model does represent the load paths through the cabinet, the deformed shape of the cabinet, and the relative stiffnesses of components.

2   Deep Door Hinge Results Correlation

An independent evaluation of deep door hinge effects was performed separately from this optimization. Due to space limitations, the details of the evaluation will not be discussed. Initial testing (of three cabinets) of the deep door hinge effects indicated that sway was increased by 36% when deep door hinges were used. The deep door hinges were evaluated in the analytical model and showed the effects to be 18%. Two additional physical tests were performed (three cabinets each) and they respectively indicated that the effect was 46% and 8%. This evaluation does not necessarily show good correlation (average of all physical testing indicate the effects are 30%, versus the models 18%). This evaluation does indicate that the structure varies enough that correlation to a single analytical model may not be possible. This evaluation also shows that physical testing without the analytical model is both misleading and expensive.

3   Optimization Results Correlation

Final testing prior to production approval was performed on standard and optimized cabinets. Seven cabinets of each were tested. The standard cabinets resulted in an average sway of 0.336" and a standard deviation of 0.015". The Optimized cabinets resulted in an average sway of 0.337" and a standard deviation of 0.015". The model indicated the optimized cabinet should perform 0.7% worse than the standard cabinet (0.338"). Section 2 above (Deep Door Hinge Results Correlation) indicates that some luck is necessary in order to achieve correlation this tight.

(9)   PROPAGATION OF OPTIMIZATION

The results from this cabinet optimization proved to be quite beneficial (hundreds of thousands of dollars in annual material savings). Due to these results, the techniques discussed in this paper are being utilized on additional product cabinet designs.

(10)   REFERENCES

1   BUREAU OF LABOR STATISTICS – Producer Price Index Revision – Current Series, Household Refrigerators and Freezers, 17.5 to 19.4 Cubic Feet, http://stats.bls.gov, 12/4/98

2   FRIEDMAN, S. MORGAN – The Inflation Calculator, http://www.westegg.com/inflation/, 12/15/98

3   ANSYS, INC. – Ansys Theory Reference, Eighth Edition, Subproblem Approximation Method, Ansys Inc., 1997

The HyperSizing Method for Structures

Craig Collier, Phil Yarrington, and Mark Pickenheim
Collier Research Corporation
2101 Executive Dr. 6th floor
Tower Box 72
Hampton, Virginia 23666

collier-research.com
e-mail: info@collier-research.com

Abstract

A practical structural optimization system specifically designed for effective engineering solutions is presented. The system, called HyperSizer™ [1], is coupled with finite element analysis (FEA). The system is based primarily on accurate engineering analyses and secondarily on discrete optimization. Its underlying method is a departure from typical finite element design sensitivity and optimization that emphasize numeric optimizers, and model based user defined constraints on strength and stability failure analyses.

HyperSizer's built-in detailed analysis capabilities, and its ease of use makes it suitable as a tool for performing automated structural analyses of any general structure. Indeed, this is the fundamental premise of the HyperSizing method. HyperSizer's ability to predict structural failure will be first presented, and then the benefits of coupling the closed form analytical capabilities with those provided with FEA. The example application is a space launch vehicle, which containing 7 assemblies, 21 optimization groups, and 203 structural components. It demonstrates how an engineer is able to provide 'real-world' expertise in the optimization process by interacting with HyperSizer for *designs on the fly*.

Introduction

Planes, rockets, automobiles, and ships require FEA to solve their 'running-loads', 'internal loads', or 'load-paths.' In essence, the vehicle FEM is first and foremost used as a 'loads' model for integrating the effects of surface pressure, temperature, and accelerated inertia into element forces and moments. It accomplishes this because the discrete shell elements accurately represent the generalized stiffness of the individual panel and beam structural components of the vehicle design. This forms the **first** premise of the HyperSizing method. That is the loads model does not need to know the actual cross sectional shapes of the panels and beams, nor their composite material layups [1]. The method is robust enough to handle panels and beams with general cross sectional shapes, including those, which are unsymmetric or unbalanced [2].

HyperSizer automates the analysis and optimization of structures by using the FEA computed 'internal' panel and beam forces and moments. These are used to check and avoid the many different types of failures that may occur within a structure [3,4]. The **second** HyperSizing premise is that once the structural component's design-to loads are accurately resolved, potential panel or beam failures can be effectively predicted with explicit, closed form methods. The simplest analogy to this capability is an automated hand stress check performed after the internal FEA loads are computed to determine if they exceed the load carrying capacity of the structural members. These non-FEA based failure analyses, which are quite sophisticated, include material strength, panel biaxial buckling, beam-column buckling, local buckling of flanges and webs, crippling of the cross section, deformation, modal frequency, etc.

A substantial challenge to automating structural analysis and optimization is 'pulling-loads.' The problem arises when many finite elements are used to represent a structural component. This is especially true if the panel has varying load from midspan to edge, or from one edge to another edge. Designing to the maximum element load could be far too conservative and result in over-weight. Buckling failure modes are more dependent on integrated type compressive load than an element peak load, which may be located at the panel's corner. A **third** HyperSizing premise is that statistics is the best way to determine the appropriate design-to load.

A **fourth**, and final HyperSizing premise is that optimization of all possible panel and beam design variables of the total structural system is best accomplished with discrete optimization [5]. This is particularly true when there are many in-service loadings subject to many diverse local level design criteria. This approach permutates panel and beam designs based on user-defined upper and lower bounds of each variable. Benefits to this approach as implemented in HyperSizer are that non-numeric optimization variables like material or structural concept can be handled, as well as discrete optimization variables such as number of plies, without the occurrence of local or false optima, and without limitation on problem size. In fact, HyperSizer is able to discretely optimize in a manner that considers material selections and panel or beam concepts in addition cross sectional dimensions, thicknesses, and layups. Using methods to accurately compute margins-of-safety for all potential failures, without depending on the user being able to derive these on his own experience, guarantees structural integrity of the selected optimum design.

The first three of these four premises are discussed briefly below, leaving the bulk of the paper devoted to the aerospace vehicle optimization example.

Generalized stiffness coupling with FEA

HyperSizer is significant due to its generality and ability to be linked accurately with planar finite element analysis (FEA). Non-linear, temperature and load dependent constitutive material data of each composite material's laminate are used to "build-up" the stiffened panel membrane, bending, and membrane-bending coupling stiffness terms and thermal coefficients. These panel data are input into the FEA program to accurately perform analysis with coarse meshed models. The method is robust enough to handle panels with general cross sectional shapes, including those, which are unsymmetric or unbalanced. Traditional methods of formulating equivalent plate panel stiffness and thermal coefficients, though intuitive, are difficult to use for a wide possibility of applications and give incorrect results for thermomechanical internal load distributions [1,2]. A technique of implementing this formulation with a single plane of shell finite elements using MSC/NASTRAN was revealed that provides accurate solutions of entire airframes or engines with coarsely meshed models, Fig. 1. These models produce accurate thermomechanical internal load distributions, Fig. 2 solved with closed form methods.

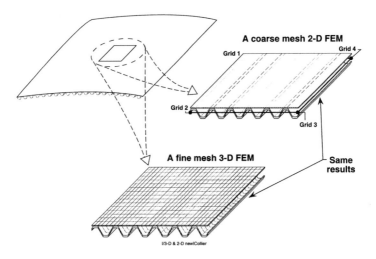

Fig. 1 Genearlized stiffnesses provide accurate FEA.

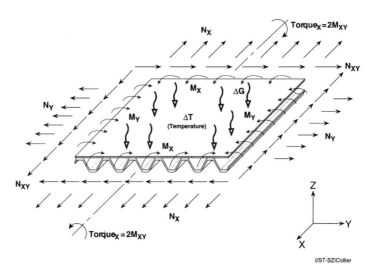

Fig. 2 These FEA produced thermomechanical load combinations can be used accurately by closed form analysis methods.

856

Closed form, physics based analysis methods

Resulting FEA solved thermomechanical forces and moments can cause many different types of failures to occur within a structure. Some of these failures can be predicted with FEA quite easily, some can be accomplished only with very discrete and finely detailed model meshes (such as local buckling of stiffened panel spans), and then others cannot be effectively accomplished with FEA (such as empirical crippling analyses). In any case, to satisfactorily achieve desired accuracy, many different types of Finite Element Models (FEMs) are usually required in addition to the 'loads model' to predict the multitude of failure possibilities.

Explicit methods can have complete knowledge of detailed design, materials, and design principles including manufacturing constraints, without being hindered by mesh density concerns. HyperSizer's robust automated structural analysis system contains an extensive list of physics based strength, stability, stiffness, and minimum frequency failure analyses as well as provisions to include user defined design criteria such as manufacturing minimum sheet gages and composite material ply layup sequences. By using detailed knowledge of the structure's design, accurate failure analyses can be performed explicitly using the forces and moments from the loads model. In fact, HyperSizer was able to predict the same failure loads, in less than a second each, for the three separate FEMs of Figs 3-5.

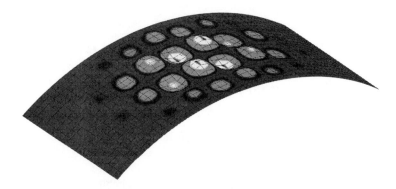

Fig. 3 Biaxial compression buckling of a cylindrical, stiffened fuselage.

857

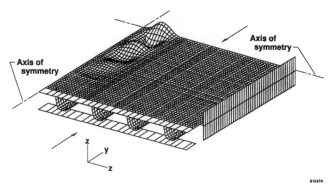

Fig. 4 Local buckling of panel. For benchmark, see page 72 of ref. 4.

Fig. 5 Local buckling of composite facesheet. See page 153 of reference 3.

FEA computed design-to loads quantified with statistical analyses

Structural analysis is performed using two primary data: applied loadings and allowable loadings. An allowable loading is due to a combination of the material's strength and the nature of the structural design such as panel concept, shape, size, etc. Reliability of a structure is defined as the probability that the allowable load is greater than the required load. Potential failure occurs when the curves overlap in the middle, Fig. 6. The ultimate question being what is the appropriate 'design-to' loading for performing a deterministic structural component analysis.

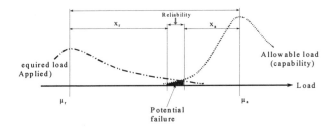

Fig. 6 A statistical approach is used for analyzing potential failure.

The "narrowness" a bell curve distribution is called Kurtosis. A large Kurtosis is desirable because of its narrow width. Unfortunately, as seen in Fig. 6, loadings sometimes have small Kurtosis, i.e. a wider curve causing a larger separation (variance) of the applied loading. The problem is one of determining acceptable levels of load or risk, which is particularly relevant to structural optimization.

$$\text{``Design-to'' loading} = \mu + K\sigma$$

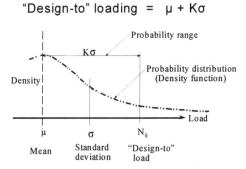

Fig. 7 The HyperSizer user can select the K standard deviation factor for determining the "Design-To" applied loading for strength analysis.

Structural analyses are typically performed using a component's peak loading without much concern given to the actual load distribution. For components with uniform loadings, i.e. narrowly varying load distributions/large Kurtosis, this approach is sufficient. However for components with widely varying load distributions, i.e. higher loading gradients, this approach becomes overly conservative. The statistical approach of HyperSizer treats the individual force components (N_x, N_y, N_{xy}, M_x, M_y, M_{xy}, Q_x, Q_y,) of each element of a structural component, in essence, as if they were a frequency distribution, or a probability histogram. In this way, the *K factor* (referred to as K sigma, such as 3σ) identified in Fig. 7 is now used to achieve the desired *confidence limit* of the component's area which is experiencing a level of load. As a result, for a one-sided distribution, a K factor equal to 1, 2, or 3 indicates 84.13, 97.72, and 99.86 % of the component's area.

Optimizing the total structural vehicle

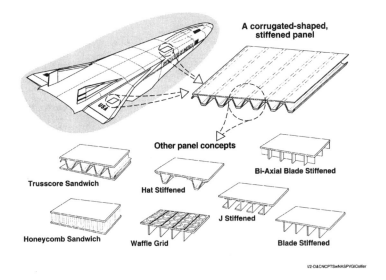

A corrugated-shaped, stiffened panel

Other panel concepts

Trusscore Sandwich

Hat Stiffened

Bi-Axial Blade Stiffened

Honeycomb Sandwich

Waffle Grid

J Stiffened

Blade Stiffened

I/2-D&CNCPTSwNASPVGlColler

Fig. 8 Discrete optimization permits the optimum selection of panel concept for all surface areas.

The ability to optimize a total structural system with all design variables is fundamental with HyperSizer, Figs 8 and 9. Since finite element sensitivity analysis and numerical optimizers are not used in this approach, there is no exponential relationship between run time and model size/number of design variables. In fact, the only model size limitation is based on practical limits of the linear static solver and pre/post processors. Run times are quick and do not have local minimum solutions that possibly occur with numerical optimization. Fast response is important for the user to be able to keep focus. Immediate feedback on optimization selections, given automatically, helps the user to interact with the solution process and to be able to stay on track with his design thoughts.

An advantage of using explicit solutions is that analyses are accomplished rapidly and can consider the multitude of failure modes and loadcases. HyperSizer's purpose is to include all possible failure modes in the assessment of a possible design. The objective is that the user should be able to depend on the software for capturing all physics based structural integrity checks. As an example taken from reference 6, page 20, in the process of a optimizing a beam, the HyperSizer user need not provide a constraint on the maximum allowable beam height to width ratio because in addition to simple bending stress criteria, HyperSizer would also investigate twisting, lateral-torsional buckling etc.

860

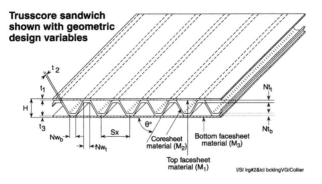

Fig. 9 All aspects of the structural design are optimized

- Panel and beam concepts
- Material selections
- Cross sectional dimensions, thicknesses and layups
- Layups are even customizable to include odd angles and ply dropoffs using an integrated composite layup builder

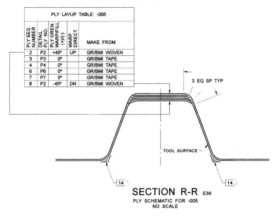

Optimization input data is easy to select. Optimization bounds can be assigned to different vehicle locations quickly, referred to as *groups*. Optimization solutions are determined for components, which are a subset of groups. *Components* are defined during the process of constructing the FEM and usually represent the smallest piece of manufacturable structure. That is structure fabricated with <u>all</u> of the same design dimensions such as stiffener spacing, panel height, web thickness, facesheet thickness, layups, etc. Visual interpretation is provided automatically for the current panel concept, component, and group.

861

About the Model

The model represents a NASA designed two-stage-to-orbit aerospace plane requiring accurate analysis capabilities to account for a complex thermomechanical environment. The integrated airframe/engine design contains a large volume of pressurized cryogenic fuel. Internal bulkheads serve as shape control members to maintain the vehicle's shape. The aeroshell is designed to be graphite/epoxy, hat-shaped stiffened panels.

Though HyperSizer can analyze and optimize FEMs as large as one million DOFs, the choice was made to build a relatively small model of approximately 30,000 degree of freedoms (DOF) for the aerospace vehicle. This allows us to take advantage of HyperSizer's unique panel and beam stiffness formulations that achieve accuracy with coarsely meshed MSC/NASTRAN FEMs.

Interaction between the engineer and the software is key to HyperSizer's design process

Engineers learn within seconds the strengths and weaknesses of their structural designs from the software's interactive reporting of margins-of-safety. Interactive 3-D graphics provide visual inspection of the structural component layout, assemblies, and drawn to scale optimum panel and beam cross sections. See Fig. 10. These features are used on the aerospace plane to quickly interpret and understand design flaws. Critical design issues were identified and resolved early in the design process, allowing ample time to perform many design trade studies. This quick and highly interactive process makes the task of saving weight easy and fun.

Conclusion

The commercially available HyperSizer™ detailed analysis and sizing optimization program, which is integrated with FEA, is described using an aerospace example. The example model is a reusable launch vehicle referred to as an aerospace plane. It contains 7 assemblies, 21 optimization groups, and 203 structural components. FEA is used for predicting internal loads. The entire plane is optimized for minimum weight with both composite and metallic materials. Structural integrity is ensured because of over 100 different failure analyses considered by HyperSizer that included strength, buckling, crippling, deformation, and frequency. Run times on a Pentium workstation ranged from two to ten minutes for the entire vehicle.

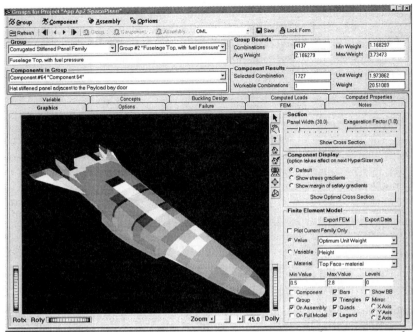

Fig. 10 Interactive display tools illustrate the computed optimum panel unit weights margins-of-safety on the assembly called 'OML'.

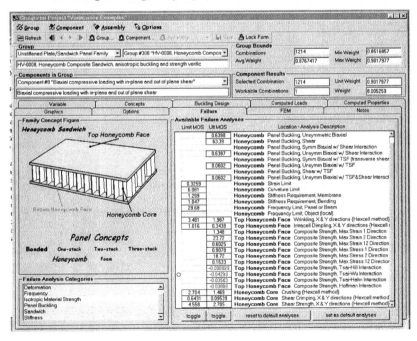

The graphical display of analysis and design results is shown to provide the engineer with a powerful insight into the structural problem, and in so doing, allows 'real-world' expertise in the optimization process. The analytical methods and general approach of this integrated tool apply to FEA users in other industries.

References

1. Collier, C.S., "Structural Analysis and Sizing of Stiffened, Metal Matrix Composite Panels for Hypersonic Vehicles", AIAA 4th International Aerospace Planes Conference, Orlando, FL, December 1-4, 1992, Paper No. AIAA-92-5015

2. Collier, C.S., "Thermoelastic Formulation of Stiffened, Unsymmetric Composite Panels for Finite Element Analysis of High Speed Aircraft", 35th AIAA/ASME/ASCE/AHS/ACS Structures, Dynamics, & Materials Conference, Hilton Head, SC, April 18-20, 1994, AIAA paper 94-1579

3. HyperSizer User's Manual, Book 1: Tutorial and Applications, Collier Research Corp., Hampton, VA, September 1998

4. HyperSizer User's Manual, Book 2: Analytical Method and Verification Examples, Collier Research Corp., Hampton, VA, September 1998

5 Formal vs. Discrete Optimization and HyperSizer, Collier Research Corp., Hampton, VA, http://collier-research.com/papers.html, August 1997

6 MSC/NASTRAN Design Sensitivity and Optimization Manual, Version 68, The MacNeal-Schwendler Corporation, Los Angeles, CA, 1994

Trademarks

HyperSizer™ is a trademark of Collier Research Corporation
NASTRAN® is a registered trademark of NASA
MSC/NASTRAN™ is an enhanced proprietary product of The MacNeal-Schwendler Corporation

Shape variable definition with C^0, C^1, and C^2 continuity functions

G. Chiandussi[1], G. Bugeda[2], E. Oñate[2]

SUMMARY

A new technique for the definition of the shape design variables in 2D and 3D optimisation problems is proposed. It can be applied to the discrete model of the analysed structure or to the original geometry without any previous knowledge of the analytical expression of the CAD defining surfaces. The proposed technique allows for defining *a priori* the surface continuity to be preserved during the geometry modification process. This capability allows for the definition of shape variables suitable for every kind of discipline involved in the optimisation process (structural analysis, fluid-dynamic analysis, crash analysis, aerodynamic analysis, etc.).

1   INTRODUCTION

The correct definition of shape variables is of the utmost importance in the layout of an optimisation problem. One of the first researchers to point out the complexity and the difficulties in the shape variable definition was M. H. Imam [1]. The design element and the shape superposition techniques proposed in [1] have been widely used till now. The former is based on the concept of design element, a macro finite element. The isoparametric shape representation, as used for individual elements, is used here to determine the isoparametric co-ordinates of any point belonging to the design element itself. The shape superposition technique concerns the possibility to superimpose two or more shapes specified in terms of model location of points on a curve or surface in varying proportions. The linear combination of these predefined shapes allows for the generation of a variety of shapes, the coefficients of the combination being the shape variables [2,7].

---

[1] Department of Mechanical Engineering, Technical University of Turin, Corso Duca degli Abruzzi 24, 10129 Torino, Italy
[2] CIMNE, International Center for Numerical Methods in Engineering, Edificio C1, Campus Norte UPC, Gran Capitán s/n, 08034 Barcelona, Spain

865

In 1984, V. Braibant and C. Fleury [3] proposed a new approach for two-dimensional shape variable definition. The design element concept has been used once more. Instead of using the shape functions of a two-dimensional finite element, blending functions typical of Bezier or B-spline curves are used to determine the co-ordinates of any point inside the design element or on its boundaries. Therefore, the shape variables are no longer the position of the nodes of an isoparametric two-dimensional element, but the points that control two families of curves whose cartesian product defines the design element [4].

The techniques above mentioned show several limitations. In the curve superposition technique difficulties arise in the selection and evaluation of the master curves or surfaces. In the design domain one only zero continuity of displacements can be guaranteed between adjacent elements. The technique proposed by Braibant and Fleury solves the lack of continuity control at the interface of two adjacent domains by introducing Bezier or B-spline interpolation functions. This approach can be easily adopted for two-dimensional problems, but difficulties arise in its extrapolation to three-dimensional ones.

The present paper proposes a new shape design variable technique. It allows for defining the shape variables directly on the geometry of the analysed structure without any previous knowledge of the analytical expression of the CAD surfaces. The proposed technique allows for an *a priori* definition of the surface continuity to be preserved during the geometry modification process. This capability allows for the definition of shape variables suitable for every kind of discipline involved in the optimisation process.

2   THE METHOD

The proposed method is able to manage two-dimensional as well as three-dimensional shape variable definition problems, the only difference being the degree of the geometrical entities to be taken in consideration. Whereas in a three-dimensional problem the geometry of the analysed structure is described using two-dimensional surfaces, in a two-dimensional problem this is obtained by using one-dimensional curves.

2.1   Surface modification

Let us consider the three-dimensional geometric model of a structure or of a fluid domain surrounding an immersed object. The geometrical model is usually defined by several patches linked together. It is possible to group the patches to define one or more macro-patches. This subdivision allows for the identification of the portion of the boundary surface of the structure that will be affected by shape modifications and the one that will be not. Very often, the mathematical definition of these patches is unknown or it is very complex and outside the scope of the numerical analysis. In these cases it is impossible to obtain the parametric equation r(u,v) of the macro-patches and, consequently, to evaluate the parametric co-ordinates describing the location of the points belonging to them.

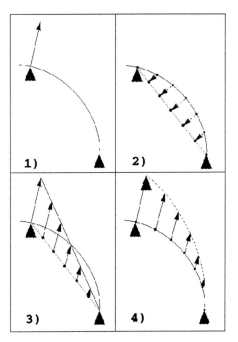

In order to obtain, at least, the approximate parametric co-ordinates of a point belonging to a macro-patch, a new patch approximating the real one is built by using the information on the co-ordinates of the vertices of the macro-patch. Figure 1 describes this process by using a two-dimensional example for the sake of simplicity. In this case the macro-patch is a curve (1) and the approximating patch is a straight line connecting the two vertices (2). The projection of every point belonging to the macro-patch over the new one along its normal can

Figure 1 The four steps of the propagation of the vertex movement on a one-dimensional macro-patch.

now be evaluated (2). The projection location over the approximating patch as well as the approximating patch equation can be transformed from the (x,y,z) reference system to a (u,v) parametric one. A set of standard interpolation functions with C^0, C^1 and C^2 continuity have been identified. Every interpolation function is described by using a parametric equation as will be seen in paragraph 2.2. The knowledge of the parametric equation of the approximating patch allows to relate it to the standard interpolation function. Consequently, it is possible to identify for every point projection on the approximating patch a point on the interpolating function describing the effect of the application of a vertex movement (3). With this approach the new position of each point of a macro-patch is defined in terms of the original geometry plus the superposition of some curved patches (4). The shape modifications are controlled in direction and magnitude by vectors applied at the vertices of the macro-patches representative of the shape variables (4). Finally, the co-ordinates (x', y', z') of each point belonging to every macro-patch after shape modification are obtained through the application of the following equation:

$$(x',y',z') = (x,y,z) + \sum_{i=1}^{nd} d_i \cdot \sum_{j=1}^{nv} v_{ij} \cdot \sum_{k=1}^{np} r_{jk}(x,y,z) \qquad (1)$$

where:

- nd is the total number of design variables,
- nv is the total number of vertices of the macro-patch,
- np is the total number of macro-patches whose one of the vertices is $\mathbf{v}_{ij}$,
- d_i is the value of the i-th design variable,
- $\mathbf{v}_{ij}$ is a vector related to the design variable i and applied at the vertex j (vector $\mathbf{v}_{ij}$ is defined in direction and magnitude),
- $\mathbf{r}_{jk}(x, y, z)$ is a function defined over the patch approximating the macro-patch k to interpolate the movement of the vertex $\mathbf{v}_{ij}$.

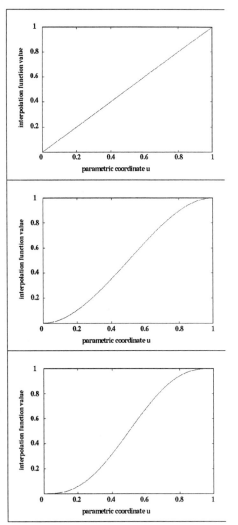

Figure 2 C^0, C^1 and C^2 mono-dimensional interpolation functions.

The modifications of the surface induced by the design variables are transformed in a displacement field of the nodes of the discrete model belonging to the macro-patches. To avoid strong distortion effects it is necessary to propagate the displacement field defined at the boundary surfaces to all internal nodes of the discrete model. The movement of the mesh points is obtained by computing the displacements of a fictitious linear elastic structure formed by the mesh elements under prescribed boundary conditions [10].

2.2   Propagation of the boundary movement

The identification of the deformed shape of a macro-patch due to the application of a movement vector to one of its vertices and the transfer of the information so obtained into a displacement field of the nodes of the discrete model belonging to it are performed together. The co-ordinates (x,y,z) of the nodes of the discrete model laying on the surface of the structure are approximated with the parametric co-ordinates (u,v) of their projection over the approximating patch. These parametric co-ordinates are used to identify the effect of the application of an imposed displacement to any vertex of the macro-patch by keeping as a reference a parametric standard surface with parametric co-ordinates (Figure 1):

$$0 \le u, v \le 1 \qquad (2)$$

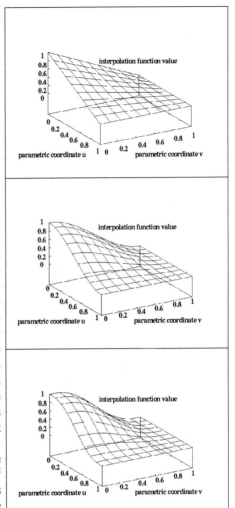

Every interpolation function r, is characterised by a unit value at one of the vertices and a null value at all the other ones. Then, for every macro-patch it is possible to identify as many interpolation functions r as the number of the vertices. The interpolation functions r are defined to keep certain continuity properties. The smoothness of the boundary surface of the structure requires continuity of the desired degree to be preserved at the boundaries along the boundaries themselves and between different macro-patches. This implies 0, 1 and 2 continuity degree respectively for the C^0, C^1 and C^2 continuity interpolation functions also across their boundaries. The requirement of unit sum (displacement of a portion of the boundary surface parallel to itself) and the consequent requirement of symmetry lead to ask for cross-boundary derivatives continuous and equal to zero.

Figure 3 C^0, C^1 and C^2 interpolation functions for the two dimensional quadrangular domain.

a)   Mono-dimensional interpolation functions

The identification of the analytical expressions of the interpolation functions in parametric co-ordinates is quite straightforward in the one-dimensional case. The C^0 interpolation function is simply a first degree polynomial in the unique parametric co-ordinate u (Figure 2):

869

$$f(u)=u \tag{3}$$

the C^1 interpolation function a third degree polynomial (Figure 2):

$$f(u)=3u^2-2u^3 \tag{4}$$

and the C^2 interpolation function a fifth degree polynomial (Figure 2):

$$f(u)=10u^3-15u^4+6u^5 \tag{5}$$

b)   Two-dimensional interpolation functions with quadrilateral surfaces

The identification of the interpolation function in the two-dimensional case is more complex. The approximating patch can be obtained as a Boolean sum of two surfaces interpolating the r(u,0) and r(u,1) boundary curves along the v direction and the r(v,0) and r(v,1) boundary curves along the u direction (Coon patch). The resulting surface patch is conveniently expressed in the matrix form by [8]:

$$r(u,v) = \begin{bmatrix} 1-u & u \end{bmatrix} \cdot \begin{bmatrix} r(0,v) \\ r(1,v) \end{bmatrix} + \begin{bmatrix} r(u,0) & r(u,1) \end{bmatrix} \cdot \begin{bmatrix} 1-v \\ v \end{bmatrix}$$

$$- \begin{bmatrix} 1-u & u \end{bmatrix} \cdot \begin{bmatrix} r(0,0) & r(0,1) \\ r(1,0) & r(1,1) \end{bmatrix} \cdot \begin{bmatrix} 1-v \\ v \end{bmatrix} \tag{6}$$

The auxiliary functions u, (1-u), v and (1-v) are called blending functions, because their effect is to blend together four separate boundary curves to give a single well-defined surface. The surface so identified has only positional continuity across patch boundaries (Figure 3).

Looking for a C^1 and C^2 continuity interpolation functions, gradient continuity needs to be preserved and the patch needs to be defined not only in terms of its boundary curves but also in terms of its cross-boundary slopes. Let us consider only the C^1 continuity interpolation function. The patch equation can be derived as above except that now it is necessary to use generalised Hermite interpolation rather than generalised linear interpolation. The resulting equation is:

$$r(u,v) = \begin{bmatrix} \varphi_0(u) & \varphi_1(u) & \psi_0(u) & \psi_1(u) \end{bmatrix} \cdot \begin{bmatrix} r(0,v) \\ r(1,v) \\ r_u(0,v) \\ r_u(1,v) \end{bmatrix} + \begin{bmatrix} r(u,0) & r(u,1) & r_v(u,0) & r_v(u,1) \end{bmatrix} \cdot$$

$$\begin{bmatrix} \varphi_0(v) \\ \varphi_1(v) \\ \psi_0(v) \\ \psi_1(v) \end{bmatrix} - \begin{bmatrix} \varphi_0(u) & \varphi_1(u) & \psi_0(u) & \psi_1(u) \end{bmatrix} \cdot \begin{bmatrix} r(0,0) & r(0,1) & r_v(0,0) & r_v(0,1) \\ r(1,0) & r(1,1) & r_v(1,0) & r_v(1,1) \\ r_u(0,0) & r_u(0,1) & r_{uv}(0,0) & r_{uv}(0,1) \\ r_u(1,0) & r_u(1,1) & r_{uv}(1,0) & r_{uv}(1,1) \end{bmatrix} \cdot \begin{bmatrix} \varphi_0(v) \\ \varphi_1(v) \\ \psi_0(v) \\ \psi_1(v) \end{bmatrix}$$

$$\tag{7}$$

The substitution of the boundary curve expressions leads to the C^1 interpolation function expression (Figure 3):

$$r(u,v) = (1-3u^2+2u^3)(1-3v^2+2v^3) \tag{8}$$

In a similar manner it is possible to obtain the C^2 continuity interpolation function shown in Figure 4. Two more blending functions are needed and the square matrix involved is of order 6x6. The patch expression is given by (Figure 4):

$$r(u,v) = (1-10u^3+15u^4-6u^5)(1-10v^3+15v^4-6v^5) \tag{9}$$

c)   Two dimensional interpolation functions; triangular surfaces

The definition of the interpolation functions for a triangular domain is more complex than the previously explained one. Following the same procedure used before it is possible to obtain the C^0 continuity interpolant for the triangle as (Figure 4):

$$r(u,v) = 1-u-v \tag{10}$$

The approach followed by Gregory [11,14] has been used to identify the interpolation function for the C^1 case. The expression of the C^1 continuity interpolation function over a triangle is given by (Figure 4):

$$r(u,v) = \frac{\alpha(u,v)r(u,0)(u+v-1)^2(2v-u+1)}{(1-u)^3} + \frac{\beta(u,v)r(0,v)(u+v-1)^2(2u-v+1)}{(1-v)^3}$$
$$+ \gamma(u,v)\left[\frac{v^2(3u+v)r(0,v)}{(u+v)^3} + \frac{u^2(3v+u)r(u,0)}{(u+v)^3} - \frac{u^2v^2(-6u+6u^2-6v+6v^2)}{(u+v)^4}\right] \tag{11}$$

where:

$$\alpha(u,v) = u^2[3-2u+6v(1-u-v)] \tag{12}$$
$$\beta(u,v) = v^2[3-2v+6u(1-u-v)] \tag{13}$$
$$\alpha(u,v)+\beta(u,v)+\gamma(u,v) = 1 \tag{14}$$

The C^2 continuity interpolation function is still missing and this will be object of future research work.

2.3   Displacement propagation inside the discrete model

The interpolation functions previously defined allow for the propagation of the displacements defined by the vectors over the macro-patches. The process takes into account only the boundaries of the discrete model of the analysed domain. To reduce the distortion effects that would take place in the areas near the boundaries of the model during the updating process operated by the optimisation code, it is necessary to propagate these boundary modifications to the internal nodes of the discrete model.

The propagation of the boundary displacements is accomplished by considering the mesh as a fictitious linear elastic structure. Different mechanical properties can be selected and assigned to each mesh element. Assigning materials with higher mechanical characteristics to the elements near the moving surfaces and materials with lower mechanical characteristics to the elements far from these surfaces, it is possible to distribute the mesh deformation more uniformly. Different laws of variation of the mechanical characteristics can be adopted following geometrical or physical criteria. The best results have been obtained using a distribution law of the Young modulus of the material depending on the strain energy [10]. Given a Poisson modulus ν and an arbitrarily selected Young modulus $\overline{E}$, the desired strain mean value $\overline{\varepsilon}$ can be obtained by a non-unifom Young modulus distribution given by:

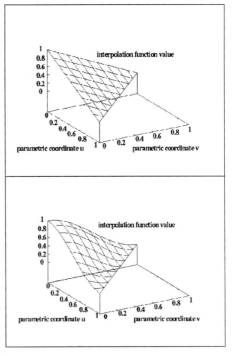

Figure 4 C^0 and C^1 interpolation functions for the two dimensional triangular domain.

$$E = \frac{\overline{E}}{\overline{\varepsilon}^2} \frac{\left(\varepsilon_1^2 + \varepsilon_2^2 + \varepsilon_3^2\right) - 2\nu\left(\varepsilon_1\varepsilon_2 + \varepsilon_2\varepsilon_3 + \varepsilon_3\varepsilon_1\right)}{3(1 - 2\nu)} \qquad (15)$$

2.4 Implementation of the method

The proposed method has been implemented in a specific computer program that takes advantage of the capabilities offered by the pre and post processor system GiD [15] developed at the International Center for Numerical Methods in Engineering (CIMNE, Barcelona). The result of the application of the program is the creation of two files for post processing purposes and one file in standard NASTRAN format containing the relationships between the design variables and the nodal displacements of the discrete model.

3   EXAMPLES

Two examples will be shown next. The first example concerns the shape variable definition of an airfoil fluid-dynamic optimisation problem. The airfoil is defined by several points connected by spline curves. The fluid domain has been defined as a rectangular box around the airfoil. Three shape variables have been defined (Figure 5). The three interpolation facilities have been used to allow for a comparison between the different results that can be obtained. Mesh modification is quite uniform and elements near the moving surfaces maintain quite the same dimension and shape as the original ones. This procedure allows for reducing element distortion effect due to surface modification during the optimisation process.

A three dimensional example concerning the bulb of a ship hull is next presented. Four variables have been defined with the C^1 interpolation functions. Figure 6 shows the resulting shape modifications corresponding to each design variable. The quality of the modified shape and the corresponding mesh for each design variable are, practically, as good as the original one.

In a complete optimisation process, all the design variables are used all together, and the shape of each design is defined through the superposition of the shape modification corresponding to each design variable using expression (1).

4   CONCLUSIONS

A new method for shape design variable definition has been proposed. It overcomes some of the pitfalls of the methods formerly proposed by several authors. It is able to manage two-dimensional as well as three-dimensional problems allowing the user to work directly on the geometrical model of the problem and simplifying the variable definition phase of an optimisation problem layout. C^0, C^1 or C^2 continuity properties of the boundary surface of the analysed structure can be preserved leading to the definition of shape variables suitable for optimisation problems in structural mechanics as well as in fluid-dynamics. Boundary surface displacements are propagated over the internal nodes of the mesh by keeping the distortion of the elements to a minimum. This allows to reduce the error introduced during the optimisation process due to the presence of highly distorted elements.

5   ACKNOWLEDGEMENTS

This work has been supported by the EU Marie Curie Grant BRMA-CT97-5761 hold by the first author. The authors want to acknowledge the support of E. N. BAZAN (Madrid) and of the International Center for Numerical Methods in Engineering (CIMNE, Barcelona).

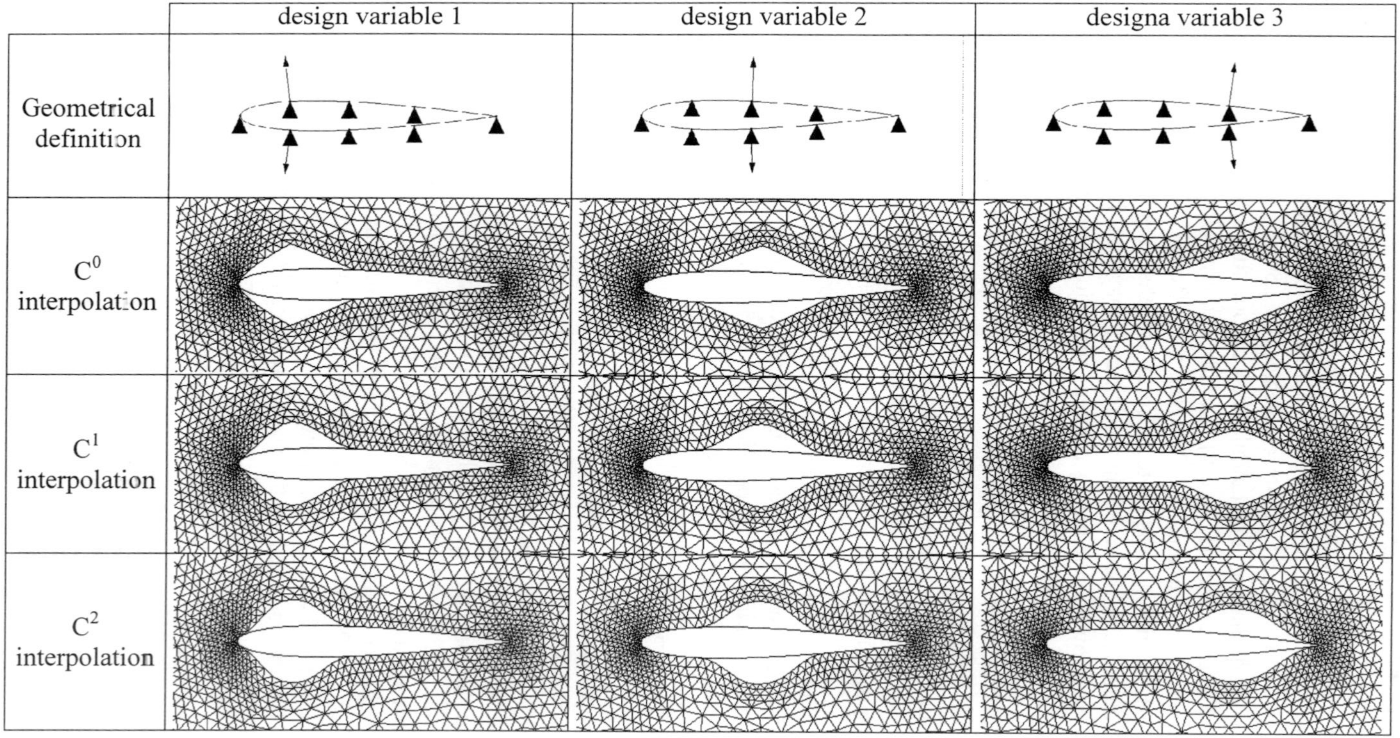

Figure 5 Local shape variables for the airfoil problem. The geometrical dfinition of the design variable is shown; the triangles define the limits of the macro-patches and the vectors the design variable. Three shape variables defined by using the three available interpolation functions are shown.

874

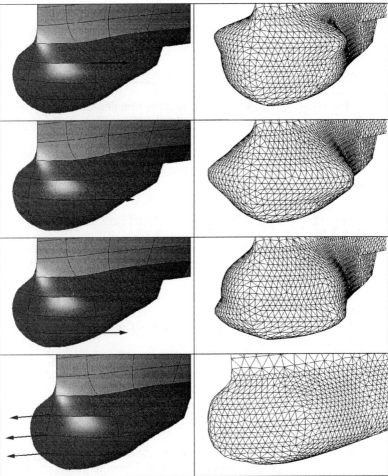

Figure 6 Symmetric surface modification due to four shape variables is shown here. On the left the geometry of the bulb and its partition in macro-patches can be observed together with the vector representative of the design variable. Tangent continuity is preserved over all the surface of the structure by means of the C^1 interpolation function used to extrapolate the vertices displacements.

REFERENCES

[1] M. H. IMAM - Three dimensional shape optimization, International Journal for Numerical Methods in Engineering, Vol. 18, pp. 661-677, 1982.

[2] G. VANDERPLAATS - Approximation concepts for numerical airfoil optimization, NASA Technical Report n. 1370, 1979.

[3] V. Braibant, C. Fleury - Shape optimal design using B-splines, Computer Methods in Applied Mechanics and Engineering, Vol. 44, pp. 247-267, 1984.

[4] V. BRAIBANT, C. FLEURY - An approximation-concepts approach to shape optimal design, Computer Methods in Applied Mechanics and Engineering, Vol. 53, pp. 119-148, 1985.

[5] R.T HAFTKA, R.V. GRANDHI - Structural shape optimisation. A survey, Computer Methods in Applied Mechanics and Engineering, Vol. 57, pp.91-106, 1986.

[6] M.K. CHARGIN, I. RASCH, R. BRUNS, D. DEUERMEYER - General shape optimisation capability Finite Elements in Analysis and Design, Vol. 7, pp. 343-354, 1991.

[7] S. ZHANG, A.P. BELEGUNDU - A systematic approach for generating velocity fields in shape optimization, Structural Optimisation, Vol. 5, pp. 84-94, 1992.

[8] G. FARIN - Curves and Surfaces for Computer Aided Geometric Design, Academic Press, 1993.

[9] R.E. BARNHILL, R.F. RIESENFELD - Computer Aided Geometric Design, Academic Press, 1974.

[10] G. CHIANDUSSI, G. BUGEDA, E. ONATE - A simple method for automatic update of finite element meshes, CIMNE Publication n. , July 1998.

[11] J.A. GREGORY - Smooth interpolation without twist constraints, Computer Aided Geometric Design, R.E. Barnhill and R.F. Riesenfeld Editors, Academic Press, 1974.

[12] R.E BARNHILL - Smooth interpolation over triangles, Computer Aided Geometric Design, R.E. Barnhill and R.F. Riesenfeld Editors, Academic Press, 1974.

[13] S.A. COONS - Surface patches and B-spline curves, Computer Aided Geometric Design, R.E. Barnhill and R.F. Riesenfeld Editors, Academic Press, 1974.

[14] R.E BARNHILL, J.A. GREGORY - Polynomial Interpolation to Boundary Data on Triangles, Mathematics of Computation, Vol. 29, N. 131, pp. 726-735, 1975.

[15] GID User manual – CIMNE, Barcelona, 1998

Topological Design of Shells Subject to Stress and Frequential Constraints

Pagnacco E., Souza de Cursi J.E.

Laboratoire de Mécanique de Rouen, UPRESA CNRS 6104
Institut National des Sciences Appliquées de Rouen
BP 8, 76801 Saint-Etienne du Rouvray, France

Abstract

The topology of shells with non constant (continuous) thicknesses is considered here. In this particular case, a dedicated characterization is more suitable for a convenient rapidity. Thus, the topology of the structure is characterized by an unknown thickness distribution and a fixed mid-surface. For the situation involving only the stresses constraints, the optimization process is based on solving the algebraic equation that characterizes the optimal thickness. For the one involving modal frequency constraints, a succession of approximate sub-problems are solved. Formulations and methods are shown to be effective for solving problems involving a large number of elements, therefore large number of design variables, even on workstations in interactive sessions.

Introduction

In the first stage of design, called "pre-design step", the designer seeks the topology of "the best" spare part subject to the imposed structural behaviors, in a design space. So, the aim of a topological optimization tool is the automatic design of mechanical pieces, in an available space, without a priori assumptions on his shape. The expected result is the overall shape of this part with the form (and number) of holes and ribs. In most of the cases, "the best" is considered as corresponding to "the lightest" spare part. The structural limitations taken into account in this step are usually the possibility of plastic restriction (to be avoided) and a frequential kind (frequential components of particular modes shapes must be avoided from some ranges). The available space is a geometrical constraint generally defined by others components in the assembly.

An additional challenge comes from the minimum time to design, due to the competitive context and the inherent functionality: the available space for designing is referred to others parts which are also in the same pre-design step; thus, modifications on a surround part could change the space for designing, and the designer must be able to quickly produce a new topology solution.

To solve this problem, the usual way is to handle binaries parameters (0/1) associated to a constant 2D thickness or a 3D mesh of the available space to denote the material absence/presence ([1]). Thus, the generation of the optimal topology leads to a problem with a large number of non-continuous parameters (typically between 1,000 to 100,000) in order to have a fine geometrical description. However, we focus on this paper on parts requested to have one small dimension compared to the two others (so considered as plates or shells) with a continuous thickness. In this case, a well suitable approach consists in characterize the part by a mid-surface and a continuous thickness distribution ρ (holes are regions where ρ is equal to zero) ([11], [8]). We assume here that the mid-surface is imposed. This approach is developed in sub-sequent sections.

Once topologies of parts are defined, a Computer Aided Design model has to be generated and parameterized. Finally, the determination of the involved parameters is performed by the optimization module of the CAD software. For more details on the global design stage proposed here, reference [9] could be consulted.

878

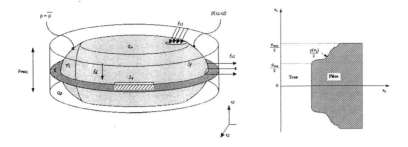

Figure 1: a) Problem Description, and b) Cross-section View of a Part.

1 Mathematical Background

In this work, we are essentially interested in the optimization of plate and shell structures widely used in the aerospace and automotive communities. We assume that the mid-surface is imposed. In the following, we will show that the part is entirely characterized by the thickness distribution, which becomes the only variable to be determined.

1.1 Geometrical model

Figure (1) illustrates the shell and the design domain. A shell is defined by a mid-surface S assumed regular and bounded. A natural description of S is obtained by using a set of curvilinear coordinates $a = (a_1, a_2)$ such that

$$S = x(\Omega) = \{x = (x_1, x_2, x_3) \in R^3 \,|\, x = x(a), a \in \Omega\} \quad (1)$$

Here, the 2D space Ω is used for defining the boundaries of the mid-surface. The design region occupied by the spare part is denoted by Q and it is defined as

$$Q = \{x + rn \in R^3 \,|\, x \in S, -\frac{\rho_{max}}{2} < r < \frac{\rho_{max}}{2}\} \quad (2)$$

where n denotes the unitary vector normal to the mid-surface S at point x, with the dimension ρ_{max}. Thus, the constraint boundary on the available design region is taken into account in implicitly manner assuming the set of mid-surface with a maximum thickness ρ_{max} is provided.

The design region can be split into two complementary sub-regions Q_0 and Q_+, the first one denoting the holes and the second one the matter, defined as:

$$Q_0 = \{x + zn \in R^3 \,|\, x \in S,\, z = 0,\, \rho(x) < \rho_{min}\} \qquad (3)$$

$$Q_+ = \{x + zn \in R^3 \,|\, x \in S,\, -\tfrac{\rho(x)}{2} \leq z \leq \tfrac{\rho(x)}{2},\, \rho_{min} \leq \rho(x)\} \quad (4)$$

with ρ a thickness distribution associated to each point x of S, and ρ_{min} a minimal thickness due to manufacturing processes which can not handle an arbitrarily small thickness in practical applications.

So, the determination of the topology is equivalent to the determination of Q_0 or its complementary Q_+: these two sets are entirely characterized by the thickness distribution ρ. Clearly, the parameter ρ defines the topology of the spare part: holes are regions where $\rho < \rho_{min}$. The topological optimization problem is therefore equivalent to the determination of the optimum distribution $\rho(x)$. In the following, we only consider the determination of the thickness distribution problem.

1.2 Mechanical model

Let forces densities f_s applied onto S_f, a subset of $\partial Q = S_u \bigcup S_f$, the Q_+ boundary. The spare part is fixed on S_u. Assuming the product $\rho\sigma$ is defined everywhere in Q(the value is zero in Q_0), the virtual work is:

$$W(u, \delta u) = \int_Q \delta\epsilon^T H\epsilon dV - \int_{S_f} \delta u^T f_s dS + \int_Q \delta u^T \mu \ddot{u} dV = 0 \quad (5)$$

$$\forall \delta u \text{ regular} \,|\, \delta u = 0 \text{ on } \partial S_u \text{ where } u = 0$$

with the stress-strain symmetric tensor H for the elastic structure such that $\sigma = H\epsilon$, and a linear strain field ϵ such as $\epsilon = \tfrac{1}{2}\left(grad(u) + grad(u)^T\right)$.

For shells, the displacement field u may be decomposed into a mid-surface component displacement u_m and a rotation of a straight section β such as: $u(x, z) = u_m(x) + z\beta(x)$ and the stress component $\sigma_{zz}(x, z)$ along the thickness is assumed to be null.

Displacement field is obtained by setting $\ddot{u} = 0$ in W. This leads to the Von Mises equivalent stress field given as $\sigma_{eq} = \left(\tfrac{3}{2}trace\left(\sigma^D\sigma^D\right)\right)^{\frac{1}{2}}$ where $\sigma^D = \sigma - \tfrac{1}{3}trace(\sigma)\mathrm{Id}$ denotes the deviator of σ. Modes shapes are obtained using $u(x, t) = \Phi(x)e^{i\omega t}$ with $f_s = 0$.

1.3 Optimization problem formulation

1.3.a The static problem

As previously observed (Section 1.1), the thickness distribution ρ characterizes the topology of the mechanical piece. Let us introduce the following problem: find ρ such that

$$\begin{cases} \min_\rho(\mu \int_S \rho dS) \\ W(u(\rho)) = 0 \quad \text{and} \quad u = 0 \text{ on } S_u \\ \sigma_{eq}(u) \leq \overline{\sigma} \\ \rho = 0 \text{ or } \rho_{min} \leq \rho \leq \rho_{max} \end{cases} \qquad (6)$$

This problem is referred in the sequel as "the static problem". We observe that the first equality corresponds to the structure equilibrium for one load case. The first inequality gives structural restrictions: the Von Mises equivalent stress field must be lower than the elastic limit $\overline{\sigma}$ in order to avoid plastic collapse. The last inequality introduces geometrical restrictions not implicitly assumed by the model: ρ_{min} is a threshold of matter underneath which a hole is assumed and ρ_{max} is a maximum value of the thickness.

We point that the mechanical restriction involves a local quantity. After discretization, it leads to a large number of restrictions (between 1,000 to 100,000) with expensive sensitivities. The class of optimality criteria method of zero-order (sensitivity less) in optimization field is well suited for this specific problem. The fundamental result due to [11] for the static problem of plane stress is:

Let $f_\mu : [0, 1] \to [0, 1]$ be a regular application such that $0 \leq f_\mu(\alpha) < 1$ if $\alpha < 1$ and $f_\mu(1) = 1$. We called ρ^* the optimal thickness, $\sigma^* = \sigma(\rho^*)$, the field of stresses associated to ρ^* and σ_{eq}^* the Von Mises equivalent stress associated to σ^*. If one solution exist and the product $\sigma^*\rho^*$ is regular enough then:

$$\rho^* = \text{proj}_{[0,\rho_{max}]}\left[\rho^* f_\mu\left(\frac{\sigma_{eq}^*}{\overline{\sigma}}\right)\right] \qquad (7)$$

Proof of this criteria does not use differential notion but the Borell-Cantelli lemma and establishes that it is necessary. This is equivalent to the *fully stressed design* criteria: $\rho^* > 0 \Rightarrow \sigma_{eq}^* = \overline{\sigma}$ and either the thickness is null ($\rho^* = 0$) or the Von Mises equivalent stress is saturated ($\sigma_{eq}^* = \overline{\sigma}$).

Approximation concepts in mechanics must be used in order to propose some convenient choice of the function f_μ. We can consider auxiliary variables such that $y = f_\mu^{-1}(\rho) = \rho^\eta$ with η a parameter. The following choice is effective in various shell problems:

$$y = \begin{cases} 0 & \text{if } \rho < \rho_{min} \\ \rho^2 & \text{if not} \end{cases} \tag{8}$$

1.3.b The frequential problem

The second optimization problem, referred as "the vibration problem", is:

$$\begin{cases} \min_\rho(\mu \int_S \rho \, dS) \\ W(\Phi_j(\rho), \omega^2(\rho)) = 0 \quad \text{and} \quad \Phi_j = 0 \text{ on } S_u \\ \omega^2(\Phi_j) \geq \varpi_j^2 \quad j = 1, ..., m \\ \rho = 0 \text{ or } \rho_{min} \leq \rho \leq \rho_{max} \end{cases} \tag{9}$$

Only the structural restriction differs from the static problem. We point that restrictions concern m *imposed* modes Φ_j such that their pulsation ω be higher than a threshold pulsation ϖ_j (typically given by an acoustic analysis).

This formulation differs from the one usual in structural optimization where restrictions concern pulsations identified by their position in the all list of pulsations in ascending order. In fact, such a formulation suffers from the following drawback: let an engineer interested in the "first mode" -referred as the lowest mode- in the initial design. The difficulty comes from the fact that this mode could be "the second one" for the optimum structure, even if its shape corresponds to "the first one" in the initial design. Thus, in this situation, the first mode swaps with the second one during the optimization iteration process, and this phenomenon is referred as the mode switching effect.

The form adopted here does not suffer from the previous phenomenon and is, in general, more straightforward for the engineer: for example, in acoustic design, the engineer is only interested in panel bending modes and no matter how is its correspondence to pulsation number. However, there is no loss of generality with this formulation since an appropriate set of modes also permits a restriction concerning the lower pulsation, if need be.

The method of resolution involves a sequence of approximated subproblems, which are constructed using the following auxiliary variable:

$$y = \begin{cases} \frac{1}{1-\eta}\rho^{1-\eta} & \text{for } \eta \neq 1 \\ \ln(\rho) & \text{if not} \end{cases} \tag{10}$$

Due to the limited number of restrictions (generally less than ten), the dual problem of this sub-problem is solved.

So, this approach using a programming method is opposed to the previous one adopted for static problems since an optimality criteria is not constructed for this specific problem. This eventually permits to handle other restrictions such a displacement limitation if need be.

Mathematical difficulties concerning this problem are still with us, as the existence of solutions and Lagrange multipliers which are closely connected to the topological properties of the set of the admissible configurations (for example, its closure or the existence of an interior). These questions will not be evoked here, since this work is mainly concerned by the numerical aspects of the problem.

2 Numerical procedures

The complete determination of the thickness distribution involves an unknown by point of the middle surface. Consequently, the first step is the discretization of the quantities involved in the optimization problem. Definition of the imposed surface is obtained by the mesh of the available space to design. Since the analysis model is a finite element model that involves the discretizing process, the thickness distribution ρ is discretized accordingly with a one-to-one correspondence by the element thickness h in n elements. This is an interesting feature since the mesh can be easily refined in order to approximate conveniently the thickness distribution. That also ensures consistency for the approximations used for both finite element model and thickness distribution. Moreover, such a discretization is sufficient because in the first design steps, we only need a rough idea of the overall shape.

In order to have an efficient method, we consider a fixed mesh during the optimization process and a procedure to treat holes. Therefore, we avoid problems involved by re-meshing. The drawback of this approach is that solving linear systems of equations or extracting resonant frequencies for re-analysis may become very expensive as the size of the model is increased, so a preconditioned conjugate gradient technique is used even in the subspace solver. In addition, global finite element matrices can be disassembled in a matrix product of geometrical and material matrices with a thickness matrix: updating the finite element model is efficient because only the thickness matrix needs to be corrected ([4]).

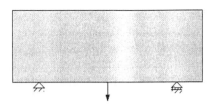

Figure 2: Design space of the second example of Mitchell trusses

For the static problem, a fixed point (zero-order) method is used to respect the optimality criteria (7) ([7]). Special care is required to take into account the maximum of two Von Mises equivalent stresses evaluated on each surface of the finite element. Typical values of time to optimize a mid-size problem of 10,000 degree of freedoms (corresponding to approximately 50,000 dofs for the equivalent 3D model) is 5 mn for about 25 iterations without special strategies previously evoked (a frontal direct solver is employed on a R10000/180 cpu).

For the vibration problem stationarity conditions of the approximated dual sub-problem are solved explicitly ([10]). Ensuring that the optimization concerns the same mode at each step is treated along iterations by a mode tracking strategy. This method is well known in structural dynamics, and called Modal Assurance Criterion (MAC). This criterion permits to find a given mode apart from the others after the model updating. For these problems, the software Matlab is used with strategies evoke before.

3 Numerical experiments

3.1 Second example of Mitchell trusses

The design space available is shown in figure (2) with boundary conditions. Dimensions are $100 \times 50 \times 10\ mm$. Force of 600 N is applied at the mid-distance (40 mm) between supports. Material is characterized by $E = 208600\,N/mm^2$, $\nu = 0.29$ and $\mu = 7800\,kg/m^3$. Maximum available stress is 20 MPa. A threshold of matter is imposed to be 0.075 mm.

Results are shown on figure (3). Final mass is 2.37 g which is the optimum mass according to formula of reference [6]. We could see that the topology is not affected by changing meshes. In addition, symmetry is respected even as such a condition is not imposed in the model.

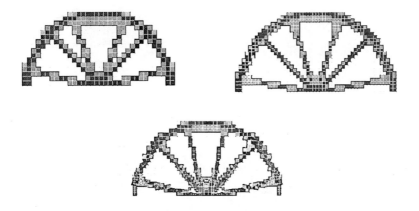

Figure 3: Topology found for different meshes

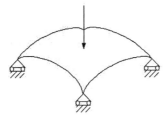

Figure 4: Geometry of the design space for the spherical shell

3.2 A spherical shell example

In figure (4) a spherical available space is shown. A segment of the shell is subjected to a load in the vertex, and vertically supported on each corners. Straight length of the segment is $5\ m$ and the spherical radius is $10\ m$. Load modulus is $1600\ N$. The minimum thickness is $10^{-3}\ mm$. Material is characterized by $E = 30000\ N/mm^2$, $\nu = 0.2$ and $\mu = 2400\ kg/m^3$ with an available stress level of $16\ MPa$. Model used a $q4\gamma$ finite element ([2]). No condition of symmetry is imposed in the model.

No checkerboard control is used for this example. After 50 iterations, an optimal topology of $152\ kg$ is found (fig. 5).

885

Figure 5: Topology for two different meshes

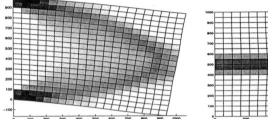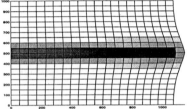

Figure 6: Ribs for: a) bending mode shape and b) extension mode shape

3.3   Plate reinforcement

3.3.a   The first bending mode shape

Space design is $1000 \times 1000 \times 10 \, mm$. Left edge could be embedded and mid-right edge have a $1000 \, kg$ mass. Minimum thickness is $1 \, mm$. Topology of ribs needs to be determined such that the mass is minimum subject to a frequency greater than $52.1 \, Hz$ for the first bending mode shape.

Material is characterized by $E = 210000 \, N/mm^2$, $\nu = 0.3$ and $\mu = 7800 \, kg/m^3$. Mesh of 20×20 finite elements implies 400 optimization parameters. Additional mass is distributed on 9 nodes. The initial constant thickness of $3.06 \, mm$ satisfies the constraint imposed on the frequency for a $23.9 \, kg$ plate mass.

Optimum mass of $20.0 \, kg$ is obtained in 4 iterations with 6 structural analysis. Optimum ribs (fig. 6.a) have a V shape beginning on vertex of embedded edge and ending to the additional mass. Results seems agree with an identical topology, and a mass lower than the $23.4 \, kg$ found in reference [3] (for a model using constant thickness ribs).

886

3.3.b The first extension mode shape

Data of Section 3.3.a are used, but restrictions are now on the first extension mode shape with a frequency greater than $100.8\ Hz$. Eight structural analysis gives the optimum mass of $15.5\ kg$ which is close to the theoretical mass of $15.0\ kg$ applying results given in [12] (remember that our model has a minimum thickness over all the design space and this is slightly different). Topology having a I shape is shown on figure 6.b).

Conclusion

The case of topological design of continuous thickness shells subjected to stress and vibration constraints is studied here. The topology is characterized by an imposed mid-surface and a thickness distribution ρ (holes are regions where ρ is equal to zero). Formulations of both static and vibration problems are kept close to the engineer problem by considering the minimum mass objective with required structural behavior. Moreover, it has been verified on a few reference examples that such formulations leads to stable solutions.

Even if results for both situations are obtained by different ways, they can be viewed as a dual method with explicit solutions, where different approximations are considered: the approximations must be global for frequencies, while they may be local for stresses. Thus, optimization processes are efficient in maintaining the number of operations as close as possible from the minimum. In addition, strategies such as preconditioned conjugate gradient technique have been used and an explicit parameterized finite element model have been developed for analysis. Consequently, a well refined mesh of the design space that would be well suited for both static and dynamic finite element model could be used to describe precisely the topology.

In conclusion, the combination of the adopted topological characterization with improved analysis and optimization strategies offers an efficient pre-design tool for the case considered.

The simultaneous optimization of the thickness distribution and the shape of the mid-surface could be an interesting extension of this work for problems involving stresses, displacements, and vibrations restrictions.

887

References

[1] ALLAIRE G., BELHACHMI Z., JOUVE F., The Homogenization Method, Numerische Mathematik 76, pp. 27-68, 1997.

[2] BATOZ J.L., DHATT G., Modélisation de Structures par Eléments finis, 3 volumes, Hermès Editeur, 1992.

[3] DUYSINX P., Optimisation Topologique : du Milieu Continu à la Structure Elastique, Thèse de Doctorat de l'Université de Liège, 1996.

[4] HEMEZ F.M., PAGNACCO E., Finite Element Disassembly for Structural Dynamics - Part II: Efficient Inverse Solvers, to published in European Journal of Finite Element, Jun 1998.

[5] MAUTE K., SCHWARZ S., RAMM E., Adaptive Topology and Shape Optimization, in World Congress on Computational Mechanics, Buenos-Aires, Argentina, 29 June-2 july 1998.

[6] MITCHELL A.G.M., The Limits of Economy of Material in Framed Structures, Phil. Mag. , 6, pp. 589-597, 1904.

[7] PAGNACCO E., SOUZA DE CURSI J.E., Minimum Mass Spare Part in 2D Elasticity, World Congress of Structural and Multidisciplinary, Goslar, Germany, 1995.

[8] PAGNACCO E., SOUZA DE CURSI J.E., Shell Topological and Shape Optimization, Vibration Case, IDMME'98, 2nd International Conference on Integrated Design and Manufacturing in Mechanical Engineering, Compiègne, France, 27-29 Mai 1998.

[9] PAGNACCO E., SOUZA DE CURSI J.E., RONDÉ-OUSTAU F., ESMINGEAUD P., Optimal Automotive Design, Fisita 98 World Congress, Paris, France, 27 Sept.-1er Oct. 1998.

[10] PAGNACCO E., Optimisation Topologique des Structures de Type Coques, Ph. D. Thesis, Université de Rouen, 1998.

[11] SOUZA DE CURSI E.J., Allégement d'une Pièce Elastique Homogène soumise à des Contraintes Planes, Research Report, No. 1/94, Laboratoire de Mécanique de Rouen, 1994.

[12] TURNER M.J., Minimum mass structures with Specified Natural Frequencies, AIAA Journal, Vol. 5, No. 3, pp. 406-412, March 1967.

888

TAKING ADVANTAGE OF USING BOTH TOPOLOGY AND SHAPE OPTIMISATION FOR PRACTICAL DESIGN

Andreas Back-Pedersen[1]

Grundfos A/S, Denmark

1 ABSTRACT

The methods of topology and shape optimisation are combined for developing an impeller hub. It is shown how these two methods fulfil different requirements during the design process

The method of topology optimisation is well suited for elaborating good basic geometries early in the design process. This method only requires input of permissible design domain and applied loads. On the other hand, the method of shape optimisation is well suited for narrow analyses later in the design process. Contrary to the method of topology optimisation, a basic geometry for shape optimisation should be created by the design engineer. This method can also take into account constraints on the geometry considering the manufacturing processes mastered by the company.

The strength of combining the methods of topology and shape optimisation is demonstrated by designing an impeller hub for a single stage centrifugal pump.

2 INTRODUCTION

Facing the daily work of the design engineer in creating new geometries for components and products, it can be realised that this creative process has traditionally been formed as a sequence of iterative loops. The design engineer creates design proposals, which are analysed by using e.g. the finite element method or by experimental tests. Based on the results of the analyses and tests, the design proposals are modified. This process is running until an acceptable geometry has been reached.

---

[1] Specialist in Solid Mechanics, M. Sc. in Engineering, Ph. D.

The modification process can be automated by extending the finite element analyses with an optimiser. But even if the design has been automated for the modification process, it still relies on the design concept created by the design engineer.

The design process outlined in the following is characterised by not relying solely on the capability of the design engineer to create conceptual designs. The emerging method of topology optimisation has been used for creating topologies. This topology optimisation method was introduced by Bendsøe and Kikuchy, [1], to solve the problem of finding the optimal material distribution. An overview of the method is given by Rozvany, Bendsøe and Kirsch, [2], and Bendsøe, [3].

At the current stage of technology, the topology optimisation algorithm is not able to restrict the topology to fit to any predefined manufacturing process. Therefore, the created topologies can often not be used directly without some interpretation by the design and manufacturing engineers. Also the fact that the used implementation creates the most stiff structure regarding the applied loads requires an interpretation when minimum stresses are wanted. However, it appears that the most stiff structure is concurrently a suitable design for minimum stresses.

The method of shape optimisation has been thoroughly described in the literature, see e.g. Haftka and Gürdal, [4]. Shape optimisation creates optimum only within the design concept created by the design engineer. Therefore, the outcome depends on the capability of the design engineer to create good design concepts. In the design process described below, the basic designs to be subjected to shape optimisation derive from the outcome of the topology optimisation.

Inspired by the created topologies, parametric finite element models are elaborated taking into account the production processes, are elaborated. These parametric models are optimised with respect to an objective function optimising the product reliability.

3 THE DESIGN PROBLEM

An impeller hub to be used in a single stage centrifugal pump must be designed. The hub should be a part of the impeller component. The working principle how the hub should fix the impeller to the shaft is predetermined on account of the connection with present product components. The fixation should be performed by friction forces caused by a taper bushing. The pump impeller including the hub must be designed for mass production, therefore, material savings and reliable manufacturing processes are key issues.

The intersected impeller without hub is shown in Fig. 1. The principle of clamping the impeller to the shaft by using a taper bushing with a predefined cone angle, is shown in Fig. 2a. The impeller hub and taper bushing are pulled together by means of a nut.

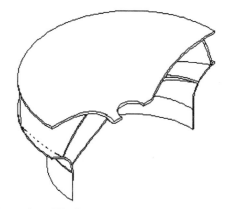

Figure 1:   The intersected impeller without hub

The hydraulic part of the impeller must not be changed, therefore, the permissible design domain for the impeller hub is restricted to be on the back side of the back plate.

The impeller hub should transfer the loads applied on the impeller to be carried by the shaft. Therefore, the loads applied are very essential for designing the impeller hub. When the impeller is mounted on the shaft by tightening the bushing, a static load is applied to the impeller hub, which will be designated as the tightening load, see Fig. 2 a. The pump impeller is placed in a volute, which has a uniform pressure distribution only at the optimal flow point. For a pump not running at that flow point, and this will often be the case, there will be a non-uniform pressure distribution on the impeller from the volute flow.

a   b

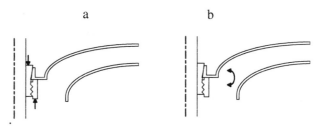

Figure 2:   Loads cases at the impeller hub. a: Clamping the pump impeller to the shaft by using a taper bushing. b: Bending moment due to non-uniform pressure distribution in volute.

891

The resulting force on the impeller can be decomposed in a bending moment component, an axial force component, a radial force component, and a torsion moment. Even for the pump running at the optimal flow point there will be an axial force component and a torsion moment on the impeller. For the actual impeller it has been found that the dominating forces are due to the tightening load and the bending load, shown in Figs. 2a and 2b, respectively.

The manufacturing process has not been pre desired. According to what will seem adequate during the design process, the manufacturing processes for the impeller hub will be chosen.

4 THE TOPOLOGY OPTIMISATION MODEL

The whole impeller has not been modelled for the conceptual design study. It has been recognised that modelling the blades and front plate does not considerably influence the behaviour of the hub region. The geometry model for the impeller therefore consists of the back plate and the permissible domain for the hub. Also the taper bushing is included and connected to the impeller by contact elements. Not modelling the blades has made it possible to use a simple axi-symmetric geometry with axi-symmetric or harmonic loads. The modelled regions are shown in Fig. 3.

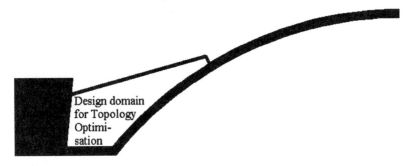

Figure 3:   Design domain for designing the impeller hub

The shape of the impeller back plate and the taper bushing must not be changed. Therefore, these regions are not subjected to changing material distribution. Only the density of the elements in the region of the hub must be changed during the topology optimisation. The amount of material for the hub is specified to be 10% of the volume of the permissible design domain.

The algorithm, which has been used for the topology optimisation, is outlined by Back-Pedersen, [5]. In the same reference, the single load case topology optimisations for all load cases are performed, including the load cases not shown in Fig. 2.

The design engineer has to design an impeller hub to resist both the tightening load as well as the bending load. This problem can be solved as a multiple load case topology optimisation. Because the tightening load and the bending moment load are static and dynamic, respectively, a suitable weight function has not been formulated. Instead topology optimisations are performed to show the transition from dominating tightening force to dominating bending moment, see Fig. 4.

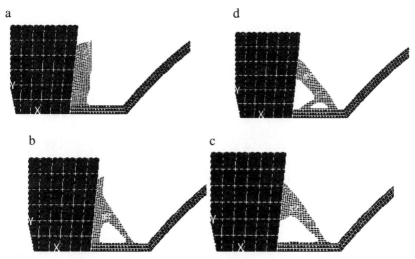

Figure 4:   Multiple load case topology optimisation showing transition b and c, from dominating tightening load, a, to dominating bending load, d. (Material densities above 0.8)

Even if the topology layouts in Fig. 4 cannot be used as final designs, they give the design engineer fundamental understanding of the behaviour of the component. They can be used as source of inspiration for creating parametric geometry models taking into consideration the manufacturing processes.

5 DESIGNING FOR MANUFACTURING

The topology layouts may inspire the team of design and manufacturing engineers to create good designs for manufacturing. This process is not automated, and the result is much depending on the creativity of the team. By combining the topologies shown in Fig. 4 with the manufacturing processes mastered by the involved company, a number of proposals for manufacturing the impeller hub turns up.

The topology shown in Fig. 4a has inspired the project team to form the impeller hub as an integrated part of the back plate, shown in Fig. 5. The manufacturing processes involved for creating this solution is sheet drawing followed by ironing to reach the pre desired height of the impeller hub. Finally a conical pressing should be performed. A narrow analysis shows that the ratio of the impeller height to the hole diameter is critical. Therefore, this solution will not be discussed further.

Figure 5:   Creating the impeller hub as an integrated part of the back plate

Another way of creating the impeller hub similar to the topology design in 4a is based on a pipe section. This solution is shown in Fig. 6a and is processed by cutting a pipe section, which is collared. The collared pipe section is spot welded on the back plate. Because of position tolerances the cone is made by revolving.

Figure 6:   Design of the impeller hub processed from a pipe section, which is collared

For dominating bending load the topology optimisation performs a stiffening member as shown in Fig. 4d. This topology has inspired the project team to create the impeller hub by sheet drawing, as shown in Fig. 7a. The impeller hub is processed by blanking a disc and deform it by sheet drawing. The drawn sheet is spot welded on the back plate, and finally the cone is made by revolving

Because sheet drawing is very attractive due to low costs, an adaptation to the topologies in Fig. 4b and 4c has been done. The outcome is shown in Fig. 7b. The tooling and processes for this solution are slightly more complex than the solution shown in Fig. 7a because two drawings are needed. However, the processes are similar

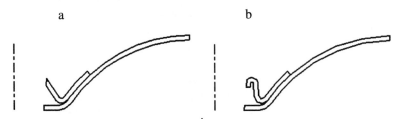

a b

Figure 7: Designs of the impeller hub processed by using sheet drawing

The topology shown in Fig. 4c has also given the idea of creating the impeller hub by using cold forging, shown in Fig. 8. The starting material is a sheet, which is firstly drawn to make the overall geometry. Afterwards, the varying thickness is created by cold forging. Finally the cone is made by revolving.

Figure 8: Design of the impeller hub based on cold forging

Lastly the ability of creating the impeller hub by using sintering and cold forging was discussed. On account of axial powder squeezing, the geometry for the impeller hub will differ from the one produced by cold forging. To obtain small thickness and narrow tolerances for spot welding, the sintering process should be followed by cold forging. The impeller hub created by using sintering and cold forging is shown in Fig. 9.

Figure 9: Design of the impeller hub using sintering and cold forging

The designs proposed above differ in the prices of starting material, tooling and processes. In table 1 price indices for starting materials as well as price indices for tooling and processes are listed for all design proposals.

Due to the price level of the starting materials the proposals A and E in Table 1 have been dismissed. For the remaining three proposals the connected geometries and manufacturing processes have been described in three parameteric finite element models. The parameters are chosen in a way corresponding to the possibilities and restrictions of the underlying manufacturing processes.

Table 1: Price indices of starting material, tooling and manufacturing processes for each impeller hub proposal

Process and design	Starting material Price Index	Tooling and processes Price Index
A	5	1
B	1	1
C	1	2
D	1.5	2
E	4	2

6 THE SHAPE OPTIMISATION PROBLEM

The three elaborated finite element models are shown in Figs. 7 and 8. The background of each parameteric finite element model has been presented, and no further details of the geometric models will be discussed here.

The two load cases, which appear for the impeller hub, are the static tightening load and the alternating bending load. The design should be dimensioned to avoid momentary break due to the tightening load and fatigue crack due to the bending load. The objective function, which is minimised, is the ratio of the alternating stress amplitude to the fatigue strength. The fatigue strength is a function of the mean tensile stress. To establish the fatigue strength as a function of mean tensile stress, the empirical relation of Goodman, [6], is used. Pre stressing due to manufacturing processes are neglected for this design study.

The modified Goodman diagram, shown in Fig. 10, is constructed using the fatigue strength under simple alternating stress without mean stress, Sa, and the tensile strength, Su.

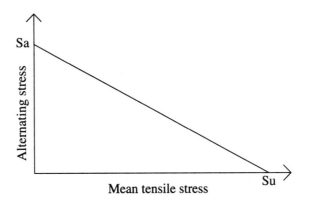

Fig. 10:   Fatigue strength as function of mean tensile stress

The optimisation loop consists of two load steps. In the first load step, giving the stress distribution due to the static tightening load, the 1st principal stress in each element, i, is found and the fatigue strength, σ_{fs}^{i}, according to the modified Goodman diagram is determined. In the second load step, giving the stress distribution due to the alternating bending load, the 1st principal stress, σ_{1a}^{i}, for each element is determined. The ratio $\sigma_{1a}^{i}/\sigma_{fs}^{i}$ is calculated and the objective function is the maximum value for this ratio. An objective function of 1 indicates failure of the impeller hub due to fatigue. The optimisation problem is formulated as

$$\min(\ \max_{i}\ (\sigma_{1a}^{i}/\sigma_{fs}^{i})) \tag{1}$$

where $1 \leq i \leq$ number of elements.

In Table 2 the objective functions for all the optimised models are listed. The objective function for each model is seen to be far below one, which means that all the elaborated designs fulfil the requirements according to structural reliability.

Table 2: Objective functions for optimised models

Model	Objective function
	0.53
	0.70
	0.49

For pointing out the best design for the impeller hub it should be kept in mind that all the designs are reliable designs. Therefore, the selection of the design for the impeller hub should be done considering manufacturing costs. The price indices in Table 1 tells that the design in Fig. 7a, manufactured by sheet drawing, has the lowest starting material costs as well as the lowest tooling and manufacturing process costs. The structural reliability together with the low costs makes this design an obvious candidate for the definitive choice.

7 CONCLUSION

The design process outlined above demonstrates the strength of combining topology and shape optimisation for developing industrial components.

It has been demonstrated how the method of topology optimisation should be used at the very beginning of the design process. Based on permissible design domain the outcome of the topology optimisations for single and multiple load cases gives the design engineer a fundamental understanding of the behaviour of the component according to the applied loads.

It has also been mentioned that the topology designs can not be used directly as final designs. At the current stage of technology, the algorithm is not able to restrict the topology to fit to any predefined manufacturing process. However, when interpreting the topology designs, these may be a comprehensive source of inspiration of physical relevant designs, which the design and manufacturing engineers should adapt to mastered manufacturing processes. Therefore, the topology design creations should be followed by a thorough analysis of possible manufacturing processes.

It has been shown that the basic behaviours of the elaborated topology designs and the possible manufacturing processes can be taken into account when creating parameteric finite element models. The parameteric finite element models can be subjected to detailed analyses of structural behaviour and to shape optimisation according to a predefined objective function.

The design process carried out shows that numerous designs fulfil the structural requirements. The definitive choice has therefore been based on prices of starting material and tooling and manufacturing costs.

8 REFERENCES

1   BENDSØE, M.P. AND KIKUCHI, N. - Generating Optimal Topologies in Structural Design Using a Homogenization Method. Comp. Meth. Appl. Mech. Eng., Vol 71, pp 197-224, 1988

2   ROZVANY, G.I.N.; BENDSØE, M.P.: KIRSCH, U. - Layout optimization of structures. Appl. Mech. Rev., Vol 48, pp 41-119, 1995

3   BENDSØE, M.P. - Optimization of structural topology, shape, and material. Springer: Berlin, Heidelberg, New York, 1995

4   HAFTKA, R.T.; GÜRDAL, Z. - Elements of Structural Optimization. Kluwer Academic Publishers: Dordrecht, Boston, London, 1992

5   BACK-PEDERSEN, A. - Designing an impeller Hub using Topology Optimization. Proc. 8th Int. ANSYS Conf. and Exhibition, Pittsburgh, Vol. 2, pp 2.81-2.90, 1998

6   FORREST, P.G. - Fatigue of Metals. Pergamon Press: Oxford, London, New York, 1962 (reprinted 1970)

CONNECTION TOPOLOGY OPTIMISATION BY EVOLUTIONARY METHODS

G.P. Steven[1], Qing Li[2] and Y.M. Xie[3]

SUMMARY

This paper presents a novel application of the recently developed Evolutionary Structural Optimisation (ESO) method for the topology, shape and/or size optimisation of structures under Multi-objective Optimisation criteria. It addresses the ongoing technical design requirement regarding the location of fasteners in a joint. A well designed multiple-fastener joint should have all fasteners involved to some extent and therefore a guiding design principle as to where to locate fasteners, be they, spot welds, rivets, bolts; should be that they all carry as close to uniform a load as possible. The joint will also be subject to multiple load cases, which represent the *corners* of the load envelope. Structural optimisation has its roots in attempting to obtain fully stressed designs for pin-jointed frames and beams under bending, however under multiple load cases any material has to be at its stress limit for at least one load case.

An initial layout for a fastened joint has every potential fastener site occupied, FEA indicated which are lowly stressed for all load cases and these are slowly eliminated. This evolutionary process is repeated with an ever increasing ratio of min to max stress until a suitable topology emerges where all the remaining fasteners have a role to play in at least one load case.

The paper presents a description of the ESO method followed by details of the specific adaptations made to ESO to accommodate the fastener topology situation. Several illustrative examples follow.

1. INTRODUCTION

Evolutionary Structural Optimisation (ESO), [16][17][10] has now established itself as a important technique for the solution of a wide range of structural optimisation problems in static, dynamic and stability analysis with

[1] Professor, Department of Aeronautical Engineering, University of Sydney, Australia.
[2] Graduate Student, Department of Aeronautical Engineering, University of Sydney, Australia.
[3] Assoc. Prof., School of the Built Environment, Victoria University of Technology, Australia.

the full range of environments both static and kinematic. Now that single structural items can have an optimum topology, shape or size distribution, then the next challenge is to look at structures with multiple components, such as the engine of a car or its suspension. To get to this stage of total structural design optimisation an important intermediate stage is to look at the optimisation of connections between components using mechanical or adhesive connections. This paper considers the optimisation of the topology of mechanical fasteners in a variety of standard jointing situations.

The fundamental idea for topology optimization [1][2] of connections as proposed here, is to model the connection between multiple components as disconnected discrete brick elements for every possible candidate fastener location. Unlike existing approaches [4][19][20] that deal with the connected components (e.g. plates) as rigid objects and connections as springs, this proposed algorithm has all relevant parts being dealt with as linearly elastic. Therefore, the structural responses of a connection system can be predicted by finite element analysis.

Previous optimization approaches to connection problems had to adopt complicated mathematical programming methods [3][13][15] such as the penalty function method, Lagrange multipliers, etc.[19][20]. This proposed approach needs no mathematical operations during the optimization process. The joint elements (e.g. brick) are set to all possible candidate locations in initial design. The objective of the optimization is to reduce the number of connections so as to cut manufacturing and assembly cost. If a connecting element is stressed at a low level, it means that the element is structurally ineffective and thus should be eliminated as part of the optimisation process. In this ESO-based algorithm, such an elimination process is carried out in an evolutionary manner. The examples in this paper give a preliminarily demonstration of the feasibility of this idea.

The approach reported in this paper does not need any mathematical operations during the optimisation process. The 3D brick elements represent all possible candidate locations in the initial design. The objective of the optimization is to reduce the number of connections so as to reduce manufacturing and assembly cost. If a connecting element is stressed at a low level relative to others (in % terms called the Rejection Ratio RR), it means such a connecting element is structurally ineffective and thus should be removed (hard-kill). Initially the iterative process of removal and re-analysis is maintained with the same low stress threshold until no other elements are eliminated, this is called a steady state. At such point the elimination relative threshold is increased (by an Evolution Rate ER%) and the elimination cycle repeated. This is the standard ESO technique as detailed in the next section.

2. EVOLUTIONARY STRUCTURAL OPIMISATION METHOD

The Evolutionary Structural Optimisation Method (ESO), follows a concept that is very simple and robust. The principles and procedures that define ESO are as follows:

1) Set up a dense FE mesh that fully covers the maximum design domain.
2) Apply all boundary constraints, loads, material, element properties, etc.
3) Specify the criteria used to optimise the structure, eg. Von Mises Stress.
4) Specify the ESO driving parameter, eg. max or mean Von Mises Stress.
5) Carry out a linear static Finite Element Analysis of the structure.
6) As the design domain is fully populated and hence over designed, there will be regions within the structure where the material is not efficiently used, ie. lightly stressed.
7) Compare the Von Mises Stress of each element with say the Max Von Mises Stress of the physical domain, those elements which are lightly stressed, for all load cases, can be removed. However, instead of removing them all at once, the Max Von Mises stress is multiplied by a Rejection Ratio factor (RR) which varies from 0 to 1, ie:

$$\sigma_{VM,e} \leq RR * \sigma_{VM,\max}$$

where:

$\sigma_{VM\,e}$ is the element mean Von Mises stress
$\sigma_{VM,Max}$ is the structure's Max Von Mises stress
RR   is the Rejection Ratio

$$RR = a_0 + a_1 SS + a_2 SS^2 + \cdots$$

where the a's are determined by experience, typically $a_0 = 0.01$, $a_1 = 0.005$, SS is the steady state number that increases by one every time. The use of RR, has the effect of damping or delaying the element removal process, such that, during each iteration only very few elements are removed, the success of the ESO process is dependant on slow removal of elements in the calculus sense.

8) If no elements were removed in step 7, it means that a local optimum and steady state has been reached. The (SS) above is incremented by 1 and steps 7 & 8 are repeated.
9) Steps 5 through 8 are repeated until all the stresses are within say 20% of the mean or maximum stress of the structure or until say only 10% of the original design domain volume remains, thus ending the ESO process for the specified design domain.

Although the driving criterion in this explanation is the Von Mises stress, any criterion can be used to drive the optimisation of a structure using the ESO method. The important aspect to understand is that no matter what criterion is used, the ESO method does not change, only the driving parameters do. This methodology can best be summarized in the flow chart of Fig. 1 which shows the set of logical steps mentioned above.

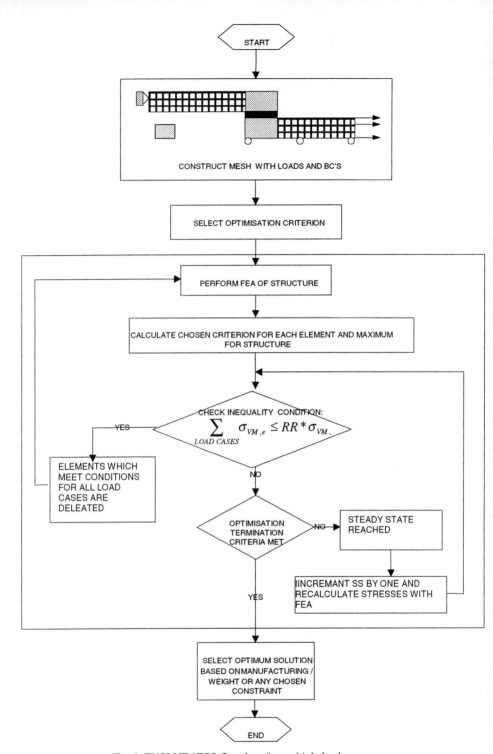

Fig. 1 EVOLVE / ESO flowchart for multiple load cases.

904

3. ILLUSTRATIVE EXAMPLES

3.1   Case 1)   overlapping structure   (ER=1%)

This first example is to connect two thick plates, Figure 2, in overlapping manner by either spot weld, rivet, pin or threaded joints. Due to the isolation of joint elements, the initial candidate connections are set to a checkerboard pattern as shown in Fig. 3a). Two moment load cases are applied at the right hand side while the left hand side of the structure is built-in, Fig. 2. Plate A and B are meshed by a mixed course and fine grid using eight node brick elements. The overlapping part is meshed with a 21 x 21 grid. Within this grid there are 100 candidate fastener locations modeled by bricks with half the thickness of the plates and half the Young's Modulus of the plates.

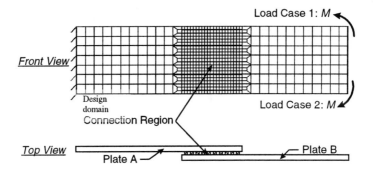

Fig. 2   Finite element model of the overlapping connection of two plates.

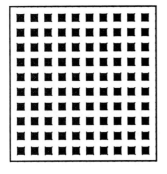

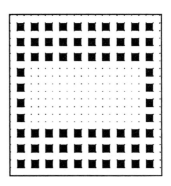

a) Initial Design   b) Steady State 1

905

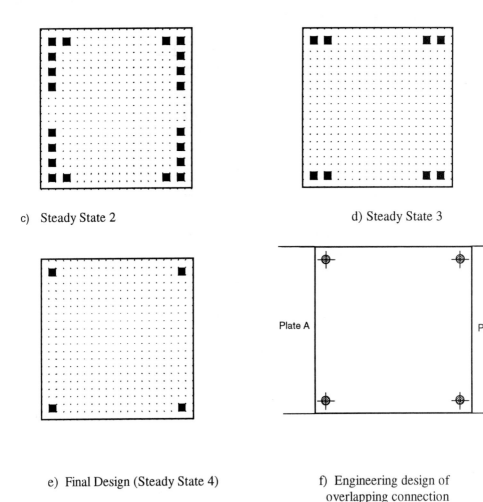

c) Steady State 2

d) Steady State 3

e) Final Design (Steady State 4)

f) Engineering design of overlapping connection

Fig. 3 Evolution process to eliminate inefficient candidate connections.

During the ESO procedure, an initial rejection ratio RRo=0 and an evolution rate ER=1% were adopted. Fig. 3a)-f) show some steady state results of evolutionary process. As the rejection ratio (RR) increase, more and more, relatively inefficient joint elements are eliminated from the candidate locations. At the end of the evolutionary process only four fasteners remain at the outmost corners. This obvious result could be argued to be some form of validation for the ESO process used here. Fig. 3f) could be the engineering drawings of practical connections where manufacturability and production considerations were being considered. If there were limits on fastener location from the edge then these are simply accommodated by not having candidate fasteners in such "no-go" areas.

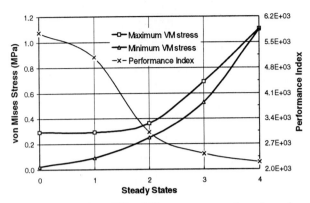

Fig. 4   Evolution histories of Von Mises stresses and performance index of
connection elements

Figure 4 shows plots of the Maximum and Minimum Von Mises stress
in the connecting brick elements. Because of the 2D constraint on the analysis
it can be said that in the connection elements this stress in dominated by the
shear in the connecting bricks. Clearly both these will increase as the number
of connections reduce, but the significant feature of the curves is that the ratio
of Min/Max moves from 0.05 to almost 1; demonstrating the potential for an
almost fully stressed design. The other curve on Fig. 3 is that for a the non-
dimensional Performance Indicator *PI,* mentioned above in Section 2.10),
[14][12] defined as $PI = \sum\limits_{loadcases} \sum\limits_{elements} \sigma_{VM}{}^e V^e / FL$, where F is a nominal load
and L a nominal length.   The lower the PI the better the structure.   Some
optimum structures, such as the classical Michell frames [9] have theoretical
values of *PI* and ESO has been shown to achieve these.

3.2   T-Shape overlapping structure (ER=1%)

For this second example the modeling region mimics a transverse
support joint that could be part of a frame structure. The initial pattern of
connection bricks in the overlap region is the same as that for the previous
example, Fig. 3a). The two loading cases are shown in Fig. 5 along with the
final layout of fasteners.

907

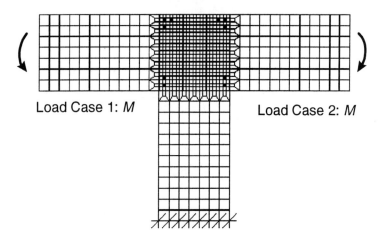

Load Case 1: *M*   Load Case 2: *M*

Fig. 5   T joint, SS=4, a design with minimum Performance Index

Figure 6 is the corresponding one to Fig. 4 showing Min and Max Von Mises stress and the performance indicator. Again the ratio of Min/Max stress tends to one at Steady State 6,. However it has a minimum at SS=4 and this corresponds to the fastener arrangement in Fig. 5. As can be seen in Fig. 5 there is an arrangement of fasteners with 3 each at top corners and 2 each at bottom corners. If the ESO process is taken further then in an effort to drive the Min/Max ratio closer to one the stress is driven up significantly, thus increasing the PI.

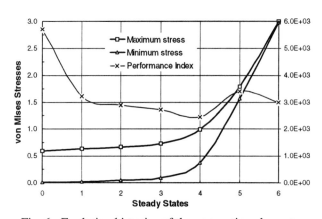

Fig. 6   Evolution histories of the connection elements

4.3 L-Shaped overlapping structure (ER=1%)

By another simple adaption of the mesh for the previous examples we can study an L shaped joint as shown in Figure 7. Two bending moments are applied and the ESO process started. Figures 7 and 8 are the corresponding

908

ones to Figures 5 and 6 of the previous example and similar observations can be made about the optimum arrangement of fasteners.

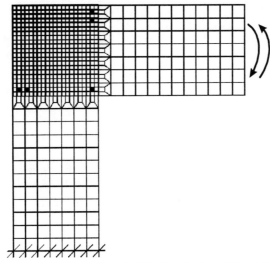

Figure 7: An L shaped joint, SS=4, a design with minimum PI.

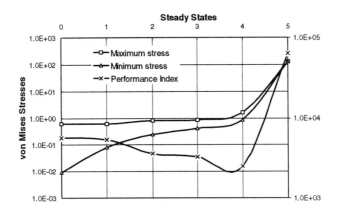

Figure 8: An L shaped joint, evolution histories of the connection elements.

4.4 T-Joint Design by Spot Weld or Threaded Fasteners

This is a design example of connection location for a T-joint structure, typical of many found in ship structures, where two plates are connected by two L shaped brackets. The connected plates and L-shaped brackets are modeled in plane strain elements (as shown in Figure 9) and are all considered as the non-design domains. Initial design of connections, 2D plate elements are also employed for modeling the joint elements as shown in Figure 9. Figure 10 shows the optimized connection positions. The connections can be either spot

909

weld or threaded fasteners. In addition, for these illustrative load cases, it could be concluded that the horizontal legs of L shape do not need so long.

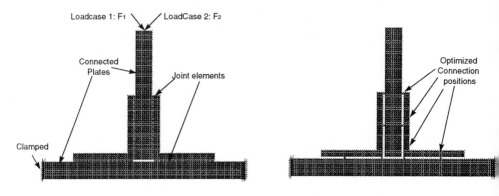

Fig. 9 Initial connection of T-Joint structure   **Fig. 10** Optimized locations of joint elements

4. CONCLUDING REMARKS

The above simple examples demonstrate the capabilities of solving connection optimization using the ESO procedure. This algorithm can be used on more complex connection problems. For instance, the design shape optimisation of adhesive bonded joints can be carried out by *nibbling* technique of ESO, [17]. Stiffness criteria [18] or both stiffness and stress may produce better results for connection topology. Further research is underway.

5. RERERENCES AND BIBLIOGRAPHY

1   BENDSOE, MP - Optimization of Structural Topology, Shape, and Material, Springer, Heidelberg, (1995).
2   BENDSOE, MP and KIKUCHI, N - Generating optimal topologies in structural design using a homogenization method. Comp. Meth. Appl. Eng. Mech., 71, 197-224, (1988).
3   HAFTKA, RT and GURDAL, Z - Elements of Structural Optimization, 3rd. edn., Kluwer Academic Publishers, Dordrecht, 1992.
4   HARRIS, HG OJALVO, IU and HOOSON, RE – Stress and deflection analysis of mechanically fastened   joints. AFLDR-TR-70-49, Wright-Patterson Air Force Base, 1970.
5   HEMP, WS - Optimum Structures, Clarendon Press, Oxford, 1973.
6   LI, Q, STEVEN, GP and XIE, YM(1997) Topology Design of Structures Subjected to Thermal Loading by Evolutionary Optimization Procedure, Proc. Of DETC'97, 1997 ASME Design Engineering Technical Conf. Paper DETC97DAC3974, Sacramento, (1997).

7   MATTHECK, C and BURKHARDT, S - A new method of structural shape optimisation based on biological growth, Int. J. Fatigue, 12(3), pp185-190, (1990).

8   MAUTE, K and RAMM, E - General Shape Optimisation-an Integrated Model for Topology and Shape Optimisation, in Rozvany, G.I.N. and Olhoff, N.(Eds), Proc 1st World Congress on Structural and Multidicipline Optimisation, Pergammon, (1995).

9   MICHELL, AGM - The limits of economy of material in frame structures. Phil. Mag. 8, 589-597, (1904),

10   QUERIN, O - ESO : Stress Based Formulation and Implementation, PhD Thesis, Aeronautical Engineering, University of Sydney, (1997).

11   QUERIN, O, STEVEN, GP and XIE, YM - Evolutionary Structural Optimisation by an Additive Algorithm, to appear, Finite Elements in Analysis and Design, (1999).

12   QUERIN, O, STEVEN, GP and XIE, YM - Development of a performance indicator for structural topology optimization, to appear, Structural Optimization, (1999).

13   ROZVANY, GIN - Structural Design via Optimality Criteria, Kluwer Academic Publishers, Dordrecht, (1989).

14   STEVEN, GP, QUERIN, O and XIE, YM - Structural Perfection:: Real or Imaginary (or just the place to FEA), Proc.NAFEMS World Congress '97, Stuttgart, 37-48, (1997).

15   VANDERPLAATS, GN - Numerical Optimization Techniques for Engineering Design: with Application, McGraw-Hill, New York, (1984).

16   XIE, YM and STEVEN, GP - A Simple Evolutionary Procedure for Structural Optimisation, Computers and Structures, 49, 885-896, (1993).

17   XIE, YM and STEVEN, GP - Evolutionary Structural Optimisation, Springer, Heidelburg, (1997).

18   CHU, DN, XIE, YM, HIRA, A and STEVEN, GP - On Various Aspects of Evolutionary Structural Optimization for Problems with Stiffness Constraints, International Journal of Applied Finite Elements and Computer Aided Engineering, Vol. 24, 197-212, (1997).

19   JIANG,T and CHIREHDAST, M – A systems approach to Structural topology optimization: Designing optimal connections. Trans ASME, 119, 40-47, (1997).

20   CHICKERMANE, H, YANG, RJ, GEA, HC and CHUANG, CH – Optimal fastener pattern design considering bearing loads, Proc ASME Design Eng.Tech.Conf., DETC'97, paper DETC97DAC3770, (1997).

Algorithm to Simulate the Construction Steps of a Cable-Stayed
Bridge by Back-Analysis

Dr F. A. N Neves
COPPE/UFRI, Brazil

ABSTRACT

The knowledge by the designer of the development tension in the cables and the flux of forces through the structure during the construction steps is very important in the design of a cable-stayed bridge.

This work deals with the study of numerical algorithm for the simulation of the construction steps of a cable-stayed bridge. Two different algorithms for the analysis of the construction steps, considering the geometrically nonlinear behaviour, and the possibility to take into account the dynamic analysis in each step of construction, are implemented.

The designer can confront the results obtained with the algorithm concerning displacements, tensions, etc. and those that will arise in situ and decide about them.

This work contains four sections, explaining: the theoretical formulation, two algorithms, the test example and, finally, a complete analysis of a three-dimensional cable-stayed bridge model, considering two and four construction steps.

Implementation of Design Sensitivity Analysis of Elastoplastic Material in Non-Linear Explicit Formulations

Dr J E Hassan, DaimlerChrysler AG, USA

Recent design requirements in automotive applications require the use of inelastic materials in an efficient way to reduce cost and cut weight. In such requirements, the elastoplastic behaviour of an automotive structure becomes a predominant factor in the design process. A typical improvement in such an environment can benefit a great deal of the sensitivity of the response of the structure to applied conditions. Measures of the sensitivity expressed in terms of the derivatives of the response, the gradient, with respect to the design parameters of interest are generally computed in the forum of design sensitivity coefficients.

A general finite element solution method for the transient dynamic response sensitivity of structures having elastoplastic material behaviour id developed. Employing a direct differentiation method, the gradient equation of motion is solved without iteration and by taking advantage of the available solution of the response. Special attention is given to material J" plasticity model that accounts for both kinematics and isotropic hardening. The method can be readily applied to other inelastic material models with analytically defined yield function and flow rule. The formulation is discretised and implemented in an explicit transient finite element code developed as a general purpose crash code for automotive applications. Typical automotive crash examples are used to demonstrate the method.

Structural Optimisation in the Design Process
of Modern Trains at Adtranz.

Per-Olof Danielsson
Structural Mechanics, Trains Division, Adtranz Sweden
E-mail: per-olof.danielsson@adtranz.se

Abstract

The need for simulation of the structural behaviour in the development of new train carbodies is large. The strongest reasons for this are that the available time for developing new trains are getting shorter and that customer demands are getting tougher, including demands on lower structural weight.

To meet these demands, the work with developing a new train has changed the way that a lot more parallel activities are present. The drawback with this is of course that if one system must undergo a big change of the design, a lot of functions could be involved and they might have to change their design too. Therefore the carbody design department must have tools to handle such changes very fast. One very powerful tool for this task is the optimisation module in MSC/Nastran. If for example one load path is cut by one system, the optimisation module of MSC/Nastran is a very powerful tool to help finding the most weight optimal design to overcome the proposed weakness.

In this paper, it is described how the optimisation module in MSC/Nastran is used in the design process of the new Oeresund Train Unit, finding the most weight optimal load paths.

VALIDATION: THE KEYSTONE TO THE BLACK ART
THE KEY TO THE BLACK BOX

C.P. Rogers[1] and S.Y. Azimi[2]

1   SUMMARY

The means of using computer-aided numerical analysis has changed. Over the past twenty years Finite Element Analysis (FEA) has moved from the fixed format entry of card images using a text editor, to point and click graphical user interfaces (GUI). Computers have changed from multi-million pound mainframes, to affordable microprocessor based PC and workstation systems. The Finite Element Method (FEM) is now widely available at affordable prices and in many enticing forms. FEA is now being employed by designers without any formal analysis background, as well as by analysts without an engineering design background. With the inclusion of FEA within Computer Aided Design and Manufacture (CAD/CAM) systems, there are instances where the FEM is being used without the conscious knowledge of the user.

Now, more than ever FEA programs are being used as black boxes, with engineers using the FEA program to expand their capability. FEA is used across industry, from mechanical through aerospace to civil, to provide design substantiation. In mechanical and aerospace it is often possible to validate the analysis by physical test in civil engineering however, this is expensive and difficult. The capability of modern FEA systems has provided the engineer with tools to solve problems that in the past have been considered intractable, and there is an increasing tendency to use FEA as the only substantiation of design. This is especially the case for the design of resistance to extreme and hazard loads for safety related structures such as civil nuclear containment, and offshore oil and gas facilities. Here, due to the scale of the structures and the one-off nature of the design, physical testing against extreme loading is difficult and in many cases, impracticable.

The liability for the validity of a design based on FEA lies with the design engineer. The FEA system is a tool, which allows the engineer to examine the product being designed in varying levels of detail. The apparent capability of the systems induces a feeling of confidence in the user, usually based on the premise that the computer is carrying out a mathematical process, which is beyond the users capability. As yet systems are not readily available that can reliably supplement the specific experience of the user, making a structural engineer into a structural dynamicist for instance. Most FEA system authors carefully

---

[1] Managing Director, CREA Consultants Limited, High Peak, UK, and member of the NAFEMS QA Working Group.
[2] Consultant, CREA Consultants Limited, High Peak, UK

include disclaimers as to the validity of results, clearly indicating that they take no responsibility for the manner in which the program is used.

It is a fair assumption that whilst most ISO9001 [1] Quality Management Systems (QMS) will demand that all software used on projects is verified, very few will demand that analyses are validated in detail. The most common procedure is to have all input data verified, checking calculations, input decks, model plots and boundary conditions, and then to carry out simple reaction force summations on output data. How many QMS procedures require the users to demonstrate that the boundary conditions are valid, that the solution scheme chosen is applicable and that the results be explained?

This paper seeks to demonstrate that validation is the key tool in the demonstration that the analysis performed is fit for purpose; that it is possible to test the operation of the black box, without an in-depth theoretical understanding of the FEM; and overall, that validation is simple engineering common sense. The discussions presented are also equally valid for other branches of numerical analysis, including Computational Fluid Dynamics (CFD) and Boundary Element Analysis. The advent of affordable high capacity computers is bringing all of these methods into the commercial domain.

2   INTRODUCTION

Over the past twenty years or so, numerical analysis, in particular the FEM has undergone a significant metamorphosis. When systems began to be commercially available in the late 70's, they were used by engineers, physicists and mathematicians who specialised in FEA. The computers available at the time were expensive to use and by today's standards; they were limited in their capacity and capability. The software of the day was also constrained by the programming technology as much as by the capacity and capability of the available computing platforms. The software users had to find a balance between arriving at a solution in a reasonable time at a reasonable cost and the accuracy of their results. Consequently, it was necessary to make gross assumptions to simplify the modelling of the structure. These assumptions were then necessarily tested on completion of the work to ensure that they were still valid and that the results presented were reasonable. This "testing" of the results also served to ensure that the program had performed reasonably. Many programming bugs were traced due to results failing to validate.

Today the working environment is significantly different. Desktop PC systems are many times more powerful than the mainframes of the 70's, storage devices such as memory and discs are relatively cheap, and programming technology bears little comparison. Today's computer user is used to advanced features such as the GUI, online manuals and real time error trapping. The GUI gives the user the same interface as mass-market products such as word-processors and spreadsheets, a feature that has significantly increased the market appeal of the systems.

In some industries, the numerical solver, particularly FEA, is now embedded in computer-aided design and manufacture (CAD/CAM) systems. There are regular discussions on the Internet relating to the use of embedded systems, especially with respect to ability of the users and software to identify errors. Another current trend is that of companies using computer software to expand their corporate experience. For instance, many of the structural design and analysis programs have added a mode-frequency extraction capability, thus giving a dynamic analysis capability. Users of these programs are now adding dynamic analysis to their portfolio, without taking on engineers experienced in the interpretation of dynamic analysis results. Similar situations occur with other expanded capabilities such as CFD and magnetostatics.

The concepts and procedures described in this paper have equal relevance in all fields of numerical analysis. The FEM is taken as the basis for this paper as this is the maturest of the common numerical analysis procedures used in engineering. The paper has been written with all numerical analysis in mind, however the examples all relate to the FEM.

3   VALIDATION AND VERIFICATION

It is useful to begin by defining the two processes by which the quality of numerical analysis should be measured, verification and validation, or V&V. The definition of and differentiation between verification and validation is a controversial subject within the numerical analysis community. This may be due to the subject crossing the boundary between computing and engineering.

The dictionary definitions of validation and verification are as follows [2]:

Valid 1. having some foundation; based on truth. 2. Legally acceptable: a valid licence. 3.a. having a legal force; effective. B. having legal authority; binding. 4. Having some force or cogency; a valid point in debate. 5. Logic. (of an inference) having premises and a conclusion so related that if the premises are true, the conclusion must be true; the conclusion will be false if one or more of the premises are false.

Validate 1. To confirm or corroborate. 2. To give legal force or official confirmation to; declare legally valid. – Validation. – Validatory.

Verification 1. Establishment of the correctness of a theory, fact, etc. 2. Evidence that provides proof of an assertion, theory, etc. 3. Law. a. a short affidavit at the end of a pleading stating the pleaders readiness to prove his assertions. b. confirmatory evidence.

Verify 1. to prove to be true; confirm; substantiate. 2. To check or determine the correctness or truth of by investigation, reference, etc. 3. Law. to add a verification to (a pleading); substantiate or confirm (an oath). – Verifiable. – Verifiableness. – Verifiability. – Verifier.

The NAFEMS Quality System Supplement[3] (QSS) [3] does not specifically define verification and validation; instead, it refers to "*software verification*" and "*validated analysis*". Extending the definitions within the QSS, the following definitions of verification and validation are offered in the numerical analysis context:

Verification is defined as the demonstration that the mathematical and logical functions necessary to analyse a physical system to an acceptable accuracy are correctly executed; and that they correctly implement the theory upon which they are based.

Validation is defined as the demonstration that the derivation and selection of mathematical models and solution algorithms to analyse a given physical system are based on sound and justifiable assumptions and approximations.

Applying these definitions to numerical analysis, the verification will demonstrate that the tool being used for the solution executes correctly. Therefore, it will produce results of an adequate accuracy given that the input has been shown to be correct. Validation of the analysis model, including such features as model discretisation and selection of the solution software and hence solution algorithms; demonstrates that the input is reasonably correct within the constraints of the prevailing technology. Thus, it is reasonable to assume that a validated model run with verified software will give results of an adequate accuracy for the system being considered. Therefore, as the premises (input) have been shown true; the

---

[3] The NAFEMS QSS is undergoing a review to be published in the first quarter of 1999.

conclusions (output) can be inferred as being true, within the bounds defined by the analysis specification.

Verification is therefore a task primarily for the developer; validation is a task primarily for the user. There are verification tasks involved in the development of numerical analysis models, such as the derivation of loading and boundary conditions, and validation tasks involved in the writing of the software, such as in the derivation of material models. It is noted that these definitions are different to those specified in the software development standards introduced by the International Standards Organisation (ISO) and others. This is because we are considering the use of software developed by others, and demonstrating that the software is being used in a reasonable manner.

If the developer cannot provide sufficient verification documentation, ISO9001 passes the requirement for providing the verification to the purchaser or user. Essentially, ISO9001 requires that all software used for production work to be adequately verified. The code does not however introduce requirements for validation of the analysis models.

In depth numerical analysis validation does not generally feature in quality assurance procedures; the most likely reason for this is that ISO9001 does not specifically call for validation in the manner that is expected for numerical analysis. It is apparent however that by considering the relevant clauses within ISO9001, the definitions given above are fully applicable. Clauses 4.4.7 and 4.4.8 read:

4.4.7 Design verification

At appropriate stages of design, design verification shall be performed to ensure that the design stage output meets the design stage input requirements. The design verification measures shall be recorded.
NOTE 10. In addition to conducting design review, design verification may include activities such as
- performing alternative calculations;
- comparing the new design with a similar proven design, if available;
- undertaking tests and demonstrations; and
- reviewing the design stage documents before release.

4.4.8 Design validation

Design validation shall be performed to ensure that product conforms to defined user needs and/or requirements.
NOTE 11. Design validation follows successful design verification.
NOTE 12. Validation is normally performed under defined operating conditions.
NOTE 13. Validation is normally performed on the final product, but may be necessary in earlier stages prior to product completion.
NOTE 14. Multiple validations may be performed if there are to be different product uses.

A QMS written to comply with the QSS would however include model validation procedures since validation is one of the underlying principals of the QSS. In general the authors of a QMS not written to comply with the QSS, will tend to require that the numerical models and results are verified (or checked) by review as if they were calculations and reports.

4   THE NEED FOR VALIDATION

Some of the reasons for validating analysis have not changed since the early days of commercial FEA. Primarily these are the need to demonstrate that the correct techniques have

922

been used, and that any assumptions in the modelling can be shown to hold once the analysis is complete. Other factors prevalent in the current usage of numerical analysis have introduced new reasons for validation; it is these that indicate that validation should take a higher profile.

One of the well-known maxims of computing and engineering is that if you increase system capacity, engineers will almost instantly find jobs that will push the new limits. Thus, if disc capacity is increased, the disc will be filled days later; and if a faster processor is acquired, it still takes as long to get a solution. The reason for this is that the engineers remove approximations from their previous work and "aim" to achieve a better resolution in their solution. Therefore, in current modelling practice we find that we make as many, if not more assumptions than in the late 70's, each of which should be tested for validity on completion of the work.

Approximation in numerical analysis is an occupational hazard. By definition, we approximate. For example in FEA, the discretisation of the structure to a Finite Element mesh, assuming steady-state dynamic response, allowing non-linear materials to act in a linear manner; these are all approximations. Some approximations are made for run-time purposes, others are to allow the results to be used along with standard design codes of practice; others approximations are introduced to simplify the analysis to help understand responses. There are many reasons to approximate; however, each approximation has a related assumption, each of which should be validated. If after the analysis has run, the modelling assumptions cannot be shown to still hold, then it is likely that the analysis will have to be declared invalid.

In the late 70's the approximations were made to reduce the problem size and thus to obtain a solution at a reasonable cost and in a reasonable time. Many gross assumptions were made, whole civil engineering structures were reduced to a few sticks, and intricate pipework systems were simplified, all in the quest to obtain an "improved" solution or to solve previously insoluble systems. These reduction techniques introduced many of the rules of thumb that are both still in use and still valid today. The assumptions were often so obviously gross that analysts had no option but to validate them.

The modern system with its GUI and large user base introduces new factors into the analysis arena. Approximations are still made, in many cases they appear to be less severe than those made in the 70's, however, it can often be seen that they are just as significant, but for different reasons. It is also the case that current practice is to introduce a large number of less significant approximations and assumptions. The cumulative effect of these assumptions can lead to more uncertainty than the small number of gross assumptions made in the early years.

The modern analysis environment has introduced a new culture to numerical analysis, one that is far removed from the culture of the late 70's. Of the reasons for validation introduced by modern programs, the so-called "Black Box Syndrome" is possibly the most important. The modern analysis system has a similar GUI to the more familiar office systems such as word-processors and spreadsheets. The GUI, which is available on most systems, hides much of the complexity of program usage. With careful programming, the GUI can be made to keep options unavailable until certain conditions have been met. The system will also trap some errors as input proceeds. This apparently helpful environment serves to raise the users confidence in the model to the extent that an assumption is made that all of the input data is correct. This confidence is then extended to the results presented, "since the system indicated that the input contained no errors, therefore the output must also be error free".

Numerical analysis systems need to serve two communities, the novice or occasional users and the power users. The novice and occasional users require a system that is capable of rigorously trapping errors and identifying potential conflicts in input data. The software developers need to attract a larger user base to increase revenue; therefore, they add the necessary features to the GUI. The addition of such user-friendly features causes a significant conflict with the power users. Power users by their very nature are using the program to its limits often to solve cutting edge engineering problems, to do this they require the ability to allow the program to run through error traps. Therefore, to enable the power users to use the systems the developers have to categorise the errors trapped. Errors that could result in poor results, but that will not cause the analysis to fail are notified to the user, but execution continues. Errors that will cause the analysis to fail are usually raised in a manner that stops execution, usually in a controlled manner.

The problem introduced by this conflict is that the novice and occasional user will assume that if the job runs to a solution, with say warnings and no errors, it is by definition valid. Where the power user will examine the error and warning messages to establish their importance to the validity of the solution, the novice or occasional user may not. Indeed such users may not be able to correctly interpret the individual and cumulative significance of the messages.

The more friendly analysis suites now available, and in some cases their low cost, has also lead to a culture among design teams and design houses where the analysis is carried out by relatively junior engineers. These engineers may not have sufficient engineering background to be able to validate results; they may not have a sufficiently detailed analysis background to be able to appreciate the weight of their modelling decisions. Validation, lead by senior engineers with a sound knowledge of the product or process being designed will provide a basis for ensuring that the analysis results are safe and appropriate.

An example of the use of numerical analysis in a developing field is that of soil-structure interaction analysis. The use of numerical techniques to analyse soil models and dynamic soil structure interaction (SSI) effects is a relatively new development in the field of geotechnical and seismic engineering. If numerical analysis is used correctly, it can provide a better understanding of soil behaviour and seismic SSI effects. However, the current state of the art is such that it is even more important to validate the results against established, often deemed conservative, methodologies. While there are many commercial FE codes available, only a few have been written specifically with SSI analysis in mind. General-purpose programs do not offer the required facilities, and much time and effort can be saved by using purpose written codes. Most explicit time-domain finite element and finite difference codes can usually incorporate any non-linear constitutive characteristics of the soil. The difficulties are not with computational procedures but with determination of the appropriate non-linear laws. Whichever method of analysis is used, it must always be remembered that the results are only as good as the input data. Therefore, all input and output requires thorough validation from a logical engineering point of view. Since the site properties are subject to variations, then it is essential that the sensitivity of the parameters used in the model are fully appreciated and an adequately credible range of properties employed in the analysis.

5   WHAT SHOULD BE VALIDATED

Validation is the demonstration that the analysis model is adequately representing the physical system being analysed. It is therefore necessary to validate all aspects of the model that influence the outcome of the analysis.

a)   All assumptions

b)   All boundary conditions

c)   Nature of loading

d)   Modelling/analysis parameters

924

e) Overall solution quality

5.1 Assumptions

Assumptions are a necessary part of numerical analysis, as a rule the more complex the analysis the more assumptions are made. Assumptions are usually designed to simplify complex features or to allow aspects of the physical system to be ignored. An important assumption that is made is the assumption that the chosen solution scheme adequately represents the physics being studied. Typical modelling assumptions, all of which would necessitate some validation, are:

a) The type of solver used, static, dynamic, etc;

b) The elements selected and their restrictions;

c) The level of idealisation, for instance the decision to use beams, shells, solids or a hybrid;

d) The type of solution, linear, non-linear, etc. (an example would be the common decision to analyse a non-linear material such as concrete using linear elements);

5.2 Boundary Conditions

The inappropriate selection of boundary conditions is one of the most common causes of modelling error. Boundary conditions include the edge fixity of the model, the means of coupling elements, (pinned beams as fixed beams for example) and the application of loading. Some boundary conditions are dictated by the analysis technique that is being applied; others are decisions that are made by the analyst. It is not always practical to model connectivity in a manner that directly represents the structural connectivity, therefore approximations are necessary. The application of boundary conditions can change between different systems and examples manuals are a good source of information on how best to apply them. One positive benefit from the introduction of standard benchmark tests such as those published by NAFEMS is that many developers now include the tests in their verification manuals. Thus, it is possible to compare how different systems expect standard details to be modelled.

5.3 Nature of Loading

The nature of loading and the manner in which the load is applied can have a significant influence on the results. In carrying out dynamic analysis using the modal response technique for instance, a displacement response spectrum would be valid for the representation of seismic loads but would not necessarily be valid for sea wave loading. Other aspects to consider are pressure loads Vs concentrated loads and pressure loads Vs inertia loads. Again, in dynamic analysis a common technique is to increase density to represent load carried by a system. In this case, it would be necessary to demonstrate that the 3D nature of this type of modelling assumption is valid.

5.4 Modelling and Analysis Parameters

Modelling Parameters cover a large and diverse set of assumptions, from the selection of week and stiff springs to represent element connectivity, to element selection, solution control switches integration techniques and time steps. Each modelling parameter may or may not have an influence on the resulting solution, in many cases only experience can indicate the best selection.

Overall Solution Quality relates to the quality of the answers produced and the quality required for the system being analysed. For example, it may well be that the assumptions made in defining the model only allow results to be derived with a confidence of $\pm15\%$, but the design shows the structure to be stressed to 95% yield. The implications of such a miss-match would need to be carefully considered.

The consideration of the overall quality can be subjective and setting strict limits can be difficult. It is here where engineering judgement and experience comes into its own as a validation tool, as it may well be necessary to balance the quality of the analysis against the stresses derived.

6   HOW TO VALIDATE MODELS

Model validation is not generally a complex process, in most instances it reduces to a combination of engineering judgement (or engineering common sense) and a series of simplified analyses or calculations. One of the most useful toolkits for validation consists of a pencil, paper and a few favourite engineering reference books.

6.1   Engineering Judgement

Engineering judgement has one major drawback, that is that it changes from engineer to engineer. This is however is not a problem if the engineering judgement is written down, thus allowing others to assess the postulated reasons for the analysis validity. The logic of the argument is clear, engineering judgement is based on engineering experience. Except in a few special circumstances, the experience of any two engineers is not going to be the same, therefore the engineering judgement calls have the potential to be different. This leads to the statement "In my judgement the analysis is valid", being both true and untrue depending on the engineer reading (or writing) the statement. However, the statement "In my judgement the analysis is valid because…" can be tested for validity by other engineers competent in the relevant field.

Validation has been defined as the demonstration that the derivation and selection of mathematical models and solution algorithms to analyse a given physical system are based on sound and justifiable assumptions and approximations. To demonstrate validity it is necessary to build arguments demonstrating that the analysis does reasonably represent the physics of the system. This is best achieved by identifying bounds for the analysis; these bounds then become the range of expected results. The bounds can be tight, that is they represent a small deviation from the result achieved, or they can be broad bounds, where the deviation can be large, sometimes even orders of magnitude. (It should be noted that the former case, tight bounding, tends towards verification.)

6.2   Hand Calculation and Simplified Models

Hand calculation is a good first step in validation, simply because the physics and mathematics have to be reduced in complexity to allow simple solution. As a simple example, a beam with partially restrained rotational end restraints falls between a beam with both ends pinned and a beam with both ends encastre. We can therefore deduce that the resulting forces, moments and deflections, (or natural frequencies), for an elastic analysis of the partially fixed beam will lie between the results for the standard cases. If a complex numerical analysis returns results between the two bounds, a certain confidence in the results can be taken. If the results from the analysis fall outside the bounds set by the simple representation, this does not signal that the results are invalid. The situation points to the possibility that other factors need to be considered shear deformation for a deep beam for instance. It is therefore possible to develop a series of simple calculations that will set reasonable bounds for the more complex analysis.

The complexity of some systems does not lend itself readily to hand calculation, though some initial hand studies always help. In these more complex cases, it is useful to build knowledge of the working of the system through simplified models. A classic example is the derivation of a non-linear transient analysis. One of the accepted analysis procedures here is to build up to the non-linear work. An initial model, it may or may not have the same level of discretisation as the final model, is built as a linear model. That is no non-linear materials, interfaces, etc. This model is then solved as a mode extraction analysis, extracting a sufficient number of modes to identify the primary modes of the structure. This tests the model for connectivity and derives "expected" mode frequencies and mode shapes for the structure. These then are the first or second estimates of the bounds for the final model. The model is then modified to be run as a linear transient analysis. It is possible to add linear features, including interfaces, and the time varying loading. This solution is tested against the mode frequency analysis to ensure that the expected frequency responses exist, and that the expected displacement responses are achieved by comparison with the mode shapes. The linear solution should be able to identify where non-linearities are likely to begin to appear. The final non-linear solution is then run. The results can be compared to both of the forgoing analyses, and the deviations after the non-linear response begins can be assessed to identify reasons for change. If the non-linear responses, or at least their onset and early stages can be linked to the elastic responses, then some of the necessary confidence can be derived.

6.3   Visual Inspection

The visual inspection of results plots is an invaluable tool in validation, especially in explaining non-linear responses. (A note of caution, 3D plots are great for impressing the client and superiors, they are of little use on their own in validation, orthogonal plots are of more use.) Plots of direct and equivalent stress, displacement, stress tensors, fluid flow, etc., all indicate the progress of the analysis. With time domain analysis, graphing various parameters on a time-scale or generating spectra is a useful means of presenting the progress of the analysis in a clear format.

6.4   Parametric and Bounding Case Analysis

Another very useful validation technique is that of parametric or bounding case analysis. Numerical analysis suites that include parametric modelling features, such as macro languages are ideal for this process. The principle is to vary parameters within the analysis to test the sensitivity of the results to that parameter. Two examples, firstly where ground conditions are modelled in a dynamic analysis of a structure subjected to seismic loading. It is difficult to derive reliable dynamic properties for materials beneath the foundations of a structure. Therefore, it is normal practice to run the analysis three times, once with the best estimate (derived) ground properties, then once each with the properties halved and doubled. (These bounds are relaxed if a good ground survey is available.) This serves two purposes, firstly it tests the response of the structure under three different though related ground conditions, and secondly, it provides three analyses, the results of which can be compared for similarity and difference.

A more common approximation is that of using stiff and week springs for internal connectivity between elements or components. The usual problem here is that the quality of the results extracted from the matrix can be influenced by the range of stiffnesses chosen, especially if the range is significant. (Some solvers will only work satisfactorily with a range of 10^8, others will begin to fail at 10^{12}, and still others will run at 10^{20}.) The test here is to vary the stiff and week spring stiffnesses by orders of magnitude to ensure that there are no unacceptable influences on the results.

927

In carrying out parametric or bounding case analysis for validation, it is important to ensure that all areas of the results are examined. Tests carried out involving the changing of spring stiffnesses used for internal model connectivity have shown failures in the balance of reaction forces long before errors in stress or deformation have become evident. Table 1 shows the results of a typical parametric study where the spring stiffnesses of connecting springs are varied. The columns relate to low-stiffness, high-stiffness, stiffness ratio summated reaction forces, X, Y and Z, the deformation and stress.

When parametric or bounding load case validation is used, there is an additional advantage that it is possible to place tolerance limits to the analysis results.

K-low	K-high	Ratio	ΣFX	ΣFY	ΣFZ	Disp	Sx
1.20×10^7	2.40×10^{13}	2.00×10^6	3.41×10^{-4}	2.98×10^{-8}	-3.75×10^6	-5.65	463
1.20×10^6	2.40×10^{14}	2.00×10^8	3.78×10^{-4}	-1.62×10^{-4}	-3.75×10^6	-5.67	449
1.20×10^5	2.40×10^{15}	2.00×10^{10}	-1.10×10^{-4}	1.79×10^{-3}	-3.75×10^6	-5.69	439
1.20×10^4	2.40×10^{16}	2.00×10^{12}	3.32×10^{-2}	1.70×10^{-2}	-3.75×10^6	-5.69	438
1.20×10^3	2.40×10^{17}	2.00×10^{14}	3.14×10^{-1}	-1.55×10^{-2}	-3.75×10^6	-5.69	438
1.20×10^2	2.40×10^{18}	2.00×10^{16}	1.78	2.24	-3.75×10^6	-5.69	437
1.20×10^1	2.40×10^{19}	2.00×10^{18}	6.18×10^1	-1.70×10^1	-3.75×10^6	-5.69	437
1.20	2.40×10^{20}	2.00×10^{20}	1.95×10^2	-5.38×10^2	-3.74×10^6	-5.67	436
1.20×10^{-1}	2.40×10^{21}	2.00×10^{22}	1.06×10^4	-5.57×10^3	-3.66×10^6	-5.52	434
1.20×10^{-2}	2.40×10^{22}	2.00×10^{24}	6.36×10^4	-2.06×10^3	-2.56×10^6	-3.98	312
1.20×10^{-3}	2.40×10^{23}	2.00×10^{26}	-4.27×10^2	4.97×10^4	-1.96×10^6	-3.77	342
1.20×10^{-4}	2.40×10^{24}	2.00×10^{28}	-2.13×10^5	-2.95×10^5	1.29×10^6	-1.58	87
1.20×10^{-5}	2.40×10^{25}	2.00×10^{30}	3.14×10^6	-1.87×10^5	-1.11×10^7	-31.4	2137

Table 1 Spring Stiffness Ratios, Reaction Force Balance, Displacement and Stress from a Parametric Sensitivity Study.

6.5   Comparison to Test and Experience Data Bases

Comparison of analysis results to physical test crosses validation with verification and care has to be taken to differentiate between the two. When comparing analysis results to physical testing from a validation standpoint, it is not necessary to show that there is an exact match to the results. For validation purposes it would for instance be necessary to show that the analysis predicts load paths, displaced shapes and stress distributions reasonably. Exact correlation between test and analysis would be a verification exercise, the value of which would have to be defined before the analysis starts.

Unlike verification, validation does not require the demonstration that the results are numerically accurate. The main objective of validation is to demonstrate that the physical process is being represented in a reasonable manner. Validation can however yield information on the numerical quality and applicability.

6.6   Peer Review

Peer review is not an alternative means of validation; it is a necessary final stage of the whole analysis process. The complete analysis validation should be reviewed by engineers with sufficient product and analytical experience to be able to identify flaws and omissions in the validation and/or the analysis itself. Ideally, the peer reviewer(s) should be independent of the design and analysis project.

THE ANOMALY

Perhaps the most important class of conditions used to validate models is that of considering anomalies in the results. Here engineering judgement is the primary tool. In carrying through validation exercises, it is likely that unexplained effects may be found. If engineering experience and common sense cannot clearly explain analysis results, then an anomaly exists. This can lead to three possible conclusions: there is an error in the modelling; a mechanism has been identified that has not been met before in the experience of the engineers working on the project; or that there is an error in the software.

The first two conditions can usually be resolved by careful checking of the model and carrying out further parametric and bounding analyses. If this does not resolve or explain the anomaly, the third situation, a program error becomes the more likely cause. The resolution of anomalies highlights the ability of validation exercises to identify and resolve modelling errors and to identify software errors.

There is therefore an argument to place validation above verification in terms of project level error detection and resolution. This does not however allow users to forego the requirements of ISO9001 with respect to the verification of software. It is still necessary to obtain or carry out formal verification exercises of purchased and in-house software.

8   CONCLUSIONS

In the early days of the commercial FEA packages, the programs were run by "experts" who were seen to work from back rooms or specialist consultancies. These experts would take the problem to be solved, build long complex looking data files (or card decks), feed them into a computer, which in turn would produce results. Weeks later, when all the validation and checking was complete, the results would be passed to the designers, often still in the form of reams of listing paper. These people were seen to be practising a black art, writing out their long incantations and somehow producing forces, moments and stresses from which a design assessment could progress. Whatever they did however, the work was all held together with the validation exercises, carried out by both the analysts to check the work and the designers, to build confidence in the work. Validation was, and still is, "the keystone to the black art". These analysts, who would today be described as power users, are still using the numerical analysis suites to the limits of the machine and software capabilities. These analysts will be making assumptions, often no less drastic than those of the 70's, and as such they will still insist on validation of their work. This is evidenced by many of the discussions on the Internet mailing lists and news groups.

In the present day, computer-aided numerical analysis is almost an everyday feature of engineering. The numerical analysis suite is a user-friendly system that has the same feel to it as a word-processor or spreadsheet program. The numerical analysis suite has to many become a black box, and engineers are content to use the systems as such. Validation is a tool that can allow the numerical analysis system to be used as a black box with a degree of safety. Validation is therefore "the key to the black box". As long as design and analysis is carried out by, or is supervised by, engineers who can fully appreciate the physics of the system being studied, the black box approach can be accepted. The concern arises when the black box results are used without detailed examination of results and physical mechanisms. It is noted that this applies to both groups of users. Power users may well carry out analysis for different industries and rely on relevant field experts to help validate results.

All numerical analysis should be validated, whatever the capabilities of the user and the software. The degree of validation should be commensurate with the target use of the results and the variability of the input data and solution scheme. Validation of analysis whether or not the system has been used as a black box demonstrates several essential attributes of the work:

929

a) That the analysis carried out has used a reasonable mathematical model of the physical system;

b) That the results are within expected bounds, or where the expected bounds are exceeded there are reasonable engineering reasons for the exceedence;

c) That the software is running in a reasonable, correct manner; and

d) That the user has demonstrated their understanding of the physical system being modelled and that they understand the correct operation of the numerical analysis system.

Validation is therefore an important tool in assuring the quality of computer-aided numerical analysis, and it is arguably more important than verification. Validation has the ability to identify the bounds of applicability of analysis results, and in cases where a model cannot be validated, to identify potential programming errors. It is also worth commenting that if a structure or component substantially substantiated by numerical analysis fails, causing loss of life or significant economic loss; the liability for the failure will almost certainly fall to the designer, not the software developer.

Validation is the same as any proof, it is possible to start with the analysis and proceed as far as is possible towards an analytical solution, then start with a known engineering solution and work backwards towards the analytical result. As long as the two meet within acceptable bounds, the validity of the analysis can be considered proven. This is of course one of the bases of forensic analysis; this again demonstrates how useful validation is as an engineering analysis tool.

There is an argument for developing a generic numerical analysis validation methodology. In defining such a methodology, it would be necessary to ensure that both power users and the novice/occasional users are fully catered for. This generic methodology, which could be a two-tier system, could then be used to derive validation procedures for analysis QMSs, procedures that will suite the environment in which users are working. The derived procedures would therefore be capable of catering for black box usage and for advanced usage. Peer review of the validation exercise, by suitably experienced and qualified engineers, preferably independent of the project, should be a pre-requisite of any numerical analysis QMS procedures.

It should always be remembered that "*a model that is too complex to validate has to be considered as invalid*".

9   REFERENCES

1   BS EN ISO9001:1994: Quality Systems – Model for Quality Assurance in Design, Development, Production, Installation and Servicing, International Standards Organisation 1994.

2   Collins Dictionary of the English Language, (Australian Edition) Wm Collins Publishers Pty. Ltd., .Sydney 1980.

3   Quality Systems Supplement to ISO9001 Relating to Finite Element Analysis in the Design and Validation of Engineering Products, Issue 2.0, NAFEMS Ltd, East Kilbride, Glasgow 1999.

TRAINING IN FINITE ELEMENT METHODOLOGY AND QUALITY

William J. Anderson

Professor of Aerospace Engineering[i] and Chairman/CEO[ii]

SUMMARY

The author has pioneered in multimedia training of FEA practioners through CD-ROM discs. This paper reviews the new CD-ROM recording method and its use in promoting FEA methodology and quality.

The Personal Professor[TM] method allows a speaker to sit comfortably at a desk and face a video camera. The speaker discusses a series of figures, and can annotate them, drawing equations, underlining, etc. The pen marks are captured on a floppy disk on a PC and are time synchronized with the speaker's words.

The method is efficient enough to allow a speaker to create an hour of finished lecture with only 2 hours in front of the camera. It basically recreates a typical live lecture as commonly done with transparencies. The approach works well for encyclopedic material that contains a lot of information in the form of sketches, lists, equations, and graphs.

Personal Professor recordings are useful for training in FEA methodology and quality. A total of 3 FEA courses have been released to date, totalling 58 hours of lectures. It is expected that technical documentation (as required by quality systems) will become the biggest usage in the long run, however.

1. MULTIMEDIA OVERVIEW

Multimedia methods involve digital recording, storage and playback of information. By using video cameras, digitizing tablets, PCs, phones, faxes, printers in digital form, all data can be handled with digital computer processing.

----

[i] The University of Michigan, Ann Arbor, MI 48109
[ii] Automated Analysis Corporation, Ann Arbor, MI, 48104.

When data are retrieved, they can be accessed in random order, in contrast to sequential viewing of video tapes, say. The viewer skips over unimportant material and navigates to the examples, problems, etc. of interest.

Information captured on compact discs and DVD discs is very secure. It is easy to make multiple copies, and send them to multiple sites. It is amazing how calming it is to have a complicated lecture or project "burned" on a CD, where it is very difficult to destroy. Multiple copies of the material make the author even more confident.

Interestingly, the largest use of a new technology such as multimedia falls into the extreme poles of pornography and Bible studies. Educational material follows as people get back to business. Technical documentation of projects will eventually emerge as a major engineering use. Current personnel turnover, and the quality system approach of documenting processes in a company will require methods such as this to enable new hires to pick up procedures quickly.

2. DEVELOPMENT OF CD LECTURES

The life cycle of multimedia lectures involves [1]:
1) preparation of material
2) recording
3) processing
4) editing
5) playback

The most time-consuming part for the author is the preparation of materials, from 10 to 30 hours per hour lecture. Recording takes about 2 hours, processing takes about 6 hours, and editing about 3 hours. The task of processing can be turned over to a computer technician. (A good high-school senior loves to work with film clips.)

The preparation of materials can be done on any computer which has a "presentation software" package. Typical candidates are MacDraw Pro(TM) Claris Draw(TM), PowerPoint (TM), Harvard Graphics(TM), and Quark Xpress(TM). The requirements for the presentation package is that it handle sketching, equations, text, sorting figures, slide show and handouts. The author prefers MacDraw Pro for all tasks except equations, done by Equation Editor(TM).

The program Quark Xpress is now dominating color printing processes, and might well be used for the multimedia "study guide" which contains all the figures. (Commercial printing houses all use it, whereas other packages are hit or miss.)

3. COMPACT DISCS

Compact discs are very common today. The 540 Mbyte storage capacity allows them to record about 5 hours of lecture with the Personal Professor method. Compact discs contain information on a spiral track, as opposed to the information contained on concentric circles on hard disk drives (note that the different spelling reminds one of this difference!)

The compact disc is made of clear plastic with a silver layer on the back side. The information is "burned" on this silver layer (the process in mass production, however, is actually a printing process). Laser light penetrates the clear plastic and is reflected or absorbed by the silver layer (Fig. 1). Typical disc players use 3 laser beams and a "voting" procedure so that an error caused by a scratch on the surface doesn't destroy the information.

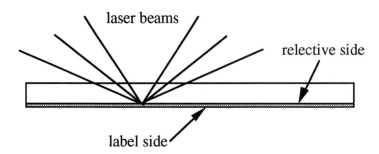

laser beams

relective side

label side

Figure 1.   Compact disc with 540 MBytes of information

4. DVD DISCS

The recent release of the DVD format discs will be a boon for those authors wishing to store entire courses on one disc. The installed base of DVD players is not great enough yet to issue training for them, but will be in about a year from now.

5. COMPRESSION ISSUES

Multimedia documents require a lot of disc storage space. The primary culprit is the video image, which must be used in small size and only sparingly. The author uses 1/16 of the monitor screen for video display, and only in video clips of 20 seconds or so. If full screen video were used without compression, one could store only a minute of video on a CD and a few minutes on a DVD disc. Audio, still figures, and annotation traces occupy relatively little disc space. Audio is recorded at 22 kbyte/sec and is not compressed.

Video compression is discussed in terms of compression ratio, i.e. the ratio of the original raw signal to the compressed information. "Lossless" compression is the type where the video can be cycled in and out of compression with no loss of resolution, and may only have a compression ratio of 2:1. "Lossy" compression is more often used, where the video will only be decompressed once and viewed, leading to compression ratios as high as 100:1.

Still figures can be compressed by JPEG compression (Joint Photographics Experts Group standard). This works within the confines of a single picture and can reach a ratio of 20:1.

Moving pictures can be compressed by MPEG compression (Moving Pictures Experts Group standard). This compression works from frame to frame, storing only the differences that arise with each new image. Because many moving pictures only have one moving part in front of a stationary backgroup, compression ratios of 100:1 can be reached.

6. PERSONAL PROFESSOR(TM) METHOD

The goal of this method was to capture large quantities of "encyclopedic" information containing movies, sketches, photographs, and annotated figures typical of university courses. Many areas of science, biology and engineering require detailed procedures done in order to accomplish a task.

A side goal was to make a method that is very easy to author. There have been some multimedia products on the market that require so much effort on the part of the author that society will never get an expert to sit down for the hundreds of hours required to record even one such lecture.

This method allows the author to sit at his/her own desk, with conventional room lighting and face a video camera. The speak talks to the camera, introducing each figure in turn, and then looks down at the figure placed on a digitizing tablet in from of him/her. With an electronic pen (Wacom or Calcomp), the author underlines, checks, draws equations and sketches, much as in a typical lecture with transparencies in front of a live audience.

This method has been taught to 6 speakers who have made recordings. The average learning time is 2 hours, with one hour spent watching a proficient author use the method.

Resources needed involve an authoring studio (a quiet room) and a processing studio. Playback is on PC or Macintosh computers, and can be done at any viewers location. Processing is the difficult step but can be purchased as a service. Licenses are being sold for the whole process, and allow an industrial documentation group to carry out the creation and processing of lecture series.

The video signal in Personal Professor is processed at 53 kbyte/second and the audio signal at 22 kbyte per second. These are low enough to allow longer lectues to be recorded, yet just high enough to provide the required resolution. These rates are still a bit high to be handled easily on the Web, yet efforts are underway to be able to send these lectures over the Web, for better distribution. The average home connection does not yet have enough speed to allow watching significant video.

A typical figure, taken from a lecture on Gaussian integration, is:

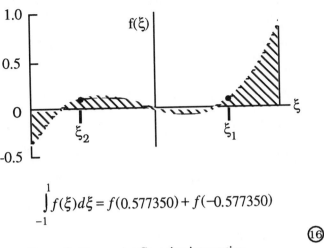

$$\int_{-1}^{1} f(\xi)d\xi = f(0.577350) + f(-0.577350)$$

(16)

Figure 2. Two-point Gaussian integration

7. EXISTING "TITLES"

Currently, there are 3 lecture series for sale which cover finite elements and acoustics:

W. J. Anderson, "Linear Static Finite Element Analysis," Automated Analysis Corporation , 1995 (30 hours of lectures with 520 page B/W study guide and 320 page text).

W. J. Anderson, "Structural Dynamic Finite Element Analysis," Automated Analysis Corporation, 1998 (17 hours of lectures with 180 page colored study guide and 210 page text).

W. J. Anderson, "Numerical Acoustics," Automated Analysis Corporation, 1997 (11 hours of lectures, 160 page colored study guide).

In addition, there is a single lecture about ergonomics (including carpal tunnel syndrome):

Armstrong, T., Ergonomics in the Workplace, Industrial and Operations Engineering, University of Michigan, 1998. (1 1/2 hours with 40 page study guide).

and a 35 hour course on the use of the Hewlett-Packard workstation operating system, used internally by Hewlett-Packard for their Navy users.

8. ANNOTATION

A novel feature of the Personal Professor approach is the use of annotation, i. e. handwritten comments made by the speaker on the still figures. Accomplishing this required some specialized software, and the process was patented by Glen Anderson [1]. The method might be familiar to football fans watching the Super Bowl, where the announcers show Xs and Os representing players, with the path of the ball drawn with a line. The football version is different in two ways--it is analog (at present) and it is not time-synchronized. The Personal Professor method was developed prior to the football usage and is more sophisticated.

9. QUALITY CONSIDERATIONS IN FEA TRAINING

To this point, we have discussed a mechanism for training FEA using anew method. The FEA methodology and quality issues are natural topicss to record this way. As a professor who has recently been exposed to quality issues, the author is now obligated to try to move toward that philosophy in the actual lectures. A series of unrelated points are mentioned below.

1) Current instruction in Universities does not include much quality information, because most professors have not been exposed to it yet.The auto industry has a strong program which can be studied [2].

2) A recent paper [3] in "Benchmark," the NAFEMS Journal, contains a well-spoken plea for FEA people to start thinking in quality terms, and suggests that Monte Carlo simulations are a good way to move from deterministic thinking to statistical thinking. This is a "must" read for all FEA people.

3) Documentation of existing processes is a major part of all quality programs. The Personal Professor method is an easy way to start doing that in your home organization.

4) Quality issues make engineers think of the design/analysis/simulate/test/build cycle from top down, rather than the current mode of bottom up. Sitting in on quality groups, and on problem solving groups can convince an instructor that the quality approach is an interesting and productive way to view problem solving.

5) The idea of a single, dominant variable (with its associated variance) causing most of the variance in the total process was originally pushed by Dorian Shainin [4].

6) Charlie Sieck of Caterpillar [5] feels that the greatest need in engineering education at presence is to account for variability in the input data, and how to account for that in the output results of simulations.

7) Because quality ideas originally were applied to production of hard goods, the philosophy seems (to the newcomer) to be more tied to driving out a few bad parts than toward increasing the mean quality of all parts (processes). This philosophy in fact will lead to better overall products in the long run, but the concept is difficult for professors to grasp. We are used to graduating students of varying ability and are more proud of the few excellent students than disturbed by the few bad students. (Parents probably have more of a natural quality orientation to their children's education than the teaching staff.) Nevertheless, teachers must consider the result of their teaching in terms of uniformly high understanding, rather than a few peaks of understanding.

8) Personnel turnover in high-tech industries seems to be accelerating. As a result, the quality approach in documenting processes is becoming more relevant. New hires have to be trained faster. Managers will sleep better if critical engineering processes are well documented.

9) The industry needs FE programs that provide error estimates. These have been "on the way" for many years, and need to materialize.

10. CONCLUSIONS

Engineers must carry out complex simulation processes to create better products. The management of engineers is getting more difficult due to personnel turnover and lack of documentation of procedures within organizations.

The Personal Professor method of video recording complex, encyclopedic material can help organizations capture the processes developed by their finest analysts and make them standard practice for succeeding generations.

Building more quality issues into the FEA training will take much more awareness on the part of instructors. We need to move from deterministic to statistical thinking, and this will take time.

REFERENCES:

1. Anderson, William J. and Anderson, Glen R., Multimedia Training and Technical Documentation, Proceedings of the 6th AIAA/ASME/ASCE/AHS/ASC Structures, Structural Dynamics and Materials Conference, New Orleans, pp. 2750-2755, April 10-12, 1995.

2. QS-9000 Quality System Requirements, 3rd Edition, Automotive Industry Action Group, (248) 358-3003, January, 1999. (142 pp)

3. Statistical Mechanical Designs Uncertainty Management in CAE via Monte Carlo Simulation, Benchmark, NAFENS, pp. 11-15, January, 1999.

4. Dorian Shainon, Quality Control Methods: Their Use in Design, Machine Design, 1953.

5. Charles Sieck, Caterpillar Corporation, private communication, 1998.

DEVELOPMENT OF A TEACHING RIG FOR IMPROVED UNDERSTANDING OF FINITE ELEMENT PROCEDURES

M J Pavier

Department of Mechanical Engineering, University of Bristol, Queens Building, University Walk, Bristol, UK, BS8 1TR

SUMMARY

An experimental rig has been designed and built for undergraduate teaching purposes to improve the understanding of important finite element concepts. In particular the rig demonstrates the role of the element stiffness matrix, the assembly of element stiffness matrices into a global stiffness matrix representing a structure and finally the solution of the global stiffness matrix.

The rig consists of a number of mechanical models of three-node triangular elements which may be connected together to form a simple structure. The structure may be constrained then loaded and the deformation of the structure measured.

In a laboratory class, the element stiffness matrix is measured by applying loads to the model element and measuring nodal displacements. A structure is then assembled by connecting a number of elements together. The deformation of the structure can be compared with that predicted by solving the global stiffness matrix assembled from the measured element stiffness matrices. The rig is intended to be used alongside a PC running a special purpose program to reduce the tedious routine of assembling matrices.

1.   INTRODUCTION

The finite element method is now taught routinely to undergraduates studying mechanical engineering. By its nature, the method is taught as a theoretical subject, which the student often finds difficult to understand. An experimental rig has been designed and built to enable the fundamental concepts of the finite element method to be demonstrated in order to improve the student's understanding of the subject.

In this paper the design of the experimental rig is presented and a technique for using the rig in a laboratory class is outlined.

2.   GENERAL PHILOSOPHY

The finite element concepts that were intended to be addressed are the element stiffness matrix, assembly of element stiffness matrices and solution of the resulting global stiffness matrix. For simplicity, a two-dimensional situation was considered.

The approach that has been taken is to represent a finite element by a mechanical spring model. The mechanical spring model can be loaded through nodes and the resulting deformation of the model element can be measured. In this way the stiffness matrix can be evaluated, providing clear understanding of the role of the stiffness matrix and the significance of individual terms in the matrix.

Once the element stiffness matrix has been measured, the next step is to assemble a structure of identical elements and then apply loads to the structure. The deformation of the complete structure can be measured and compared with that predicted by assembling the global stiffness matrix for the structure from the measured element matrices, then solving the global matrix.

In order to measure the element stiffness matrix easily, a model representing a three-node triangular plane element was used. An isosceles geometry was chosen so that the element can be assumed symmetric requiring fewer stiffness measurements to be made.

The experimental rig is intended to be used alongside a PC running a special purpose finite element programme, removing most of the tedious aspects of entering and manipulating stiffness matrices.

3.   DETAILS OF RIG

The mechanical model of the triangular element is shown in Figure 1. The model consists of three spring assemblies running parallel to the edges of the element. To allow elements to fit side by side, the spring assemblies are joined at intermediate nodes, within the nodes used to join elements together. Provided the intermediate nodes are close enough to the real nodes and the deformation of the element is small, errors using this approach can be minimised.

To provide the student with the impression that the element represents a triangular area of material, the element may be covered by a rubber membrane to hide the details of the mechanism.

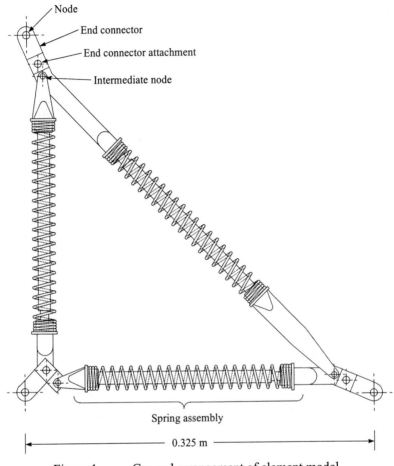

Figure 1   General arrangement of element model

The stiffness matrix of the element can be evaluated to be

$$
\mathbf{K} = k
\begin{bmatrix}
1 & 0 & 1 & 0 & 0 & 0 \\
0 & 1 & 0 & 0 & 0 & 1 \\
1 & 0 & 1+\alpha/2 & \alpha/2 & -\alpha/2 & -\alpha/2 \\
0 & 0 & \alpha/2 & \alpha/2 & -\alpha/2 & -\alpha/2 \\
0 & 0 & -\alpha/2 & -\alpha/2 & \alpha/2 & \alpha/2 \\
0 & 1 & -\alpha/2 & -\alpha/2 & \alpha/2 & 1+\alpha/2
\end{bmatrix}
\tag{1}
$$

where k is the stiffness of the two springs running along the edges at $90°$ to each other and $k\alpha$ is the stiffness of the remaining spring. This stiffness matrix can be compared with that for a three-node triangular plane stress element of the same geometry with Young's modulus E, Poisson's ratio v and thickness t:

$$
\mathbf{K} = \frac{Et}{2(1-v^2)}
\begin{bmatrix}
1+\dfrac{1-v}{2} & v+\dfrac{1-v}{2} & -1 & -\dfrac{1-v}{2} & -\dfrac{1-v}{2} & -v \\
v+\dfrac{1-v}{2} & 1+\dfrac{1-v}{2} & -v & -\dfrac{1-v}{2} & -\dfrac{1-v}{2} & -1 \\
-1 & -v & 1 & 0 & 0 & v \\
-\dfrac{1-v}{2} & -\dfrac{1-v}{2} & 0 & \dfrac{1-v}{2} & \dfrac{1-v}{2} & 0 \\
-\dfrac{1-v}{2} & -\dfrac{1-v}{2} & 0 & \dfrac{1-v}{2} & \dfrac{1-v}{2} & 0 \\
-v & -1 & v & 0 & 0 & 1
\end{bmatrix}
\tag{2}
$$

It can be appreciated that the two matrices cannot be made similar, therefore the mechanical spring model can never truly represent the behaviour of a three-node triangular plane stress element. Nevertheless, it is not considered that this point is likely to be realised by an undergraduate carrying out the experiment and in any case does not effect the principle of the approach. For simplicity, the three springs for each element were chosen to be the same.

Figure 2 shows the details of one of the spring assemblies for the element. Each spring assembly is essentially the same, except for the components used to connect the assemblies together.

TOP VIEW (spring in place)

SIDE VIEW (spring removed)

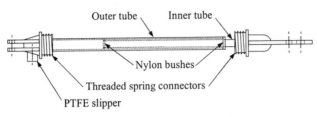

Figure 2   Details of spring assembly

The assembly is constructed of aluminium components, apart from the spring, the two nylon bushes and the PTFE slipper. The slipper enables the elements to slide easily on the surface of the base plate of the rig. The spring itself is wound specially with the coils at each end closed together. The spring can be screwed on to the threaded spring connectors allowing both tension and compression to be supported by each spring assembly.

Figure 3 shows details of a typical end connector. The end connectors attach to the elements allowing the elements to be connected together using a simple pin. A range of connectors has been made to allow a number of elements to be connected to the same node.

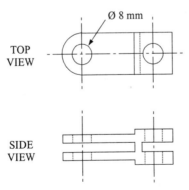

Figure 3   Details of end connector

A structure of four elements assembled on the base plate of the rig can be seen in Figure 4. Nodes of the structure can be constrained from moving in two directions using the pinned constraint block which can be fitted to the base plate in a number of different positions. A node can also be prevented from moving in one direction only using the sliding constraint block. Again, the sliding constraint block may be fitted to the base plate in different positions.

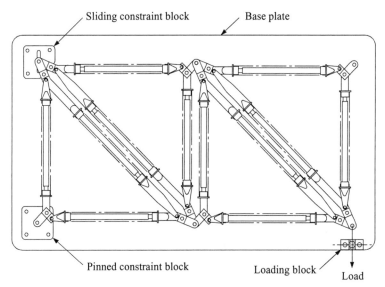

Figure 4   General arrangement of teaching rig

Details of the pinned and sliding constraint blocks are shown in Figure 5. The sliding constraint block includes a pin incorporating two miniature ball bearings to reduce friction.

4.   USE OF THE RIG IN A LABORATORY CLASS

The experimental rig is intended to be used in a laboratory class situation using the following procedure:

(i)   One element is placed on the base plate and loads applied to its nodes in a systematic manner so as to evaluate the terms in the stiffness matrix. It can be seen from equation (1) that many of the terms are equal, therefore the number of stiffness terms that require measuring is limited. Measuring terms on the leading diagonal is straightforward. Two nodes are constrained completely using the pinned constraint blocks. The remaining node is constrained to move in one direction only using the sliding constraint block. Load is applied to this node and the resulting displacement measured. Measuring off-diagonal terms is not as simple. One node is completely constrained and the other two nodes constrained to move in one direction only. Load is applied to one of these nodes and then a sufficient load is applied to the other node to reduce its displacement to zero.

(ii) As stiffness terms are evaluated they are typed into a matrix displayed on a PC running a special purpose program. The program will know which terms in the stiffness matrix are identical and will automatically set all such terms once one has been typed in.

(iii) A structure is then assembled on the base plate such as the one shown in Figure 4. The structure is constrained in some way to eliminate rigid body motion and then loads applied. The displacements of the nodes of the structure are measured.

(iv) A finite element analysis of the structure is then carried out by assembling the stiffness matrices for each element into the global stiffness matrix and then solving the resulting set of equations.

(v) Finally, the measured displacements of nodes in the structure may be compared with those predicted by the computer programme.

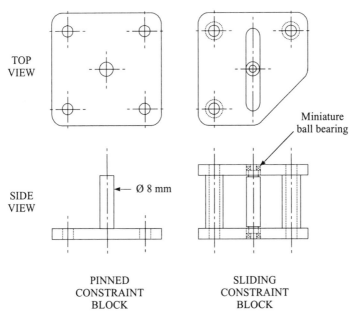

Figure 5   Details of pinned and sliding constraint blocks

F.E.A. OF BARS STRUCTURES USING MATHCAD

M.H. Tierean, M.C. Tofan, I.A. Goia

"Transilvania" University of Brasov, Romania

Abstract: The paper will present two F.E.M. applications, developed for truss and frame structures, using MathCAD. There are presented the basic elements of the programs and the drawings of distorted structures. The structure bars are checking at the traction, bending (in the case of frame structures) and at the buckling, considering the elasticity of the nodes. The results attain were checked with ANSYS.

1. INTRODUCTION

Finite element analysis (F.E.A.) is in the present the numerical method most used to the calculus of complex structures. There are many commercial programs for F.E.A. realised by well know companies. This soft requires strong hard capabilities and specialised users.

There are thousand of papers that describe F.E.A. applications, one of there realised with individual soft, others with commercial programs. We consider useful to present two simple F.E.A. applications, realised with MathCAD, for the truss and frame structures.

Because there are structures realised from straight bars, the meshing is easy, the element boundaries being the successive junctions.

2. CALCULATION PROGRAM FOR TRUSS STRUCTURES

As application is considered the parabolic truss structure of a bridge (figure 1). The forces are $Q_1 = 2 \cdot 10^5$N şi $Q_2 = 10^5$N.

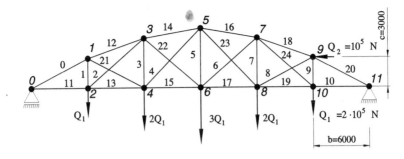

Figure 1. The parabolic truss structure

The calculation program start with a draw section of the structure. The impute data are the number of bars, the horizontal dimension "b" and height of first pillar "c". The structure drawing is generating by successive covering of the nodes by the position vector (figure 2).

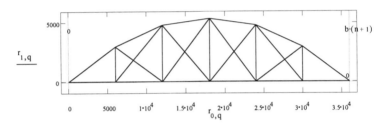

Figure 2. The structure drawing

Next, a nodal junction matrix (I) will build, which defines the elements of the structure. In this matrix the column number is the same with the element number, the first line shows the start node and the second line the arrival node. Thus the column 0 defines the 0 bar, the start node being 0 and the arrival node 1. The last column defines the 24 bar the start node being 7 and the arrival node 10.

$$I = \begin{bmatrix} 0 & 1 & 2 & 3 & 4 & 5 & 6 & 7 & 8 & 9 & 10 & 0 & 1 & 2 & 3 & 4 & 5 & 6 & 7 & 8 & 9 & 1 & 3 & 5 & 7 \\ 1 & 2 & 3 & 4 & 5 & 6 & 7 & 8 & 9 & 10 & 11 & 2 & 3 & 4 & 5 & 6 & 7 & 8 & 9 & 10 & 11 & 4 & 6 & 8 & 10 \end{bmatrix}$$

In the next step is generate the expansive incidence matrix (SE), which has the same number for the lines and for the structure nodes; the number of columns is the same with the number of its elements. If the node is the start node, the correspondent matrix value is 1, if is the arrival node the value is -1.

$$SE_{\left(I_{0,z}\right),z} := 1 \qquad SE_{\left(I_{1,z}\right),z} := -1 \qquad (1)$$

948

Additional, with these definitions (nodal junction and expansive incidence matrix) the elements are toggled in plan by polar co-ordinates. For the assemblage of the rigidity matrix (K), the calculation of the rotated matrix of incidence (SER) is necessary, trough the multiplication of the incidence matrix (SE) and the rotation matrix (ROT), and for the local stiffness matrix (KL) also.

$$SER_{2\cdot i + u, 2\cdot z + v} = SE_{i,z} \cdot ROT(z)_{u,v}$$

$$\text{(2)}$$

$$KL_{2\cdot z + u, 2\cdot z + v} = KLOC_{u,v} \cdot \left(\frac{E \cdot A_z}{l_z} \right) \qquad K = SER \cdot KL \cdot SER^T$$

After this step, the fixing structure will be attain and, the outer forces are introduces in nodes. Every node corresponded with two values: the first refers on X force and the second refers on Y force. After that the nodal

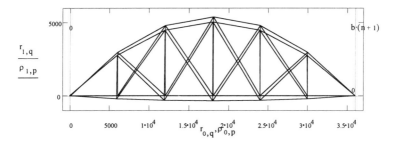

Figure 3. The deformed truss structure

displacements are calculated (two for each node) by multiplying the inverse of stiffness matrix with the outer forces vector. These nodal displacements draw the distorted structure (figure 3), amplified by 100.

The forces, on the bar direction, in the starting node, are determined by projecting the nodal distortions in the bar local base, amplified by their rigidity.

$$FL_z = -ROT(z)^{<0>} \cdot DEF^{<z>}$$

$$F^{<z>} = -ROT(z) \cdot KLOC \cdot ROT(z)^T \cdot DEF^{<z>} \qquad \text{(3)}$$

$$DEF^{<z>} = \begin{bmatrix} \delta_{2\cdot I_{1,z}} - \delta_{2\cdot I_{0,z}} \\ \delta_{2\cdot I_{1,z}+1} - \delta_{2\cdot I_{0,z}+1} \end{bmatrix} \cdot E \cdot \frac{A_z}{l_z}$$

949

After the nodal equilibrium and the calculation error verification, (in the first occurrence the error is less than 10^{-6}, and in the second occurrence the error is less than 10^{-7}), the structure bars are checking at the traction and at the buckling. The results attain were checked with ANSYS 386/ED.

2. CALCULATION PROGRAM FOR FRAME STRUCTURES

As application is considered the frame structure (figure 4), loaded by the forces $Q_2 = Q_6 = 25 \cdot 10^3 N$, $Q_3 = Q_4 = 10^4 N$ and the bending moments $M_3 = M_4 = 10^4 Nm$.

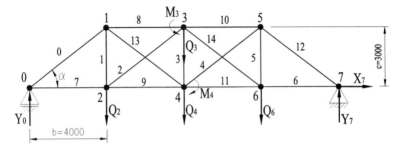

Figure 4. The frame structure

As in the previous chapter, the calculation program starts with the drawing of the structure, the impute data being the number of bars, the horizontal "b" and the vertical dimension "c".

Because in this case are three displacements the rotation matrix is

$$ROT(z) = \begin{bmatrix} \cos\left(\alpha_z\right) & -\sin\left(\alpha_z\right) & 0 \\ \sin\left(\alpha_z\right) & \cos\left(\alpha_z\right) & 0 \\ 0 & 0 & 1 \end{bmatrix} \tag{4}$$

and by multiplying with the translation matrix (because of bending) result the roto-translation matrix.

$$RT(z) = ROT(z) \cdot Td(z)^T \tag{5}$$

The components of rotated matrix of incidence (SER) are the same as the elements of rotation matrix, in the case of starting node and same as roto-translation matrix, in the case of arrival node.

$$SER_{3 \cdot i + u, 3 \cdot z + v} = if\left(SE_{i,z} > 0, SE_{i,z} \cdot ROT(z)_{u,v}, SE_{i,z} \cdot RT(z)_{u,v}\right) \tag{6}$$

950

The stiffness matrix results by projecting the free elements in the incidence space of the structure (transformation of the transformation).

$$K11(z) := \begin{bmatrix} \dfrac{E \cdot A_z}{l_z} & 0 & 0 \\[2ex] 0 & \dfrac{12 \cdot E \cdot Iz_z}{\left(l_z\right)^3} & \dfrac{6 \cdot E \cdot Iz_z}{\left(l_z\right)^2} \\[2ex] 0 & \dfrac{6 \cdot E \cdot Iz_z}{\left(l_z\right)^2} & \dfrac{4 \cdot E \cdot Iz_z}{l_z} \end{bmatrix}$$

$$KL_{3 \cdot z + u, 3 \cdot z + v} := K11(z)_{u,v} \qquad K := SER \cdot KL \cdot SER^T \tag{7}$$

After the fixing of the structure, the outer forces are introduces in nodes. Every node corresponds with three values: the first refers on X force, the second refers on Y force and the third to the bending moment M.

After that step the nodal displacements are calculated (three for each node) by multiplication of the inverse of stiffness matrix with the outer forces vector. These nodal displacements draw the distorted structure (figure 5), using cubic Hermite interpolation.

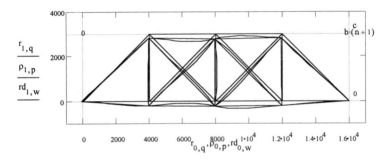

Figure 5. The deformed frame structure

For the calculus of forces is necessary the displacements imagine in the global reference system.

$$TD(z) := ROT(z) \cdot Td(z) \cdot ROT(z)^T \tag{8}$$

It is also necessary to calculate the strain matrix in the start (δEF) and in the arrival node (ΔEF). With these matrix are calculated the force matrix in

951

the local (FL) and in the global reference system (F), in the start (i) and in the arrival node (j).

$$\delta EF^{<z>} = TD(z) \cdot \begin{bmatrix} \beta_{3 \cdot I_{1,z}} \\ \beta_{3 \cdot I_{1,z+1}} \\ \beta_{3 \cdot I_{1,z+2}} \end{bmatrix} - \begin{bmatrix} \beta_{3 \cdot I_{0,z}} \\ \beta_{3 \cdot I_{0,z+1}} \\ \beta_{3 \cdot I_{0,z+2}} \end{bmatrix}$$

$$\Delta EF^{<z>} = -TD(z)^{(-1)} \cdot \delta EF^{<z>} \tag{9}$$

$$Fi^{<z>} = ROT(z) \cdot FLi^{<z>} \qquad Fj^{<z>} = -RT(z) \cdot K11(z) \cdot RT(z)^T \cdot \Delta EF^{<z>}$$

$$FLi^{<z>} = -K11(z) \cdot ROT(z)^T \cdot \delta EF^{<z>} \qquad FLj^{<z>} = -Td(z)^T \cdot FLi^{<z>}$$

After the bars' equilibrium checking (less than 10^{-6}), nodal equilibrium (less than 10^{-8}) and the calculating error (less than 10^{-9}), the structure bars are checking at the traction, bending and buckling, considering the elasticity of the nodes. The results attain were checked with ANSYS.

3. CONCLUSIONS

The presented programs give the deformed structures, the displacement values and the bars stresses, in the local and in the global reference system. Using the computer based on the Intel 80468 DX/66MHz processor, with 16MB RAM, the computing time is 4s for the first program and 9s in the second case.

Due to the simplicity and the higher velocity, the program is easily to implement in any type of calculation for the structure bars avoiding the laboriously methodology of the drawing, introducing of the data and calculation, which claimed by commercial programs f or F.E.A.

REFERENCES

[1] TOFAN, M.C., GOIA, I., TIEREAN, M.H., ULEA, M. - Deformatele structurilor (Structure's strains), Editura Lux Libris, Brasov, Romania, 1995.
[2] TIEREAN, M.H. - Contribuţii la optimizarea bazata pe fiabilitate a structurilor de rezistenţă–Teză de doctorat (Contributions to the optimisation on the basis of reliability of strength structures–Ph.D. Thesis), "Transilvania" University of Brasov, Romania, 1995.
[3] MATHSOFT, Inc. - MathCAD, Version 4.0, User's Guide, Cambridge, MA., 1993.

THE DEVELOPMENT OF QUALITY SYSTEM METRICS USED IN A SOFTWARE ENVIRONMENT

William J. Bryan
Quality Assurance Manager
ANSYS, Inc.

1. ABSTRACT

According to research, the mere knowledge that a performance measurement process exists will actually cause performance to improve. Studies have also proven that feedback derived from measurement (if implemented properly) can improve performance. In addition, public awareness of performance adds to its improvement. Realizing these performance characteristics, ANSYS, Inc. developed and implemented a set of quality system metrics to improve quality system processes used to develop ANSYS software products.

A set of twelve high level metrics in the areas of Customer Satisfaction, Product and Process, and Sales and Marketing have been implemented. Some of these metrics are based on real time customer feedback obtained through Internet accessible customer surveys. Others rely on more conventional measurements such as error reports. These metrics, which are made available to all employees via the ANSYS Intranet, are used to measure the success of the software quality process and drive ANSYS's quality improvements.

The paper will present the process of defining, developing, accepting and implementing the high level metrics which drive the quality system improvement process. These metrics, which are reviewed at top level management meetings, have become an integral part of the decision making process. Insights gained from these metrics, combined with process knowledge gathered from other sources, are used to conduct business. The metrics and their delivery system have given value added information to the decision making process at ANSYS.

2. BACKGROUND

In 1993 ANSYS, Inc. expanded its quality system to meet the requirements of the International Quality Standard ISO 9001. Previously, the quality system was established to meet the applicable requirements of the U.S. Nuclear Regulatory Commission rules and regulations Title 10, Chapter 1, Code of Federal Regulations, Part 50, Appendix B and supplemental requirements of NQA-1 subpart 2.7, Quality Assurance Requirements for Computer Software. Many ANSYS users are in the field of electrical nuclear power generation. To meet these customer's needs, ANSYS' initial quality system was set up to meet the nuclear industry's quality assurance requirements.

The ISO 9001 standard applies to commercial software products produced and services rendered by ANSYS, Inc. As part of the requirements of ISO 9001, a Quality System Steering Committee was established in 1993 comprising the CEO, department managers, and the quality assurance manager. To establish the new quality system, all required activities, as specified in the ISO 9001 standard, were documented and our quality manual modified. In 1994, a registrar was selected and our company was registered and certified to be ISO 9001 compliant in May of 1994. The registration was easily achieved because of the commitment of our management and employees to quality. The registration assessment by our registrar went very well with no corrective actions being assigned, a first with the two lead auditors assigned.

In 1994, ANSYS had established quality metrics in some areas of the company. However, a high-level set of metrics were not part of our quality system at that time. In early 1997, the ANSYS Quality System Steering Committee began the development of a concise on-going set of high-level metrics for managing product and service quality. These metrics would allow management to have information for early insight of potential problems as well as information to support decisions. For these metrics to become effective, they had to become an integral part of the decision making process and be reviewed in future Quality System Steering Committee meetings. Insights gained from metrics would be merged with process knowledge gathered from other sources in the conduct of business. A goal was established to develop and implement a set of high level metrics by the end of 1997.

3. METRICS DEVELOPMENT PLAN AND REQUIREMENTS

A plan was first developed which defined goals and objectives for metric development. The plan consisted of six items:

954

- Define goals and objectives in terms of software product/service and software management processes.
- Ask process owners, product, and release managers to define what goals and objectives were set.
- Select metrics based on data needed to answer the questions, which were posed.
- Define data collection processes and recording mechanisms.
- Define data analysis, goals and objectives, and feedback process.
- Implement metrics as part of a closed loop system which provides current and historical information to management.

Requirements for the metrics were agreed on and a list distributed to all process owners and product managers. The metric requirements are:

- Metrics must be understandable.
- Metrics must be easy to gather.
- Metrics should be highly leveraged, permitting significant improvements to be made.
- Metrics must be timely.
- Metrics must give proper incentive for process improvement.
- Metrics must be homogenous throughout the processes.

In initial meetings with product managers, six areas of major importance were identified. These were; product reliability, meeting schedule, error free product, repeat audit findings, product yield, and customer satisfaction. From these initial meetings and discussion within the Quality System Steering Committee (QSSC) a set of concise metrics were proposed. Once these metrics were adopted by the QSSC, they were made available on the company's intranet.

Next, process owners and product managers were requested to establish goals for the metrics. These goals along with feedback mechanisms were then proposed to the Quality System Steering Committee. After approval the goals were reported with the metrics on the company's intranet. This process then became the vehicle for establishing 1998 Quality Goals.

The concise system of metrics were complete in November, 1997. The following section presents definitions of each of the twelve high level metrics and typical data for these three areas: Customer Satisfaction, Product and Process, and Sales and Marketing.

4. HIGH LEVEL QUALITY SYSTEM METRICS

1. Customer Satisfaction Metrics

Customer Satisfaction - Products -- The metric measures our customers' overall satisfaction with our products.

The data comes from an on-line customer satisfaction survey. This high-level metric is a combined weighted average over all products for six areas of satisfaction. (Categories measured are product, quality, performance, ease of use, reliability/accuracy, and capability/features.) It is computed by directly averaging the response for product with the average response for quality, performance, ease of use, reliability/accuracy, and capability/features. (Since the product response should include all these areas, this process validates the customer's product responses and adds confidences to the measure.) Customers rate our products from poor (shows as 20%) to excellent (shows as 100%) in the survey. All high level metric ordinate scales were developed by using the 100% level as excellent. This philosophy has been shown to be better understood and accepted by employees. Dates on the abscissa are shown by year with the current year presented by quarter. Evaluation of individual product families' satisfaction ratings can be reviewed by drilling down one level of detail. Individual product ratings can be viewed by drilling down two levels of detail. Figure 1 shows a representative Customer Satisfaction - Products Metric.

Customer Satisfaction - Services -- The metric measures our customers satisfaction with our services.

The data comes from the on-line customer satisfaction survey. This high-level metric is a combined average weighted equally for the three areas of:
- Technical Support (a summation of 4 categories)
- Training (a summation of 5 categories)
- Business Operations (a summation of 3 categories)

It is computed by directly averaging all responses for these service areas together. Customers rate our services from poor (shows as 20%) to excellent (shows as 100%) in the survey. Categories for which Technical Support is rated are Technical Support, Professionalism, Responsiveness, and Expertise. Categories for which Training is rated are Courses, Course Availability, Course Content, Instructor's Expertise, and Training Facilities. Categories for which Business Operations is rated are Product Delivery, Contract Licensing, and Accounting/Invoicing. Each categories' response can be determined for both ANSYS, Inc. and ANSYS Support Distributors (ASD). Evaluation of ANSYS, Inc. and ASD performance can be reviewed by drilling down one level of detail.

Evaluation of individual service areas can be reviewed by drilling down two levels of detail, and categories of the areas can be reviewed by drilling down three levels of detail. Figure 1 shows a representative Customer Satisfaction - Services Metric.

Product Usability -- The metric correctly measures our customer's evaluation of our product's ease of use.

The data comes from the on-line customer satisfaction survey. This high-level metric is a combined, weighted average over all products. It is computed by directly averaging all responses for each product together. Customers rate our products from poor (shows as 20%) to excellent (shows as 100%) in the survey. The evaluation of individual product families' usability can be reviewed by drilling down one level of detail. Figure 1 shows a representative Product Usability Metric.

Customer Responsiveness -- The metric directly measures our responsiveness to our customers.

This metric is a combined average, weighted equally for four areas for the quarter. It includes the percent of customer requests to Technical Support closed within two business days, the percent of Technical Support requests to Product Creation requests closed within five days, the percent of complete, valid orders filled (from time of order entry into InfoOrder to shipment out the door) by Business Operations within three days, and the percent of complaints closed within one month. The evaluation of each of this metric's components can be reviewed by drilling down one level of detail. Figure 1 shows a representative Customer Responsiveness Metric.

ANSYS Support Distributor Performance -- The metric is a composite indicator which measures the combined performance aspects of our ASDs.

Data for this metric is a combination of two objective measures of ASD performance and three measures taken from our customers' evaluation of ASD performance as rated in the on-line Customer Satisfaction Survey. The five measures are:

- Percent of ASDs certified per quarter
- Percent of ASDs without complaints registered (from the customer complaint system) per quarter
- Customer Satisfaction Survey results for administrative services
- Customer Satisfaction Survey results for technical support services
- Customer Satisfaction Survey results for training

Responses to our on-line customer satisfaction survey are obtained when customers rate our ASD services from poor (shows as 20%) to excellent (shows as 100%) in the survey. Categories for which ASD Technical Support is rated are Technical Support, Professionalism, Responsiveness and Expertise. Categories for which ASD Training is rated are Courses, Course Availability, Course Content, Instructor's Expertise, and Training Facilities. Categories for which ASD Business Operations is rated are Product Delivery, Contract Licensing, and Accounting/Invoicing. Evaluation of each aspect of the metric can be reviewed by drilling down one level of detail, and categories of the service areas can be reviewed by drilling down two levels of detail. Figure 1 shows a representative ASD Performance Metric.

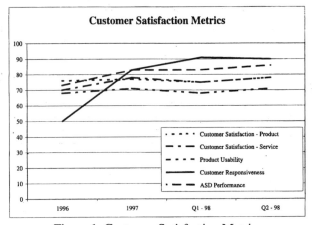

Figure 1   Customer Satisfaction Metric

2. Product and Process Metrics

Product Reliability - Surveyed -- The metric directly measures our customers' evaluation of our products' reliability and accuracy.

The data comes from the on-line customer satisfaction survey. This high level metric is a combined, weighted average over all products. It is computed by directly averaging all responses for each product together. Customers rate our products from poor (shows as 20%) to excellent (shows as 100%) in the survey. Evaluation of individual product families' reliability can be reviewed by drilling down one level of detail. Figure 5 shows a representative Product Reliability - Surveyed Metric.

Product Reliability - Measured -- The metric measures the total number of unique errors reported per task in all released software products per time.

958

For consistency in comparing these a baseline of our average customer satisfaction rating for product and reliability was used. The number of tasks used to normalize the error count is determined at the beginning of each quarter, while errors are counted at the end. The yearly metric is determined from the end of year data. Evaluation of other aspects of specific products' error rates and un-normalized error data can be reviewed by drilling down levels of detail. Figure 5 shows a representative Product Reliability - Measured Metric.

a. Error Metrics

Software quality can be defined using various measures; but ultimately, customers perceive it by the incidence of errors in the software. That is why error identification is so important in the software development process. At ANSYS, error metrics also establish readiness of the software to proceed in development, to be reviewed, to be reworked, or to be released.

Studies of software development have shown that defect removal costs roughly double with each stage of development [Ref 1]. This is due in part to the fact that errors related to the early stages, (i.e., requirements definition and design specifications) will cause many errors in later development. Therefore, at ANSYS every effort is made to identify and correct errors early in the development cycle. Additional production metrics have been chosen to encourage the early identification and correction of errors. The sooner errors can be detected, the lower the cost to correct them and the lower their impact on the schedule.

Comparisons of similar product software errors tracked during development are good indicators of schedule risk. For example, if cumulative errors found are low compared to errors in similar products at the same time in the development cycle, then testing may be lacking. This could impact the end of the development cycle when all critical errors must be corrected. The rate at which errors are opened, resolved, and closed are good indicators of schedule risk. If the rate at which errors are identified exceeds the rate which errors are being closed, a problem is occurring that will impact schedule.

If error discovery and resolution is keeping time, then the error correction process is working. However, little effort may be being expended to discover errors. At ANSYS, we try to understand both the data and what the data represents by historic comparisons. By correlating these metrics to performance milestones, indications of impact on schedule can be determined and resources can be shifted to maintain development commitments.

b. Unreleased Product Errors

Early identification and correction of errors are performed in a constructive way at ANSYS. Early detection of errors is encouraged, and the cumulative number of errors is evaluated. The number of errors being corrected is compared to those which have been resolved and closed. (The difference between resolved and closed is a check by an independent reviewer to confirm that the error has indeed been corrected.) Also, the cumulative number of open errors in unreleased products is compared with the cumulative time required to resolve and close the errors. The graph of the cumulative number of unreleased errors versus time helps determine when a product is ready for release. As the rate reduces, the slope decreases, which gives an indication that the product is approaching release, see figure 2.

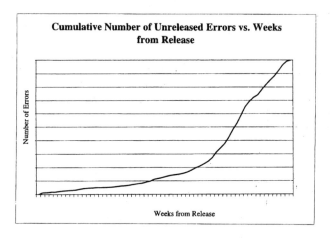

Figure 2 Unreleased Error Detection Rate

Approximately six months before release, error discovery begins to slow and reliability data collection is initiated. Total time of ANSYS product use and number of errors encountered are measured and charted. A reliability value of one error encountered per month of ANSYS product use by an average user is targeted for product release, figure 3.

Other indications can be obtained from this data. For instance, the error discovery rate can be tracked to determine the acceptability of products at a particular stage of the product creation cycle. When analyzing the discovery rate, one needs to correlate the rate with the types of efforts being managed. A declining number of reported errors may be caused by a change in effort from testing to correcting errors. The correlation of product creation resources to the discovery rate is also evaluated. Analyzing the discovery rate can also indicate whether or not errors are being identified faster than they are being resolved and closed.

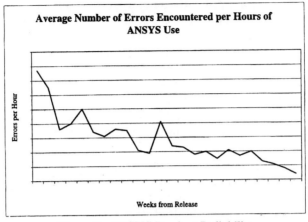

Figure 3 ANSYS Product Reliability

Understanding both the data and what the data represents is very important. It is suggested that correlation of error data figures to measures such as milestone performance and effort expenditure can help managers predict schedule changes.

c. Released Product Errors

The same types of comparisons used in analyzing unreleased product errors are used for released errors. However, there are some additions. For example, a running average of errors identified in the released product is maintained for both the total errors reported and those reported by customers. The unaveraged data is also charted and analyzed using statistical process controlled techniques. Figure 4 shows a product's representative data.

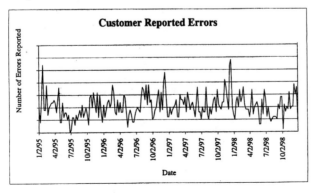

Figure 4 Customer Reported Errors

961

Process variation has two causes; those that are natural and inherent to the process and common to all measurements, and those that can be identified and prevented. The rate at which errors are identified in a released product by customers is considered a stable process, one that is in statistical control. The underlying distributions of its measurable characteristics are consistent over time, and the results are predictable within limits and can be used to predict future results if the development process does not change. As the number of weekly errors reported is charted, if the number goes above the Upper Control Limit (UCL) or below the Lower Control Limit (LCL), it indicates that the process has been altered in some way and the reason is investigated.

Error root cause evaluations are also performed on randomly selected customer identified errors. These evaluations are used to drive the quality improvement process at ANSYS, Inc. This action is also prescribed in DoD-Std-2168 [Ref 2] which states that a formal defect prevention program should be established that empowers developers and software testers to analyze the causes of defects and enact improvements to their own local development process. This empowerment helps prevent future defect insertion and enhances the detection process.

Meeting Release Schedule -- The metric assesses our ability to plan product releases and deliver to the plan.

This includes major development milestones (completed in each quarter) for all products, weighted equally. If we meet or beat a given major milestone, based on the milestone dates committed to at low level design, the measure is 100%. Missed milestones are determined by subtracting the total number of business days by which a product milestone is missed from 100. This measure is based only on milestones after the design phase is completed. Figure 5 shows a representative Meeting Release Schedule Metric.

Repeat Audit Findings and Corrective Action Requests -- The metric measures the effectiveness of our quality system's corrective and preventive action systems.

The Repeat Audit Findings and Repeat Corrective Action Requests (CARs) Metric is determined by adding the number of repeat audit findings and repeat corrective actions occurring from internal auditing, external audits, and employee findings over the year. The timeframe which would constitute a repeat finding is three-years. Figure 5 shows a representative Repeat Audit Findings and CARs Metric.

Product Yield -- The metric measures the revenue yield of released software product lines per developer.

The metric is determined by dividing the total product line revenues per quarter by the total number of product developers. It is a weighted average over all product lines. Evaluation of individual product line yields can be reviewed by drilling down one level of detail. Figure 5 shows a representative Product Yield Metric.

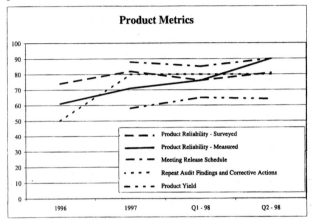

Figure 5 Product and Process Metric

3. Sales and Marketing Metrics

ANSYS Support Distributor Satisfaction with ANSYS -- The metric is measured by ASDs' responses to questions twice a year.

This high level metric is an average of the ASD responses for seven areas of satisfaction. Categories measured are business operations, authorization files/license keys, new releases, global and GSA accounts, services, sales and marketing, and communications. Figure 6 shows a representative ASD Satisfaction with ANSYS metric.

Advertising and Marketing Effectiveness -- The metric is measured by ASDs' responses to questions for four Sales/Marketing areas twice a year.

Categories measured are marketing literature, marketing communications, market analysis, and market service. Figure 6 shows a representative Advertising and Marketing Effectiveness metric.

963

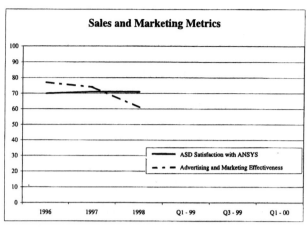

Figure 6 Sales and Marketing Metric

5. CONCLUSIONS

The twelve high level metrics presented are reviewed at top level management meetings and have become an integral part of the decision making process. Insights gained from these high level metrics, combined with process knowledge gathered from other sources, are used to conduct business. The high level metrics and their delivery system have given value added information to the decision making process at ANSYS, Inc.

6. REFERENCES

1. USA, Department of the Air Force, Software Technology Support Center - Guidelines for Successful Acquisition and Management of Software Intensive Systems: Weapon Systems Command and Control Systems Management Information Systems, Version 1.1, February 1995, Volumes 1 & 2.
2. USA, Department of Defense Standard 2168.
3. Humphrey, Watt S - Managing the Software Process, Software Engineering Institute, Addison-Wesley Publishing Company, Reading, Massachusetts, 1990.
4. Jones, Capers - Applied Software Measurement, McGraw Hill, New York, 1991.
5. Crosby, Philip B. - Quality Improvement through Defect Prevention, Philip Crosby Associates, Inc., 1985.

Evaluation and Comparison of Plate Finite Elements in ANSYS

S. Boedo[i]

ABSTRACT

Plate elements have become a standard feature in most of the commercially available finite element software packages available today, yet the seemingly endless variety of element formulations can provide a source of uneasiness and confusion to the practical engineer. This paper evaluates ANSYS plate element formulations in the analysis of a symmetrically loaded pre-tensioned clamped circular plate and in the analysis of the nonaxial bending of a clamped ring plate subjected to an edge point load. These problems were chosen for comparison to recently available analytical solutions and for their continuing academic and industrial interest.

1   INTRODUCTION

The classical treatment of plates and shells was well established before the advent of modern computation, and hence, it is not surprising that these structures were a natural choice in early applications of the finite element method. What is surprising has been the unexpectedly long effort in obtaining reliable plate and shell elements, especially when compared with the development period devoted to three-dimensional solid elements. Various compatibility- and convergence-related formulation difficulties [1] have inevitably led to the introduction of seemingly innumerable plate and shell elements, which in turn has left a source of uneasiness and confusion to the practical engineer as to which element is appropriate. Moreover, as commercially available software programs found in industry need to provide analysis capabilities for a large class of generic solid mechanics problems, they can only devote a limited set of validation case studies to plate and shell elements.

---

[i] Staff Engineer, Borg-Warner Automotive, Ithaca, NY 14850.

This paper addresses the performance of plate finite elements available in the commercial program ANSYS 5.4 in light of recent analytical work in this area. In particular, an investigation of stress and deflection induced in a uniformly loaded pre-tensioned clamped plate and in the nonaxial bending of a clamped ring plate will be presented here.

2   CLAMPED PRE-TENSIONED CIRCULAR PLATE

The first case study concerns the deflection and stress induced in a clamped circular plate with an initial in-plane tension and subjected to a uniform surface pressure. This problem is of considerable contemporary interest to the microelectronics industry for the design of miniature pressure sensor devices. Even in the absence of initial tension, the plate deforms into a non-developable surface [2], and, thus, in-plane membrane forces will become a contributing factor as the plate deflections become large compared with the plate thickness. A recent semi-analytical solution by Sheplak and Dugundji [3] provides stress and strain distributions from the coupled effects of in-plane tension and bending, and they also provide further validation of large deflection plate studies discussed elsewhere [2].

Figure 1 shows a clamped circular plate with radius a and thickness h ($h/a < 1/25$ [2]) subjected to an initial in-plane radial tension $N_r = N_0$ (per unit length) and a subsequent uniform pressure load p_0. Figure 2 shows the corresponding finite element models employed in ANSYS. The model is comprised of an assemblage of one-dimensional axisymmetric plate elements (SHELL51). For the validation studies in this paper, we are concerned with stress and deflection at the center of plate, but as in [3], we are required to make the mesh finer near the plate boundary so as to capture edge-zone bending behavior near the clamped edge. Each element node has three degrees of freedom—translation in the x- and y-directions and rotation about the z-axis. Following [3], the location x_i of system node i along the x axis is given by

$$x_i = a\left[1 - \frac{(\beta + 1) - (\beta - 1)\left(\dfrac{\beta + 1}{\beta - 1}\right)^{\frac{i-1}{N-1}}}{\left(\dfrac{\beta + 1}{\beta - 1}\right)^{\frac{i-1}{N-1}} + 1}\right]$$

where the number of nodes N is an odd number and where β is a stretching parameter which clusters more points near the plate boundary as β approaches

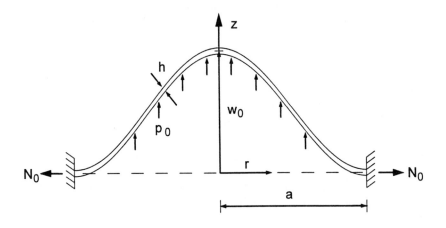

Fig. 1. Clamped circular plate with axisymmetric load.

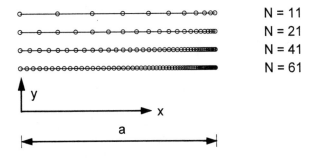

Fig. 2. Finite element representations of clamped
circular plate with axisymmetric load.

1. For the meshes in Figure 2, β is fixed to 1.05 in accordance with [3] but the number of finite element nodes used here is much smaller than the number of grid points used in their finite difference solution of the problem.

We first solve the clamped plate problem with zero initial in-plane tension. At node 1, the x-displacement and z-rotation degrees of freedom are restrained from symmetry, and at node N, all three degrees of freedom are restrained. A uniform pressure is then applied to each of the elements. Within the ANSYS 5.4 solution module, the nonlinear geometry option is selected (NLGEOM,ON) along with an adaptive solution control option (SOLCON,ON). The number of load substeps are initially set to 20 (NSUBST,20) which can change via the adaptive load stepping feature.

Employing nomenclature in [3], dimensionless plate deflection W, tension parameter k, loading parameter P, and mid-plane radial stress S_r are defined by

$$W = w/h \qquad k = \frac{a}{h}\left(\frac{12(1-v^2)N_0}{Eh}\right)^{1/2}$$

$$P = \frac{p_0 a^4}{Eh^4} \qquad S_r = \frac{\tilde{N}_r a^2}{Eh^3}$$

in terms of dimensional plate deflection w, Young's modulus E, and Poisson's ratio v. The $\tilde{N}_r$ term above refers to the incremental change in radial in-plane tension from that imposed initially. For the zero initial tension case ($k = 0$), the incremental and actual in-plane tensions are identical. The nodal y-displacements obtained from the ANSYS solution correspond to the dimensional plate deflection w in [3].

With $k = 0$ corresponding to zero initial in-plane tension, Figures 3 and 4 show the effect of pressure load P on the deflection W_0 and mid-plane stress $S_r(0)$ at the center of the plate. Using mesh densities $N = 11$ through $N = 61$, it is observed that the ANSYS finite element results at the plate center converge to values which are quite close to the curves obtained in [3] throughout the specified dimensionless pressure load range.

Next, we solve the pre-tensioned clamped plate problem in two steps. First, an in-plane radial tension is applied to the plate boundary, and the resulting radial deflection is calculated. Second, the plate is clamped at the specified radial displacement on the boundary and a pressure load is applied. In ANSYS, this is carried out in the first load step by first fixing the x-

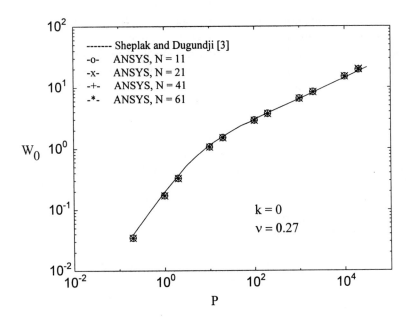

Fig. 3. Center deflection of clamped circular plate
with axisymmetric load and zero initial tension.

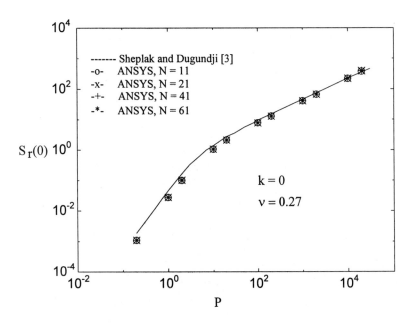

Fig. 4. Center mid-plane stress of clamped circular plate
with axisymmetric load and zero initial tension.

displacement and z-rotation degrees of freedom at node 1, fixing the y-displacement and z-rotation degrees of freedom at node N, and applying a point load (per unit length) in the x direction at node N. The x-displacements are then calculated. In the second load step, the x-displacement at node N is fixed to its previously computed value and a pressure load is applied as in the zero initial tension case.

With k = 50 and P < 10000, Figures 5 and 6 show that the converged finite element results for deflection and mid-plane stress at the center of the plate obtained from ANSYS agree well with those obtained in [3]. For loads P > 10000, stress and deflections predicted by ANSYS are somewhat larger than those predicted in [3], yet the trends are qualitatively similar. Note that $S_r(0)$ shown in Figure 6 refers to the incremental mid-plane stress induced due to pressure loading from the initial pre-stressed state.

3   NONAXIAL BENDING OF A CLAMPED RING PLATE

The second case study is concerned with the deflection and stress distribution in a circular ring plate of uniform thickness h and flexural modulus D clamped along its inner radius b and subjected to a point load P on its free outer radius a, as shown in Figure 7. This problem continues to be referenced in current design handbooks [4] as it forms the kernel for arbitrary edge loading distributions for practical engineering structures. Following the usual assumptions of thin-plate, small-deflection Kirchhoff theory, the problem was originally solved by Reissner [5] and reprinted elsewhere [2]. Recently, Boedo and Prantil [6] presented a corrected analytical solution to the problem, where it was discovered that the correct maximum radial stress can be much as 10% greater than that reported previously.

Figure 8 shows half-model finite element representations of the clamped ring plate for two different plate aspect ratios λ = b/a. The mesh is divided into N uniform circumferential divisions and M non-uniform radial divisions. The radial element length increases by a fixed scale factor to produce a nearly 1:1 aspect ratio for those elements which lie along inner and outer plate radii. The meshes shown in Figure 8 employ four-noded ANSYS plate elements (SHELL63) with six degrees of freedom at each node; however, an eight-noded ANSYS plate element with mid-side nodes (SHELL93) will also be evaluated with the same mesh discretization procedure.

With ν = 0.3, Tables 1 and 2 show the effect of plate aspect ratio and mesh density on the maximum radial bending stress and maximum axial deflection. The maximum radial stress is located at r = b, θ = 0 and the maximum axial deflection is at the point of load application. It is observed that the maximum axial deflection obtained using ANSYS is nearly identical

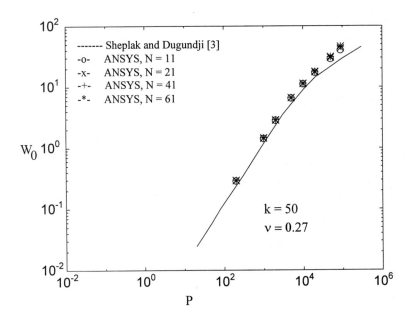

Fig. 5. Center deflection of clamped circular plate
with axisymmetric load and non-zero initial tension.

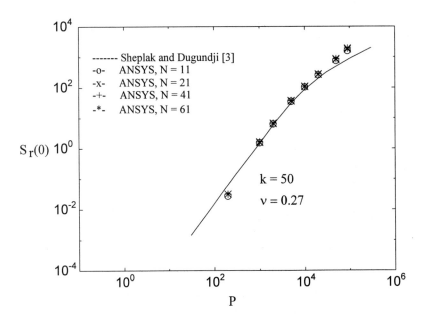

Fig. 6. Center mid-plane stress of clamped circular plate
with axisymmetric load and non-zero initial tension.

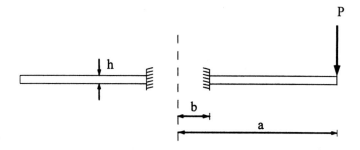

Fig. 7. Clamped ring plate with edge point load.

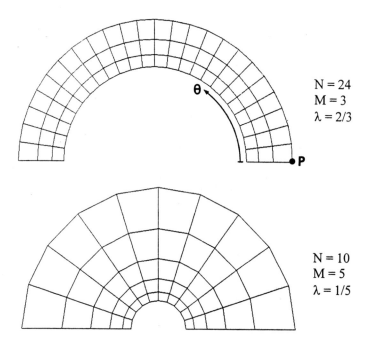

N = 24
M = 3
λ = 2/3

N = 10
M = 5
λ = 1/5

Fig. 8. Finite element representations of clamped ring plate with edge point load (top view).

972

Table 1. Maximum stress and deflection, $\lambda = 2/3$, $\nu = 0.3$.

mesh (NxM)	max $\sigma_r h^2 /P$ 4-noded elements	8-noded elements	max $wD/(Pa^2)$ 4-noded elements $(x\ 10^{-2})$	8-noded elements $(x\ 10^{-2})$
24x3	3.9362	4.2198	2.2309	2.2613
40x5	4.0161	4.2099	2.2389	2.2671
80x10	4.0945	4.2179	2.2471	2.2663
120x15	4.1308	4.2212	2.2499	2.2661
160x20	4.1513	4.2226	2.2511	2.2663
200x25	4.1643	4.2232	2.2518	2.2664
240x30	4.1734	4.2236	2.2522	2.2664
Analytical [6]	4.2229		2.2520	
Analytical [4]	4.25			
Analytical [5]	4.249			

Table 2. Maximum stress and deflection, $\lambda = 1/5$, $\nu = 0.3$.

mesh (NxM)	max $\sigma_r h^2 /P$ 4-noded elements	8-noded elements	max $wD/(Pa^2)$ 4-noded elements	8-noded elements
10x5	8.5941	9.5187	0.1794	0.1792
20x10	8.9089	9.6159	0.1790	0.1794
40x20	9.2179	9.6510	0.1790	0.1793
80x40	9.4230	9.6626	0.1791	0.1793
120x61	9.4998	9.6649	0.1791	0.1792
Analytical [6]	9.6666		0.1791	
Analytical [4]	8.8			

Table 3. Maximum radial stress, $\lambda = 2/3$, $\nu = 0.3$, NxM = 240x30.

	$\max \sigma_r h^2 / P$	
element	r/a = 5/6	r/a = 5/6
type	$\theta = 0$	$\theta = \pi/2$
4-noded	1.5509	0.0032
8-noded	1.5584	0.0032
Analytical [6]	1.546	0.0032
Analytical [5]	2.776	0.1356

to the corrected analytical solution [6], even with coarse meshes and four-noded elements. It also appears that the maximum radial stress computed with four-noded elements is converging slowly to the corrected analytical solution [6], while even coarse meshes with eight-noded elements are quite close to the analytical solution. Table 3 shows maximum radial stress for $\lambda = 2/3$ at selected interior points also agrees well with the corrected analytical solution [6] and differs appreciably from the original solution [2], [5].

It is also very important to observe that the converged maximum radial stress for $\lambda = 1/5$ computed using ANSYS agrees very well with the corrected analytical solution [6] and thus differs by approximately 10% from analytical results reported previously [4]. Without the corrected solution, one could erroneously conclude that the ANSYS plate elements will converge to the wrong result as the mesh is refined.

4   CONCLUSONS

Two case studies have shown that the plate element formulations incorporated in the ANSYS 5.4 software package give acceptable results for symmetrically loaded stress-stiffened circular plates and nonaxially loaded Kirchhoff ring plates when compared with analytical and semi-analytical solution approaches. Moreover, the finite element results give greater confidence in the correctness of recent analytical solutions provided by Sheplak and Dugundji [3] and Boedo and Prantil [6].

974

5 REFERENCES

1 ZIENKIEWICZ, O C, TAYLOR, R L – The Finite Element Method, 4th ed., McGraw-Hill, New York, 1991.

2 TIMOSHENKO, S, WOINOWSKY-KRIEGER, S – Theory of Plates and Shells, 2nd ed., McGraw-Hill, New York, 1959.

3 SHEPLAK, M, DUGUNDJI, J – Large deflections of clamped circular plates under initial tension and transitions to membrane behavior. ASME Journal of Applied Mechanics, Vol. 65, pp. 107-115, 1998.

4 YOUNG, W C – Roark's Formulas for Stress and Strain, 6th ed., McGraw-Hill, New York, 1989.

5 REISSNER, H – Über die unsymmetrische biegung dünner kreisringplatte. Ing.-Arch., Vol. 1, pp. 72-83, 1929.

6 BOEDO, S, PRANTIL, V C – Corrected solution of clamped ring plate with edge point load. ASCE Journal of Engineering Mechanics, Vol. 124, No. 6, pp. 696-697, 1998.

Benchmarking the behaviour of crushed salt -
The CEC BAMBUS project

U. Heemann[1], S. Heusermann[1], N. C. Knowles[2]
[1] Fed. Inst. for Geosciences and Natural Resources, Hannover, Germany
[2] W S Atkins Science & Technology, Epsom, U K

SUMMARY

For over two decades research studies have been carried out into the geomechanical behaviour of rock salt to investigate its suitability as a host material in which to construct repositories for radioactive waste. However, only recently has attention been directed on the behaviour of crushed salt. This is foreseen as a suitable backfill material with which to encapsulate the waste casks and canisters after placing them in the repository drifts and boreholes.

Crushed salt is somewhat unusual material in that it exhibits a strongly non-linear behaviour as it compacts under the combined influences of heat and lithostatic pressure and ultimately assumes the properties of virgin rock salt. Conceptual numerical models to account for this behaviour are well advanced but their reliability and robustness for general application are untested. Accordingly, an international benchmark exercise has been carried out to compare finite element based predictions of crushed salt's behaviour.

The paper describes the exercise in broad terms and presents details of the final stage. The key results are discussed and some preliminary conclusions are attempted.

1   INTRODUCTION

Within the frame of the BAMBUS project (Backfill and Material Behaviour in Underground Salt Repositories) the project "Comparative study on crushed salt (CS)^2" was an international benchmark exercise dealing with the numerical simulation of the behaviour of crushed salt. It forms part of a comprehensive ongoing research program into the use of crushed salt as a potential backfill and seal material for the disposal of heat-producing radioactive waste in deep geological salt formations. The broad objectives were to improve confidence in computer predictions of the behaviour of the fill material under the combined influences of heat and lithostatic pressure.

The geomechanical behaviour of crushed salt is not well understood and the constitutive models which are at the centre of existing calculations are based on a small number of laboratory experiments. Experiments to investigate the viability of crushed salt as a backfill material show that, directly after placing, the backfill porosity is about 35 %. Its permeability is thus high and its sealing capacity is very low. However, as the surrounding host rock converges the backfill is compacted and its voids close up until ultimately its properties, except for grain dimension, are assumed to become indistinguishable from those of the host rock. Drift convergence and backfill compaction are accelerated by the higher temperatures resulting from the heat producing waste. Backfill thermal conductivity increases with increasing density. These interactions between temperature, thermal conductivity, pressure, drift closure and permeability have to be faithfully modelled if predictions of the backfill performance are to be robust. The CS^2 project was designed to investigate systematically the issues involved by means of a series of benchmarks of increasing complexity.

Acronym	Name	Location	Role
BGR	Federal Institute for Geosciences and Natural Resources	Hannover, Germany	Coordinator Participant and Experimentalist
CIMNE	Department of Geotechnical Engineering, Universita Polytechnica de Catalonia	Barcelona, Spain	Participant, subcontractor to ENRESA
NRG	Netherlands Energy Research Foundation	Petten, Netherlands	Participant
ENRESA	Empresa Nacional de Residuos Radiactivos	Madrid, Spain	Observer, sponsor
FZK/INE	Forschungzentrum Karlsruhe	Karlsruhe, Germany	Participant, Experimentalist
GRS	Gesellschaft für Anlagen- und Reaktorsicherheit	Braunschweig, Germany	Experimentalist, sponsor
DBE	Deutsche Gesellschaft zum Bau und Betrieb von Endlagern fur Abfallstoffe	Peine, Germany	Participant, subcontractor to GRS
G.3S	Groupement pour l'etude des Structure Souterraines de Stockage, Ecole Polytechnique	Palaiseau, France	Participant
W S Atkins	W S Atkins Science & Technology	Epsom, U K	Facilitator

Table 1: Participants and Roles

978

In this context "benchmark" refers to a common analytical problem which is defined and agreed by all participants. Each participant subsequently analyses it independently using whatever constitutive model and computer code is deemed appropriate. The results are then compared. Emphasis is placed on the collective performance of all participants and the conclusions that can be drawn from this overall perspective. No undue attention is paid to individual performance and participants and codes are not ranked in any way.

The exercise commenced in April 1996 and completed in January 1999. Nine organisations from five European countries took part (Table 1).

The organisational methodology followed was similar to that of the successful COSA and INTERCLAY exercises [1,2]. A key feature was the series of regular 2-day plenary meetings held approximately every 6 months. These provided the vehicle for the dissemination of ideas and exchange of experience. They were conducted under the aegis of an independent chairman/facilitator (W S Atkins) with the intent of ensuring a balanced and impartial discussion of the issues involved. The plenary meetings were also the forum for reporting and reviewing progress achieved and planning subsequent activity.

2 CONSTITUTIVE MODELS

Participants embraced a number of different approaches in the constitutive models of rock salt. One approach, adopted by FZK, DBE and in a slightly modified form by NRG, leans heavily on the existing work of Hein [3] and describes the inelastic compaction in terms of a conventional Arrhenius type creep model augmented by a term which is a function of the porosity and hydrostatic stress. Korthaus [4] subsequently modified this model to include dependence on deviatoric stress.

A rather different, physically inspired approach of CIMNE is based on the deformation mechanisms of grains of salt with idealised geometry [5]. This leads to a classical flow rule potential function for the inelastic deformation which incorporates both compaction and shearing.

BGR made use of a model based mainly on phenomenological assumptions. It is an improvement of the hydrostatic model of Zhang [6].

A further approach used by G.3S is to fit the compaction test data to a conventional soil-mechanics Cam Clay model in which a hardening function is introduced to account for the dependence of compaction rate on porosity[7].

It is difficult, if not impossible, to reconcile the detailed theoretical bases of these different approaches and comparison is only meaningful via benchmarking.

3 PROJECT ORGANISATION

The exercise was split into three distinct stages:

Stage 1: Benchmark computations on simple, hypothetical problems to compare codes against each other and to investigate the behaviour of the various constitutive models. This stage involves essentially only 'verification' of the codes.

Stage 2: Comparative calculations against small-scale laboratory tests with the intention of validating the constitutive models over a limited range of physical parameters.

Stage 3: Modelling of in-situ behaviour - in this case the TSDE experiment (Thermal Simulation of Drift Emplacement) at the Asse research facility in Germany - using the constitutive models developed in Stage 2.

Stages 1 and 2 have been reported elsewhere [8]. The present paper therefore concentrates on Stage 3 and provides only summary details of the first two stages.

3.1 Stage 1

Two benchmark problems were successfully addressed in Stage 1. Both were hypothetical problems and served as verification tests of the constitutive models and the codes comprising them. **Benchmark 1.1** was an idealisation of an oedometer test on dry crushed salt. The problem is effectively one-dimensional. **Benchmark 1.2** can be viewed as an extension of the first benchmark; the oedometer is replaced by a borehole in rock salt containing crushed salt backfill. The problem can again be treated as one dimensional and involves interaction between the host rock salt and the crushed salt. In both benchmarks interest centres on the ability of the constitutive models to predict the compaction of the crushed salt under isothermal conditions.

The benchmarks provided a good basis on which to compare the attributes of the various constitutive models. All models were in good broad agreement - after initial difficulties over terminology had been removed.

3.2   Stage 2

In Stage 2 a further two benchmarks were tackled: **Benchmark 2.1** was based on a set of multi-stage compaction tests on crushed salt under isothermal hydrostatic loading. These tests were carried out as part of the benchmark exercise and were partially motivated by the desire to resolve the apparently contradictory results of earlier experiments: oedometer tests at BGR with deformation control had indicated stress exponents of the order of 14, increasing with compaction rate. However, similar tests at FZK under constant stress had found substantially lower exponents in the range of 5 -7. The same tests were therefore carried out by BGR, GRS and FZK using different testing apparatus but on material from the same source. Concurrently with the tests, participants were asked to predict the measured behaviour. Existing experimental data from BGR and FZK were supplied for participants to fit to their constitutive models.

The three sets of experimental results did not agree well. Although all showed the same form, plots of the predicted void ratio with time differed by a substantial 'shift' of up to a factor of 2. The differences have not been fully explained, but several reasons have been investigated including the effect of slightly different ambient temperatures and several possible error sources in the measurement procedures. Whatever reason, the results illustrate the uncertainty associated with experiments at laboratory scale.

Predictions of the behaviour were rather more consistent. The differences between the results obtained using the BGR data set and those using the FZK data set was not as great as had been expected and was of the same order as those due to the different constitutive models. In isolation all predictions were plausible.

Notwithstanding the distinct differences in the three sets of experimental results, it was agreed that the FZK/INE data set would be used for deriving the model parameters in the finite-element calculations of Benchmark 3.1.

Benchmark 2.2 was an extension of Benchmark 1.2 to introduce non-isothermal effects by including a heat source. The physical basis is a heater in a borehole backfilled with crushed salt. Predictions of the evolution of temperature, porosity, stress etc. were required. Once again both BGR and FZK data sets were supplied but since the problem was hypothetical no particular significance was attached to either.

In a strict sense there is coupling between the thermal and mechanical effects which necessitate fully coupled calculations. Few codes had this capability. Nevertheless the problem was successfully tackled by all participants, although with some ingenuity involving iterations and/or restarts.

Generally, the results were all acceptably consistent; some differences were observed but were attributed to the use of different strain measures.

3.3 Stage 3

Stage 3 comprised two benchmarks, of which only **Benchmark 3.1** is reported here. This benchmark is based on the ongoing TSDE test in the Asse research facility in Germany. The idealised problem is depicted in Figure 1.

In reality the TSDE comprises two parallel drifts with a distance of about 15 m apart(from axis to axis) and 800 m below the surface, in a disused salt mine. Cylindrical heaters (of a size and heat output corresponding to casks containing heat-producing radioactive waste) are placed centrally on the floor of the drift which is then back-filled with crushed salt. The backfill is placed by "slinger" and there is some evidence that this results in an uneven distribution of salt grains and some variation in the "as placed" porosity. Another important feature is that there are only three heaters in each drift and the behaviour of the backfill can then be expected to be affected by heat flow along the drift. It will be apparent that there are considerable complexities in modelling the in-situ behaviour faithfully and it was decided to base the calculations on a symmetrical two-dimensional model for reasons of computational simplicity - in the full expectation that this would result in over-prediction of the thermo-mechanical effects. A number of other simplifying assumptions were also agreed.

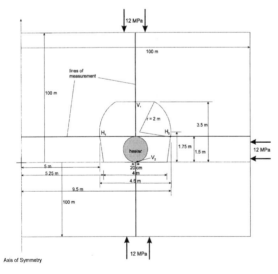

Figure 1: Specification of Benchmark 3.1

4 RESULTS OF BENCHMARK 3.1

The specification for Benchmark 3.1 called for temperature, stress, strain and void ratio for three points of time along a horizontal and a vertical line through the drift to be compared with experimental data. Horizontal and vertical closure was also compared with measurements. Model parameters were to be based on existing FZK experimental data.

982

In general the results on Benchmark 3.1 agree very well and are consistent with expectations. Differences between the participants stem in part from the different FE-meshing and the different post processing (e.g. automatic smoothing of results). Some typical results are shown in Figures 2 - 5. Within this paper only the results along the vertical line of measurement are displayed.

It should be noted that the problem is dominated by the thermal response of the host rock salt over most of the time of interest. This governs closure of the drift and thus compaction of the backfill. Due to the two-dimensional modelling of heat transport this effect of volume expansion is much more pronounced than in the real TSDE test. Thus measured and predicted temperatures are in reasonable agreement only for about 1 year. After that, due to heat transfer in the longitudinal direction the predicted temperature is significantly higher than measured. Accordingly, the measured closure over 10 years is much less than calculated (Fig. 2). The calculated model thus corresponds much better to a very long drift with heaters.

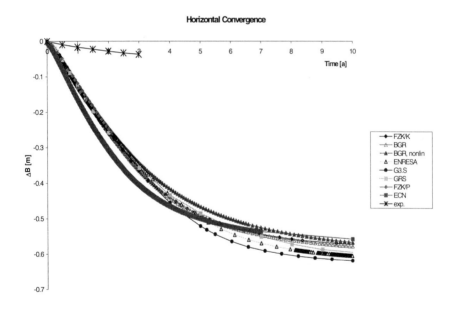

Figure 2: BM 3.1, horizontal convergence as a function of time.

The development of void ratio and the scatter of the results can be seen in Figures 3 and 4. Some participants (ENRESA/CIMNE, GRS/DBE) reached zero or negative (G.3S) void ratio. The negative void ratio shows the need for some further code improvement.

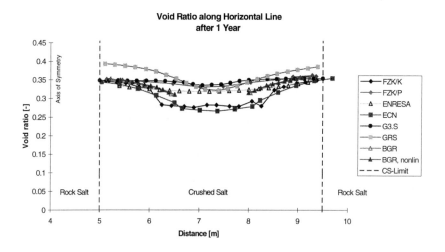

Figure 3:BM Void Ratio after 1 year

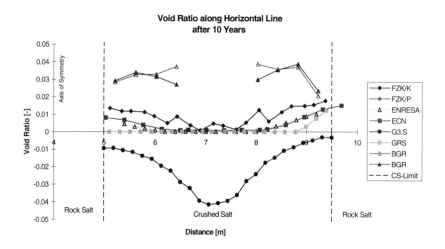

Figure 4: Void Ratio after 10 Years

There is a distinct lower amount of compaction for small porosities in case of the BGR-model. The good agreement of the two curves of BGR shows the relatively small influence of the effect of geometric non-linearities. Figure. 5 is typical of the good agreement of calculated stresses.

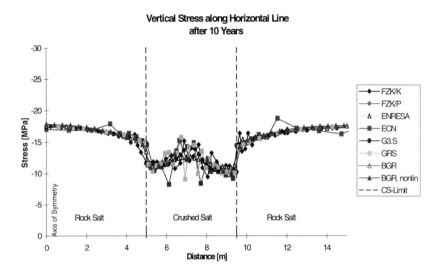

Figure 5: Vertical Stress after 10 years

5 CONCLUSIONS

The exercise provides considerable insight into the state of the art with respect to the ability to predict the thermo-mechanical behaviour of crushed salt. Using a variety of constitutive models all participants were able to make meaningful calculations for a range of scenarios relevant to nuclear waste repositories, including the coupling of the thermal and mechanical behaviour and the interaction between the host rock and the crushed salt backfill.

Differences in spatial and time discretisation and other details of the numerical methods seem to have a minor influence on the quality of the results. Although few codes exhibited some numerical difficulties affecting the stability and speed of solution, these deficiencies did not preclude satisfactory answers being obtained. Code verification no longer appears to present any particular difficulty. However at the present time it appears that fully three dimensional calculations are impractical.

A major difficulty appears to be the lack of robust experimental data with which to calibrate the constitutive models. The tests carried out as part of Benchmark 2.1 highlight the difficulty in getting such data and the inevitable uncertainty therefore associated with any uncorroborated data set.

A continuing issue is the control of human error. The exercise revealed a number of instances of mistakes arising from misinterpretation of the specification, transcription errors and confusion over terminology.

The benchmark highlighted again the need to use consistent definitions of physical quantities and for high quality measurements to validate the predictions.

In the future, work is necessary to investigate further (both experimentally and theoretically) the influence of grain-size distribution and to provide a more robust foundation for the more complex three-dimensional scenarios associated with repository safety cases.

ACKNOWLEDGEMENTS

The CS^2 benchmark exercise was funded by the German Federal Ministry for Education, Science, Research and Technology (BMBF) and co-sponsored by the European Commission within the framework of the BAMBUS project (Backfill and Material Behaviour in Underground Salt repositories).

REFERENCES

1   LOWE, M.J.S. & KNOWLES, N.C. - COSA II: Further Benchmark Exercises to Compare Geomechanical Computer Codes For Salt. EC Report 12135 EN, Brussels 1989.

2   JEFFRIES, R.M. - Interclay II: A Co-ordinated Benchmark Exercise On The Rheology Of Clays. EC Report EUR 16204 EN, Brussels 1995.

3   HEIN, H.J. - Ein Stoffgesetz zur Beschreibung des thermomechanischen Verhaltens von Salzgranulat. Doctoral thesis, RWTH Aachen, Germany 1991.

4   KORTHAUS, E. - Consolidation and Deviatoric Deformation Behaviour of Dry Crushed Salt at Temperatures up to 150 °C. The Mechanical Behaviour of Salt, Ed. Aubertin, M. & Hardy, H.R., Trans Tech Publ., Clausthal-Zellerfeld, Germany 1998.

5   OLIVELLA, S. - Non-isothermal multiphase flow of brine and gas in saline media. Tesis Doctoral, Departamento de Ingeniería del terreno, UPC, Barcelona 1995.

6   ZHANG, C.L, HEEMANN, U, SCHMIDT, M, STAUPENDAHL, G. Constititive Model for Description of Compaction Behaviour of Crushed Salt Backfill Proc. ISRM Int. Symp. EUROCK'99 1999

7   SCHOFIELD, A. & WROTH, C.P. - Critical State Soil Mechanics. McGraw Hill, New York 1968.

8   HEEMANN, U., HEUSERMANN, S. & KNOWLES, N.C. - Current Status of the Benchmark Exercise „Comparative Study on Crushed Salt (CS^2)". Proc. EC Cluster Seminar In-Situ Testing in Underground Research Laboratories for Radioactive Waste Disposal, pp. 215-236, Alden Biesen, Belgium 1997.

3D VISUALIZATION AND ANIMATION OF LARGE FE MODELS

Dr. Ketil Aamnes
Managing Director, ViewTech ASA

SUMMARY

Numerical analysis is an important part of a design cycle, and decisions are made based on the results from these calculations. The main objective for numerical analysis is to verify a design, and to investigate the effects of changing material properties and model dimensions. Good analysis tools are required, and in most design cases more than one analysis tool is used. There are a number of specialized analysis tools available on the market to help engineers to analyze different phenomena.

A crucial step in any design cycle is to visualize the results from the numerical computations. Referring to Albert Einstein: If you can't picture it, you can't understand it. A visualization tool should help the engineer to understand the results from the computation, and enable him or her to communicate the results in a form that is well-suited and understandable for decision-makers. The use of 3D graphics and animation is important to communicate results from dynamic FE analysis.

This paper describes the use of the compact and effective 3D visualization program GLview, which is developed by ViewTech ASA. The main focus will address visualization and animation of large FE models on Windows NT/95/98. The paper will also present the use of GLview as a verification tool on the next generation of web-based technical reports.

ViewTech is a spin-off company from SINTEF Applied Mathematics. SINTEF is the largest independent research organization in Scandinavia and performs contract research and development for industry and the public sector in technological areas.

INTRODUCTION

One of the major research areas at SINTEF Applied Mathematics is presentation tools and methods for engineering analysis. Effective tools for scientific visualization is demanding to be able to understand complex dynamic results. As a result of this ongoing research [1-5], the GLview program was developed as a general tool for visualization of scientific data. The main reason for developing a new visualization program GLview, starting in 1992, was to meet the requirements from our clients in their effort to understand and communicate results from numerical experiments. ViewTech ASA was established in 1995 to commercialize the visualization program GLview.

GLview is today used in a variety of application areas (Figure 1) like structural engineering (LS-DYNA, FORGE3D, SESAM [6], Abaqus, Ansys, Nastran, Usfos and FEDEM), Computational Fluid Dynamics (Phoenics, CFDesign), Basin modeling (Petroflow3D, SEMI) and Marine Hydrodynamics (SWAN, MASHIMO, WAMIT).

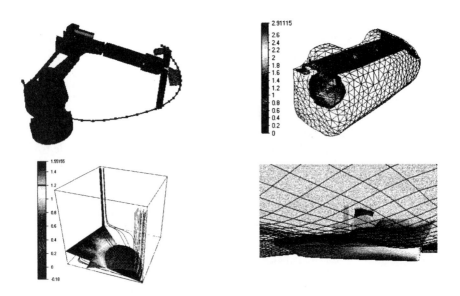

Figure 1 Application areas (robotics, metal forging, CFD, hydrodynamics)

990

The development of GLview was originally done on high performance Silicon Graphics workstations utilizing the Graphics Library [7] from Silicon Graphics. When the graphics standard OpenGL was defined, all the graphics in GLview was converted to OpenGL. An important milestone in our development process was when Microsoft supported OpenGL on Windows NT in 1994. Today GLview is available on all the major Unix platforms (SGI, HP, IBM RS6000, DEC and SUN) in addition to Windows NT/95/98. The Unix version has a Motif user interface, while the PC version has a Windows user interface (Figure 2). The look end feel for the Unix version and Windows version is the same. GLview handles the lack of binary compatibility between Unix and Windows. Therefore, the Windows version of GLview can be used to visualize results generated on a Unix platform.

The degree of high performance in GLview is dependent on the graphics hardware on the actual computer. Graphics hardware that supports OpenGL is available for most Unix workstations and Pentium based PCs running Windows NT/95/98. If OpenGL is not supported in hardware, most platforms support software rendering. The picture on the screen look the same, but the graphics speed is much slower if no hardware is present to help rendering OpenGL graphics.

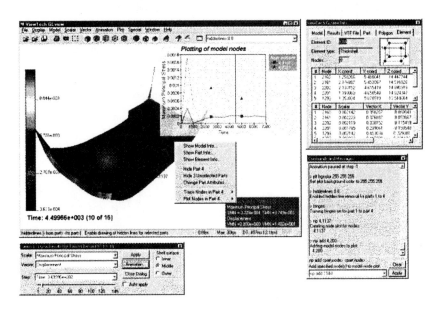

Figure 2 GLview graphical user interface.

High performance graphics is demanded if the user need to inspect large FE models, and crucial for interactive animation. For visualization of time dependent results, animation is a key way to get an overview of the results. Details about the behavior of specific nodes or elements in the model can be tabulated or presented in a 2D graph (Figure 3).

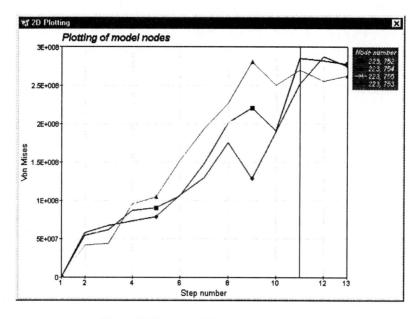

Figure 3 Plotting of time dependent scalar results.

MAIN BENEFITS

The three main benefits in using GLview to explore analysis results are:
- Ease of use
- Interactive animation
- High performance graphics

GLview is easy and effective to use with an intuitive "look and feel". The user interface has a combination of menu/mouse control and powerful commands. A high level scripting language is available allowing the expert user to speed up the process of visualizing data. While beginners tend to use menus and dialogs to interact with the program, more experienced users control the visualization through a mixture of script files and menus. Some users even control the visualization from another program sending commands to GLview through Unix pipe or DDE on Windows.

992

ADVANCED INTERACTIVE ANIMATION

Advanced interactive animation is an important aspect of GLview that enables handling of large data files effectively and ensures animation of time-dependent data. Results data can be animated and visualized on the model. GLview was designed as an engineering tool for animation of numerical results.

The interactive animation capability of GLview enables the user to interact with the animation:
- Pan/rotate/zoom the model
- Run the animation stepwise forward and backward or in cycle
- Pause the animation to investigate the model in a specific step during animation
- Adjust the speed of the animation (set a delay between each step)
- Change display attributes "on the fly"
- Change scaling factor for vector results (e.g. displacements, velocity and acceleration)
- Change the legend range for scalar results (e.g. stresses and strains)
- Enable/disable different parts of the model
- Tabulate results for specific nodes (through picking) for the current step in the animation
- XY-plot of scalar results for selected nodes (time history)
- Trace lines representing displacements for specific nodes (time history movements)
- Create cutting planes and iso-surfaces for models containing volume elements

SUPPORT FOR MULTIPLE ANALYSIS CODES

Customers using multiple application programs can benefit from using one compact and effective tool to explore and communicate results - Understanding by Visualization. The direct reading capabilities in GLview enable fast access to analysis results with no need to convert data before loading. Supported analysis files can be loaded as "primary files", while script-files containing attribute settings can be loaded as "secondary files".

For cross-platform environments, the Windows and Unix version of GLview can load the same binary data files. Most organizations use Windows as a platform to write reports, and the Windows version support cut/paste of images.

For direct reading of analysis results from Ansys, Nastran, Abaqus and I-DEAS Universal files, ViewTech utilize the VdmTools toolkit from Visual Kinematics Inc. This is a C++ object library that is integrated in GLview.

GLview also support results stored in a proprietary file format, which is well suited to store results for in-house analysis codes.

The current version of GLview support direct reading (Figure 4) of results from the following analysis codes:

- MSC Nastran
- Abaqus
- Ansys
- LS-DYNA
- I-DEAS Universal files
- SESAM (OEM with DNV)
- Forge3D (OEM with Transvalor S.A.)
- Petroflow3D (OEM with IES GmbH)
- CFDesign (OEM with BRNI Inc.)
- FEDEM (OEM with Summit Systems AS)
- SEMI (OEM with SINTEF)

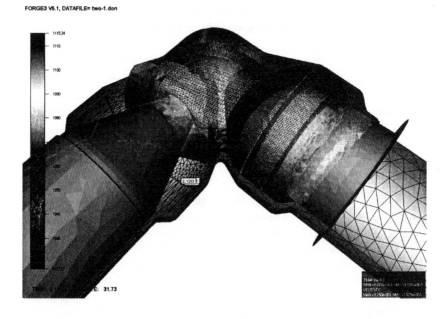

Figure 4 Postprocessing of Forge3D results with adaptive mesh.

In addition to the interactive animation capabilities, GLview support creation of MPEG movies (Figure 5). MPEG is a compressed format, and it is well suited as a communication link to share results on a CD-ROM or via

Internet. GLview also support direct creation of AVI files (Windows version only) and animated gif-files. Animated gif is supported by Web-browsers, and is well suited to promote animations on web-sites.

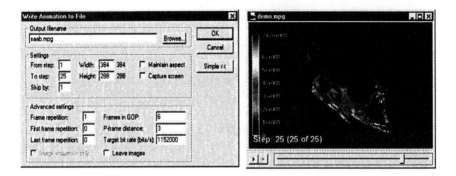

Figure 5 Direct creation of MPEG, AVI and animated gif files.

INTEGRATION WITH LS-DYNA

The integration with LS-DYNA is very tight (Figure 6). GLview supports direct reading of LS-DYNA output files. As soon as a d3plot file or time history file is loaded, the model appears on the screen in the initial state. A dialog will pop up and invite the user to start interacting with the results.

ViewTech have put a lot of effort into this integration, which have full support for adaptive mesh and eroded elements. An extensive 2D plotting capability is available to show time history data. Cutting planes and iso-surfaces can be calculated and displayed for models that contain volume elements.

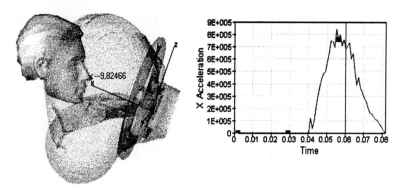

Figure 6 Integration with LS-DYNA.

Binary LS-DYNA output files created on the Unix platform can be visualized with both the Unix and Windows NT/95/98 version. Necessary "byte-swapping" is handled within GLview. This is an important feature in a heterogeneous environment, allowing engineers on Unix workstations and PCs to access the same results. By using the Windows version, it is very easy to cut images from GLview and paste them into your favorite documentation program (e.g. Word). Images can be printed directly or stored to a bitmap file (BMP, TIFF, GIF, RGB, or PPM).

VERIFICATION OF ANALYSIS RESULTS

GLview is designed as an engineering tool and presentation tool for verification of analysis results. A model in GLview can be divided in different parts or objects. Typically, each material model in the FE analysis code is treated as a separate part. A whole range of display options is available for each part in the model (Figure 7):

- Model viewed as shaded surface/lines/points/outline/hidden line
- Overlaid mesh lines showing the finite element mesh
- Filled contours or fringes for scalar results (textured, plain or shaded)
- Texture and environment mapping for realistic rendering
- Deformed model display
- Vector arrows
- Outline mesh

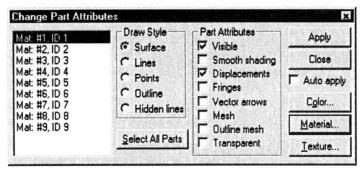

Figure 7 Part attributes

The main capabilities in GLview for verification of results are:
- Animation of time-dependent results (scalar and vector results)
- Extraction of element results through picking
- Extensive 2D plotting of time-dependent results
- Trace lines of vector results (node tracing)
- "Ride view" functionality to fix the eye-point to or relative to a moving node
- Rigid body transformation
- Output images in BMP, TIFF, GIF, RGB, PPM and MPEG format
- Output model to VRML format
- Cutting planes and iso-surfaces
- Particle tracing for CFD results
- Mirror model along x, y and z-axis
- Stereo view support for real 3D visualization

A customized version of GLview is developed to run in a CAVE environment on Silicon Graphics hardware. The software has been successfully implemented at the computer lab of the Oil Company Norsk Hydro in Bergen, Norway. This is a Virtual Reality environment, where the user can inspect analysis results in "real 3D".

HIGH PERFORMANCE VISUALIZATION

Silicon Graphics is well known for high performance graphics. They have been the driving forces in this arena for many years and they still hold the number one position in terms of graphics performance. Realistic 3D graphics is now also driven by the game industry. A number of inexpensive 3D graphics accelerators are available on PCs, but unfortunately many of these boards currently supports accelerated 3D only on low resolution, or on low depth (Z-buffer).

The main players for good graphics performance on Windows NT are Intergraph, HP and recently SGI. Their graphics boards support OpenGL acceleration on engineering workstations. Although GLview perform well on a standard Pentium PC without any graphics accelerator, these machines gives a performance boost of about 6-20 times for interactive animation.

THE NEXT GENERATION OF TECHNICAL REPORTS

Technical reports are most often printed on paper containing text and pictures from the visualization tool. The next generation of technical reports will be stored on a CD-ROM in the html-format, allowing an internet/intranet browser to be used to explore the content. It will contain text, graphics, animated MPEG sequences and analysis results.

The main benefit of storing analysis results as part of the technical report is to allow the reader to explore the results that are not already presented as pictures or animated sequences. Relevant questions to address could be:
- How does the 3D-model look from the other side?
- What about the stress distribution in another time step than already presented in the report?
- How does the analysis results compare to the tests in the lab?
- How does the analysis results compare to the results of a new modified design?

A CD-ROM can hold 650MB of data. This is in most cases enough to store the important results from a technical analysis. The benefit of storing the results in GLview-formatted files is that the results can be explored using an effective and user friendly tool on both Unix and Windows NT/95/98. Decision-makers can access the data using a common tool for analysis results.

ViewTech have developed a program named ViewVT that is well suited as a frontend for displaying analysis results on a CD-ROM. This tool is part of the GLview delivery, and the user need to set up files to describe the individual models that can be accessed by pressing one of the buttons in the ViewVT panel (Figure 8). The panel can show the results in GLview, play a MPEG file or display an html-file in a web-browser. A web page can be created to give a detailed description of the given model and results.

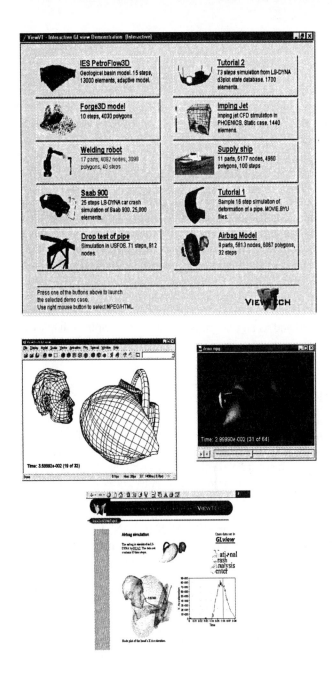

Figure 8 Using ViewVT for easy access to analysis results.

CONCLUSIONS

GLview is well suited as a high performance visualization tool for LS-DYNA results on Unix and Windows NT/95/98. Support for multiple analysis codes and direct reading of analysis results, make it easy to verify, compare and present results. ViewTech will continue to add functionality based on user requirements. The program is designed as a compact and effective engineering animation tool, which utilize the graphics standard OpenGL for high performance 3D graphics.

GLview is well suited as a visualization tool for the engineer and decision-maker on the next generation of technical reports stored on CD-ROM. By using the ViewVT program provided by ViewTech, clients can easily view information about analysis using the combination of GLview display, MPEG/AVI/animated GIF playback and html-reports.

GLview has been used in a wide range of application areas, and will be developed further to address new applications both as a standalone visualization program, and as an integrated window for graphics display of analysis results.

REFERENCES

1. Aamnes K.: "Visualization of results from mechanical engineering analysis, Ph.D. Thesis, Division of Structural Engineering, The Norwegian Institute of Technology, Trondheim, 1990.
2. Aamnes K., V. Vesterheim, and T. Rolvaag: "Simultaneous simulation and visualization using DGL, STF10 A92018, SINTEF, 1992.
3. Aamnes K., and S. Gronaas: "Visualization of results from meteorological simulations on CRAY, STF10 A92019, SINTEF, 1992.
4. Rolvag T. and Ketil Aamnes: "FEDEM User's Guide, Finite Element Dynamics of Elastic Mechanisms", SINTEF Production Engineering, Trondheim 1991.
5. Mathisen K.M et al.: Interactive-Adaptive Nonlinear Structural Analysis, Visualization 93, Trondheim, 1993.
6. Det Norske Veritas Software, Hovik, Norway. SESAM Users Manual, SESAM System, 1997.
7. McLendon P.: Graphics Library - Programming Guide, Silicon Graphics, Mountain View, CA, 1991.

Maximizing the Efficiency and Benefits of a Quality System for Software Engineering Through the Use of CASE and CAST tools.

Kenneth G. Podlaszewski
Quality Assurance Engineer
ANSYS, Inc.

1. ABSTRACT

Quality systems that encompass all aspects of software development and delivery have long been considered instrumental in providing quality software. Advances in desktop computing and available Computer Aided Software Engineering (CASE) and Computer Aided Software Testing (CAST) tools has made it possible to go from a minimal, expensive and sometimes agonizing infrastructure required by the software engineering quality system to a much more advanced, useful and symbiotic infrastructure that facilitates as well as supports the quality system. This paper will investigate the use of CASE and CAST tools to support and facilitate software engineering quality systems. Practical cases from the quality system in use at ANSYS, Inc. will be used to demonstrate how such tools can improve the quality of software, increase cost effectiveness in developing, delivering and servicing software and how tools can serve as a catalyst for continuous improvement of a quality system.

2. THE QUALITY SYSTEM

What is a quality system? Volumes of literature have been written that attempt to answer this question. Certainly an entire paper could be dedicated to this subject alone but for the purpose of this paper, a quality system is defined as the defined processes and specific, defined quality goals that are used within an organization to produce and supply products. Fundamentally, the basic goal of the quality system is to provide for the production and delivery of quality products to the customer, which meet their needs and expectations. Ideally, the quality system should evolve over time in a way that improves the processes that are in use thus adding value to the products and services that are provided.

The mission of ANSYS, Inc. (formerly Swanson Analysis Systems, Inc.) is to provide Computer Aided Engineering (CAE) solutions to the engineering community. For nearly 30 years, ANSYS, Inc has been performing this function and has been a recognized leader in its field. From its earliest days, ANSYS, Inc. has led the industry with its software quality assurance. The quality system that has evolved over this time and that is currently in use at ANSYS, Inc. is an ISO 9001 certified quality system[1]. The scope of Ansys' ISO 9000 certification includes the design, development, production and support of commercial and customized software products and training for engineering analysis and design and the provision of quality assurance and verification services for software products.

3. HISTORY OF COMPUTING AND CASE/CAST TOOLS AT ANSYS

Before detailing the use of CASE and CAST to facilitate and support a quality system for software engineering it is important to briefly look at the history of computing, and CASE and CAST tools use at ANSYS, Inc.

In the early days of CAE, the ANSYS program provided basic thermal and static finite element analysis. The program has evolved to automate many complex engineering tasks such as optimization, multi-field analysis, complex dynamics, fluid flow analysis and crash simulation, just to name a few. Increasing the software's capabilities and sheer size, while at the same time maintaining a high level of quality, has been possible only through substantial improvements in, and increased use of, CASE and CAST tools. The use of such tools has also been instrumental in the evolution of ANSYS, Inc.'s quality system. The tools now in use go beyond merely assisting developers in creating software to provide additional capabilities for managing, monitoring and reporting within the tool itself.

A look at how computing as well as CASE and CAST tool availability has increased over the years at ANSYS, Inc. and how it has been put to use, demonstrates the significance of the use of CASE and CAST tools in the development of a complex computer program such as the ANSYS program. Figure 1 gives a historical view of the relative estimated total computing power, in terms of CPU clock speed, in use at ANSYS, Inc. and shows that the estimated available CPU has increased over 100 fold since 1987.

[1] ISO 9000 registrar – Underwriters Laboratories Inc.

Relative Total Available CPU (estimated)

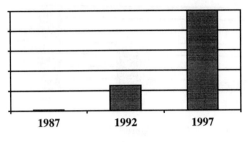

Figure 1.

Figure 2 shows how the available CPU has been put to use at ANSYS, Inc. over time.

Use of Total Computing Resources

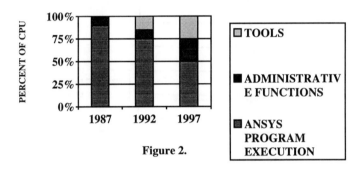

Figure 2.

It is clear that in terms of CPU use, the amount of CASE and CAST tool use has increased dramatically since 1987. It should be noted that that since 1992, when a nontrivial number of tools were in use, CASE and CAST tool use has increased by a factor of almost 7.

The amount of non-comment source code added in an ANSYS release increased from 1987 to 1997 by a factor of approximately 17. At the same time, the number of developers increased only by a factor of approximately 2.5. Figure 3. shows the relative increase over time in the amount of new code introduced per developer. Figure 4 shows the number of new lines of code added to each of the last three releases of the ANSYS program and the error incidence rate. Not only has there been an increase in code production; there has been a decrease in the error incidence rate. Thus it can certainly be argued that the increase in CASE and CAST tools usage has had a direct positive impact on the production of lines of source code as well as the quality of the software.

NEW LINES OF CODE MANAGED PER DEVELOPER

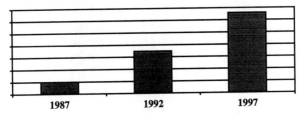

1987	1992	1997

Figure 3.

Number of new lines of non-comment code and Error Incidence Rate

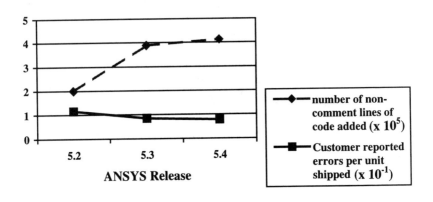

ANSYS Release

Figure 4.

4. USE OF CASE/CAST TO SUPPORT THE QUALITY SYSTEM

Although it is evident that the use of CASE and CAST tools directly benefits the developers in their efforts to produce new code, a secondary but significant benefit of the use of these tools is how they can specifically support and enhance the software engineering quality system. Since many of the available tools now not only provide basic functionality but address many of the needs inherent to a good quality system, with proper planning and foresight, tools can be chosen, created or customized to accomplish this with a minimal expenditure in resource. In fact, some of the activities that a quality system should encompass can only be accomplished in a cost-effective manner when provisions are made for the tools to support them.

These activities include, but are certainly not limited to, project management, process control and monitoring, the use of metrics for gauging program quality and program testing and validation. A certain level of quality control is always necessary and a good software engineering quality system will provide for this as needed. However, it is a well-known fact that the return on investment for quality assurance is much greater than the return on investment for quality control, so a good quality system should also focus on quality assurance as well. The proper use of CASE and CAST tools can be critical in this endeavor.

Now lets take a look at some of the specific tools in use at ANSYS, Inc and how they support the quality system. Although a few of the tools are "off the shelf" tools that have had little or no modification, most of these tools are highly customized "off the shelf tools" or have been developed at ANSYS, Inc. In many cases, the individual tools have been made to interact among themselves in order to provide support for the quality system. Although the following list of tools is not a comprehensive list of the tools that support the quality system at ANSYS, Inc., it does highlight those that provide substantial support to the quality system. It should be noted that one of the most valuable ways in which many of these tools support the quality system is to provide for cost effective generation of metrics that are used both for defining or monitoring progress towards quality goals and evaluating the system for process improvement purposes.

1. Corporate Intranet, including on-line quality system documentation

The corporate Internet is not actually a CASE/CAST tool in and of itself but it does support the quality system by providing much of the communication for the development, testing and quality planning that is part of the quality system. The most important support that the Intranet lends to the quality system is communication to employees of quality system documentation. The quality system documentation within the Intranet is a "paperless" system. It has been developed to maximize the employees' ability to access and make use of the quality system documentation through the use of links between related parts of the documentation and built in search capabilities. In fact, customized online training has been developed within this system to provide necessary quality system training to some new employees that is specific to their job function.

2. Development project management system

The development project management system has been developed to truly be a CASE tool. Its primary function is to manage and control the complete program development process from initial requirements definition to verification testing to final approval for release. It is highly customized to

1005

ensure that many of the steps of the process are fully complied with, including proper independent review or approval where it required. This tool is also an invaluable tool for communication between all persons involved in software product creation. It has been developed to provide support for monitoring of the development process and for supporting quality metrics creation. It is instrumental in planning for and monitoring progress towards quality goals that are defined in terms of milestones. It has also been developed to ease quality assurance's task in conducting audits required before product release.

3. Software bug tracking tool

The software bug tracking tool is a powerful tool that has been highly customized to support the software error handling requirements of the quality system. It supports the quality system by ensuring, among other things, that proper independent review and verification of error corrections is performed. This tool has also been tailored to control processes that provide for satisfying regulatory requirements as well. In addition to minimizing the overhead work for developers with regard to communicating, resolving and testing program defects, the tool is instrumental in providing quality metrics. Such quality metrics are used to monitor progress towards release criteria quality goals such as program reliability and error incidence rate. This tool interacts with the test case management system to provide for compliance with the quality system requirements for testing of error corrections.

4. Call tracking tool for technical support

This tool's obvious function is to support the quality system in the area of product technical support. It has been customized to provide a level of process control within the technical support process as well as to track and record technical support correspondences. Another important way that this tool supports the quality assurance system is to provide data for quality metrics.

5. Software configuration management system

The software configuration management system is a highly specialized system. It is truly a CASE tool that facilitates the quality system. Not only does this system provide for efficiency and cost effectiveness in changing and adding source code and handling the associated record keeping that is necessary, it also automates a lot of the fundamental activity that is part of the quality system. This system is really a set of tools that interact closely together to control the process of modifying the program baseline and the process of testing code changes and additions. It includes code analyzers that check for coding that is prone to be unstable across systems and coding that contains known invalid or problematic constructs.

The system also interacts with the test case management system and the automatic stability/regression testing system as well as other tools to qualify code changes and additions before they become part of the baseline code. It does this by automatically running tests and verifying their results to insure that the changed code performs as expected and does not damage other parts of the baseline that could be affected. Should this system detect any anomaly, it rejects the code change until the developer resolves the problem. Among the benefits of this system is detection of many programming errors early on while the coding is still fresh in the developer's mind. This system has provided for significant progress towards "designing" errors out of the software ahead of time, as opposed to testing errors out after development is complete.

6. Test case management system

The test case management system is a CAST tool that provides for efficient access, control and handling of test cases by software developers and testers. It facilitates the quality system by providing process control and record keeping for histories and signatures for reviews and approvals. This tool is also designed to interact closely with the software configuration management system as noted earlier. Data is obtained from this system for metrics and reports that are used to evaluate and monitor testing activity.

7. Automatic stability/regression testing system

The automatic stability/regression testing system CAST tool performs automated execution of test cases and comparison of results to certified results. This tool interacts very closely with the test case management system. It facilitates the quality system by automatically performing many of the required testing activities and producing the necessary reports and records. This system has been evolved to where human intervention has been essentially reduced to cases where testing detects a problem with the software. Data from this system is collected on an on-going basis to measure progress towards quality goals allowing constant monitoring and appropriate adjustment to programming and testing resource allocation.

8. Customer error notification tools

The customer error notification tools interact very closely with the software bug tracking tool. These tools automate much of the work required to provide notification of serious software defects to customers. They provide for the creation of reports that are distributed to customers in various forms including automatic emails and automatic posting to the Company's web page. These tools directly support this vital part of the quality system.

9. Data collection and metrics tools

The collection of data and production of metrics is a very important part of the quality system. A variety of tools are used to support this function and thus they support the quality system. Metrics are used for various purposes including, process monitoring, project management and evaluating progress towards meeting defined quality goals. A large suite of tools has been developed over time to interact with many of the CASE/CAST tools to extract the necessary data and produce the necessary metrics. In many cases, the tools have been automated to the where metrics are provided to appropriate parties automatically without human intervention using distribution channels such as email and the company Intranet. This automation has made it possible to produce a large number of metrics in support of the quality system in a cost-effective manner. In some cases where metrics are generated manually, tools such as databases and spreadsheets are used to reduce the effort involved.

10. Other tools

Other tools such as html editors, graphics tools, flowcharting tools and documentation control tools are used to increase the efficiency of various tasks that directly support the quality system. These tools play an important part in increasing personal productivity in many quality assurance functions.

5. SUPPORT OF ANSYS'S QUALITY SYTEM – THEN VS. NOW

It is possible to evaluate the positive impact that the development and use of these tools can have with respect to a quality system and the production of software by comparing some of the processes in place now at ANSYS, Inc. to similar processes that were used in the past. In doing this comparison, the benefits of these tools become apparent. For many of the comparisons presented below, the benefits are significant and obvious and warrant no quantification. In some of the cases, the benefits have been quantified to show their significance.

a. General improvements at ANSYS, Inc. attributable to tools

Since 1987 many general improvements have resulted from the increased use of tools to support and facilitate the quality system. Some of these improvements are detailed here.

Process related communications and transactions are now almost completely electronic and are virtually instantaneous. For example, with the tools currently in use, the hand-off of a software bug from the initiator to the appropriate developer now occurs within minutes. In the past, when paper systems were used, such hand-offs took a day or more. Because most processes involve a number of such hand-offs, the total turnaround time for many processes has been substantially reduced.

Process guidance and provisions to ensure quality system or regulatory compliance have been programmed into tools where it is warranted. In many cases, it is no longer necessary for employees to remember "what do I do" or "how do I do it" since the tools guide them through it. This allows them more time to concentrate on the work at hand rather than the details of the process.

The process improvement function has been made more efficient as a result of the increased use of tools. Process improvement efforts can now be focused and monitored more frequently at a much lower cost. This has significantly increased the return on investment for the process improvement effort.

b. Specific improvements at ANSYS, Inc. attributable to tools

Tremendous benefits have been gained through the use of tools in the area of system dependency prevention and detection. Code checkin tools now detect system dependencies before code is placed into the baseline. These tools incorporate code analyzers and automatic multi-platform compiles to detect system dependencies and prevent them from being placed in the baseline. Stability checking tools now allow developers to automatically verify that a code change has the desired affect (or no affect) on tests that exercise the code before the changed code is placed into the baseline. Thus errors that are detectable in existing test cases are now found and fixed by developers very quickly. Because of this, the error's impact is limited and its correction is less costly. It was not feasible make this improvement to the process until the tools to support this effort were available. Without the tools, this activity was prohibited by the amount of resource required to carry it out. Figure 5. quantifies the reduction in the effort required to remove software operating system dependencies which is one of the benefits of this process improvement.

Effort to Remove System Dependencies

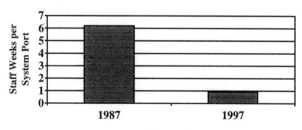

Figure 5.

1009

When the effort expended to remove system dependencies is normalized by the amount of code that is changed, as in figure 6., the dramatic benefit of this process improvement becomes clear.

Effort to Remove System Dependencies
Normalized by Amount of Code Changed

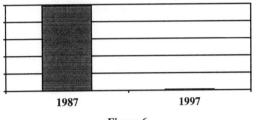

Figure 6.

Test set maintenance and regression testing processes have also benefited substantially from the use of tools and are now more efficient. Test case results and their review and approval are now handled electronically. Regression testing tools have been developed to provide for more efficient regression activity much earlier in the development process. Regression tools and test set maintenance tools are set up to be integrated into the code checkin tools as well as to be accessed independently. Furthermore, expertise in a cumbersome test set maintenance procedure is no longer required and thus developers are able, with very little additional effort, to participate in the regression process. This has allowed the code change process to be improved to provide for regression testing to occur in conjunction with code checkin rather than at a later time as a separate process. An added benefit is that this has moved the review of test case output changes up in the process to a point where the associated code change is fresher in the developer's mind. As can be seen in figures 7 and 8 the cost of regression testing has been reduced substantially as a result of this tool based process improvement.

Regression Test Set Size

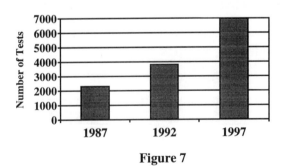

Figure 7

Regression Testing Effort
Normalized by Test Set Size

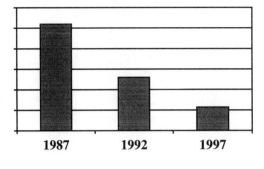

1987   1992   1997

Figure 8.

Another benefit of this improved process is the increased frequency at which test cases can be run. In 1987, the entire set of 2300 tests was run on average one to two times per month on one operating system. Currently, the entire set of 4300 tests is run approximately 30 times per week across 17 different computers and operating systems.

As was stated earlier, metrics are a very important part of the quality system. They are instrumental in gauging program quality, process monitoring, defining and monitoring quality goals and in quality system improvement. As a result, they are critical to improving a quality system to be proactive with focus on quality assurance rather than be reactive with focus on quality control. Record keeping and collection of desired data is now performed automatically by tools where possible rather than manually through the processing of paper. The reduction in required resource is obvious. This, coupled with the increased automation of metrics generation through tools, allows for the collection of a much larger number of metrics at a much higher frequency. Additionally, desired data can now be gathered much more reliably and with much less involvement from the process participant. Figure 9. demonstrates how the number of metrics and their use has increased over time. This increase can be attributed mainly to the increased use of tools.

Number of Software Quality Metrics Gathered

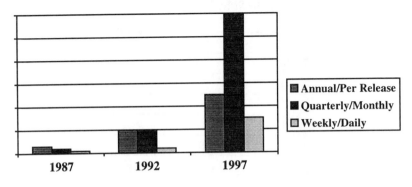

Figure 9.

6. CONCLUSIONS

It has been shown that the use of increasingly available CASE and CAST tools can considerably improve cost effectiveness and efficiency in software production and support processes. It has also been demonstrated that such tools can also be a used to support and improve a quality system for software engineering. Intelligent use of these tools can be of ever-increasing benefit to quality systems and to cost effectiveness in developing, delivering and servicing high quality software.

Bryce Gardner
Automated Analysis Corporation
USA

ABSTRACT

on

"The Prediction of the Acoustics Performance of Foam Absorbative Materials"

This paper presents an overview of the relevant theories and phenomena involved in acoustic effects in porous foam materials. It then shows how modern predictive methods correlate with test cases on material specimens.

Finally, the paper describes some real-world applications of the methodology, including aiding the design of vehicle, interior trim, and guiding the development of acoustic linings in ducts.

FATIGUE DESIGN OF FORCEMATED (COLD-EXPANDED BUSHING) LUGS SUBJECTED TO AXIAL LOAD

Alec Keith Hitchman
Associate Engineer, Fatigue Technology Inc.
Seattle, Washington USA

SUMMARY

The following paper details a method developed to facilitate the preliminary design of axially loaded lugs with respect to fatigue. Using the finite element procedure outlined in this manuscript, an analytical method is developed for the fatigue design of lugs with bushings installed by the Fatigue Technology Inc. proprietary ForceMate® process. This process radically expands an initially clearance fit bushing into a hole at high interference, while simultaneously imparting residual compressive stresses into the surrounding material. Fatigue life improvement of 3 to 1 or better is achieved typically.

The finite element analysis is performed on a ForceMated bushing and lug (using MSC/PATRAN and MSC/ABAQUS) to determine the appropriate stress concentration, for a given axial load, at those locations where lug failure generally occurs: net section tension, bearing and shear. These finite element results are used to define an analytical method for estimation of the fatigue life of the ForceMated lug. An integral part of this method is the determination of a series of stress concentration factors that can be referenced in the preliminary lug design process. Verification of the lug fatigue design is presented, demonstrating the fatigue design method developed herein. Results generated by this analytical procedure are compared to actual ForceMate fatigue test data.

1 INTRODUCTION

Structural design text presents standard methods for the design of lugs subject to axial, transverse or oblique static load for a set of general lug configurations. A good example is given by Niu [1]. These lug design methods use "efficiency factors" to account for stress concentration effects resulting from the loading applied to the lug. The "efficiency factors" are based on both a theoretical basis and actual test result comparisons.

The lug is considered to possibly fail in any one of three modes: (a) net section tension, (b) shear tearout, with the assumption that all loading is transmitted on "40-degree planes," and (c) bearing. As (b) and (c) can be considered a single mode of failure, we are left with considering the two modes of failure, as indicated in Figure 1.

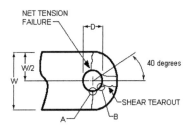

Figure 1
Lug Net Section and Shear Tear-Out Failure Due to Axial Loading

When considering the design of lugs subjected to fatigue type loading, a method such as this is inadequate, as this method is based only on static test data. Furthermore, fatigue failures may occur at stress levels that do not result in local lug yielding, although the static loading tests used to calibrate the method typically result in considerable yielding in the lug.

In an attempt to address these issues and extend the current semi-empirical method to account for repeated service loads, Tsang presented, in a paper published in the Journal of Aircraft [2], a method whereby finite element analyses of axially loaded lugs were used to obtain new stress correction factors. These stress correction factors replace the "efficiency factors" of the static load method, while taking into account the multi-axial state of stress at each of the two failure locations.

As an extension to the work performed by Tsang, this manuscript presents new stress concentration factors determined by finite element models of lugs with ForceMate bushings installed. The ForceMate bushing installation method used to extend the fatigue life of structure is described below.

2   FORCEMATE PROCESS

The ForceMate process involves simultaneous cold expansion of the hole and installation of an interference fit bushing. The process provides a significant improvement in the fatigue life of bushed holes in metallic structure, as well as reducing bushing installation cost. The fatigue life improvement is provided by creation of residual compressive stresses in the parent structure, as well as a reduction in the applied cyclic stress range caused by the interference fit bushing.

As shown in Figure 2, a specially sized and internally lubricated bushing is placed on an expansion mandrel. The mandrel is attached to a hydraulic puller unit, and the mandrel/bushing assembly placed in the hole. The puller is activated, drawing the expansion mandrel through the bushing, simultaneously expanding it into the hole at a high interference and cold expanding the surrounding material.

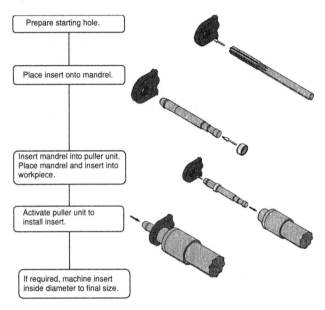

Figure 2
ForceMate Process Steps

The amount of bushing/mandrel interference is known as the applied expansion (Ia), and is determined by the following expression:

$$Ia = \frac{OD_{mandrel} - ID_{bush}}{ID_{bush}} \times 100\% \tag{1}$$

where Ia is applied expansion, in percent, $OD_{mandrel}$ is the mandrel major diameter, and ID_{bush} is the bushing inside diameter. Applied expansion, while dependent on bushing/parent materials and bushing size, typically ranges from 2 to 6%.

The ForceMate process has advantages over other bushing installation methods, including: (a) a higher interference than can be obtained by traditional bushing installation (press- or shrink-fit) methods, (b) no cryogenic materials required to shrink the bushing, and (c) little risk of galling or scraping off of typical bushing coatings [3]. Bushings of various configurations and sizes can be installed using the ForceMate process, including flanged bushings and bushings with fretting-resistant epoxy coatings (FTI's BlueCoat™). Numerous military and civil aircraft applications are in service, which use a broad range of bushing/parent material combinations, and over a wide range of bushing sizes installed in a variety of material thicknesses.

While FTI has developed ForceMate bushings and installation tooling corresponding to standard National Aerospace Standard (NAS) bushings, most new potential ForceMate applications are of a non-standard size or material combination. Even the most preliminary of ForceMate designs often require physical static and fatigue testing, and, as computational capabilities have increased in recent past, time-intensive finite element models are often required to determine the residual stresses, amount of interference, and differences in applied expansion and bushing material.

Therefore, this paper presents a method whereby preliminary design of ForceMated lugs subjected to axial cyclic loading can be performed. The purpose is to present a method whereby the effects of ForceMate for fatigue life enhancement can be ascertained during the preliminary lug design process, thereby streamlining the more detailed phases of design work.

3   FINITE ELEMENT METHODOLOGY

The finite element models for this study were created using MSC/PATRAN v7.5 and MSC/ABAQUS v5.6 finite element software. All analyses performed were two-dimensional, using plane stress elements with either reduced integration (to reduce computation time, where allowable) or standard integration, isoparametric, quadrilateral elements. Both bushing and lug materials were modeled as non-linear, isotropic hardening materials; with stress/strain curves obtained from Mil-Hdbk-5G [4]. The bushing is made from 17-4 Ph stainless steel, heat-treated to condition H1025. The lug was modeled as 7075-T73 aluminum plate. Use of the isotropic hardening material resulted in some difficulty as will be seen. Also, as the bushing/lug material combination is a crucial aspect of how the ForceMate process performs, the analysis results contained herein can be viewed as applicable only to the specific materials that were modeled. The pin used to apply axial load to the bush/lug assembly was modeled as linear, high-strength steel (AMS 4340), again per Mil-Hdbk-5G. To reduce model size, the pin was generated as a "ring" of material two elements deep, with a rigid multi-point constraint

(MPC) between the pin interior nodes and the pin center. ABAQUS "Contact Pairs," with initial clearance, were defined between the bushing and lug, as well as between the bushing and pin. Boundary constraints consisted as a rigid fixity along the lug face, as shown, concentrated load applied at pin MPC master node, and a uniform radial displacement to the bushing inside diameter (modeling the effect of the expansion mandrel). The finite element analysis modeled both 2 and 4% applied expansion to the ForceMate bushing inside diameter (see Figure 3).

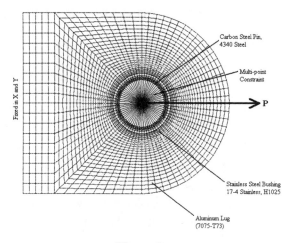

Figure 3
Typical Finite Element Model Mesh

The analysis was conducted as a multi-step, non-linear analysis. The steps for each analysis are shown in Table 1. Due to the ease of element removal and reintroduction, it was decided to include the post-ForceMate reaming step. While ForceMate can be performed to result in a required final bushing hole size without reaming, most ForceMate bushings are reamed after installation.

Analysis Step	Step Description
1	Installation of ForceMate bushing by prescribed radial displacement
2	Relaxation of bushing (no loads applied)
3	Removal of bushing elements constituting ream
4	Reactivation of pin elements and bushing contact (via displacement)
5	Application of load via pin Multi Point Constraint (MPC)

Table 1
Typical Analysis Steps

The models were constructed of a general lug geometry, chosen to best represent a wide range of width-to-diameter (W/D) ratios, as well as being of a size that is directly comparable to FTI test data [5]. Bushing sizes were FTI's standard ForceMate bushings and the W/D evaluated were determined by the standard system dimensions and the initial lug geometry. Information regarding initial lug hole diameters and final bushing inside diameters can be found in FTI's Engineering Handbook EH-3 [6].

Width, W, of the lug was held constant at 75.44 mm (2.97 inches), with a constant thickness, t, of 12.70 mm (0.50 inch). A total of 10 bushing sizes were evaluated, with W/D ratios from 9.49 to 2.26. Axial load applied was 85.63 kN (19250 lbf), giving a gross section stress of 89.38 GPa (12.93 ksi).

4 FINITE ELEMENT MODEL CONVERGENCE CRITERIA

Any model of the ForceMate process is, by necessity, highly non-linear and must consider elastic-plastic interaction between the bushing and the parent material. It is often difficult to use those convergence criteria that are usually applicable to linear elastic finite elements. Strain gradients are typically high in the yielded regions surrounding the hole, and in the bush itself. Adequate mesh refinement is required to avoid severe mesh distortion. Use of reduced integration elements resulted in hour-glassing in some cases, thus standard integration elements were used for some of the models analyzed as part of this effort. A good criteria for overall model accuracy is how the finite element model compares with test data in the particulars of final bushing inside diameter (before ream) and bushing/lug interference, as determined by the contact pressure between the bushing outside diameter and the lug inside diameter. All of the models shown indicated interference to be about 10 to 20% greater than physical test, and bushing final inside diameter (before ream) to be about 4 to 5% greater than indicated in physical testing.

These differences can be attributed to the use of the isotropic hardening model in defining material response. The isotropic hardening model, while easy to define from typical uniaxial stress/strain data such as that found in Mil-Hdbk-5G, does not take into account the phenomena known as Bauschinger's Effect, the reduction of material yield strength after load reversal. Strictly speaking, the material yield surface may grow in stress/strain space, but must remain centered about the origin. Most engineering metallics do not behave this way, and display some amount of yield surface growth and translation; i.e., following both an isotropic and kinematic hardening rule. While MSC/ABAQUS allows for a so-called combined hardening material definition, the material law must be carefully calibrated from strain-controlled cyclic testing of the materials, and this test data was not available. ForceMate can be viewed as applying nearly one full stress cycle to the bush (and often the parent material): yielding in tension during mandrel engagement, followed by reverse yielding when mandrel is withdrawn.

Using the isotropic hardening material response, knowing its limitations, it is clear that the increased contact pressure (and thus, interference) are the result of the bushing resisting plastic collapse after the mandrel has been removed. This may result in an increase in residual Von Mises stress in the lug, which might affect the fatigue design analysis, as described below.

5 RESULTS DISCUSSION

To obtain the applicable stress concentration factors for each of the W/D ratios and levels of applied expansion, in the same method as for a non-ForceMated lug, the Von Mises stress during load application was obtained from each model at the node corresponding to the two modes of failure, shear (bearing), and net section. Stress concentration factors are defined as:

$$K_{tn} = \frac{F_{VA} A_{tn}}{P} \qquad (2)$$

$$K_{br} = \frac{F_{VB} A_{br}}{P} \qquad (3)$$

where F_{VA} and F_{VB} are the finite element Von Mises stresses at points A and B, respectively (see Figure 1). P is the axial load applied, and A_{tn} and A_{br} are the tensile and shear-bearing areas, as determined by:

$$A_{tn} = (W - D)t \qquad (4)$$

$$A_{br} = Dt \qquad (5)$$

The values of K_{tn} and K_{br} for varying W/D ratios and levels of applied expansion are shown in Figures 4 and 5.

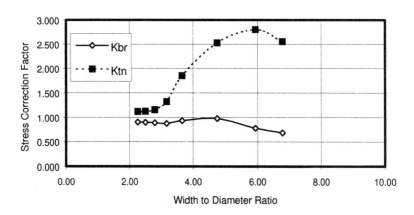

Figure 4
Relationship Between Stress Correction Factor and W/D Ratio
2% Applied Expansion

1021

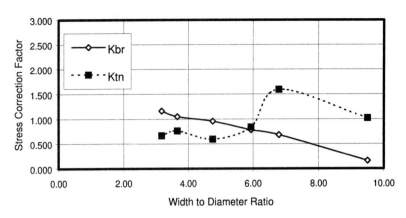

Figure 5
Relationship Between Stress Correction Factor and W/D Ratio
4% Applied Expansion

In addition, we wish to look at stress correction factors while considering the reduction in cyclic stress amplitude (R ratio) due to the ForceMate process. To determine these stress correction factors, it is necessary to determine the residual Von Mises stress in the lug prior to application of axial load. This is achieved by querying the finite element model for the nodal Von Mises stress after the ForceMate bushing has been installed and reamed, but before the axial load is applied. The new equations, for R-ratio reduction stress correction factors are:

$$K_{tn} = \frac{|F_{VA} - f_{VA}|A_{tn}}{P} \tag{6}$$

$$K_{br} = \frac{|F_{VB} - f_{VB}|A_{br}}{P} \tag{7}$$

where f_{VA} and f_{VB} are the residual Von Mises stresses present after ForceMate, at locations A and B in Figure 1. Notice that the absolute value of the difference in residual and loaded Von Mises stresses are used, since, by its definition, Von Mises stress is always a positive value. Also note that if no residual stresses are present, then Equations 6 and 7 reduce to the basic stress correction factor Equations 2 and 3. The variations of these values of K_{tn} and K_{br} for varying W/D ratios and applied expansions are shown in Figures 6 and 7.

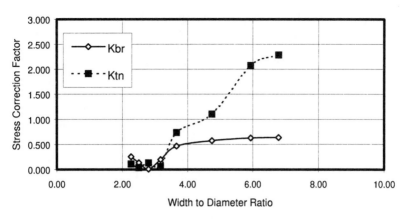

Figure 6
Relationship Between R-Ratio Reduction Stress Correction Factor and W/D Ratio - 2% Applied Expansion

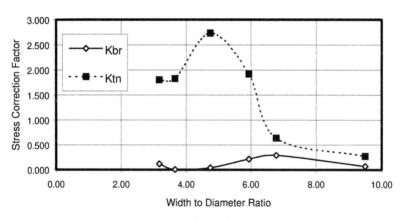

Figure 7
Relationship Between R-Ratio Reduction Stress Correction Factor and W/D Ratio - 4% Applied Expansion

6 FATIGUE ENHANCEMENT ESTIMATION ANDCORRELATION

As an example of how the new stress correction factors may be used in practice, using the methodology developed by Tsang, we will now proceed to estimate the fatigue life improvement offered by ForceMate over a similarly sized shrink-fit type of bushing. As mentioned previously, since the stress correction factors determined above are specific to the material combination studied, we are required to consider a stainless steel bushing, of type 17-4 Ph, heat-treated to condition H1025. The lug into which both bushings are to be installed is 7075-T73 plate. To allow correlation with existing fatigue test data [5], we chose to investigate bushings installed into a lug whose width, W, is

1023

75.44mm (2.97 inches), and with a thickness, t, of 12.70mm (0.50 inch). The bushings will be of standard NAS type for receiving a 25.40mm (1.00 inch) pin, which requires a lug starting hole diameter, D, of 30.17mm (1.1878 inches). This equates to a W/D ratio of 2.5. If we assume a ForceMate bushing installed with 2% applied expansion, and a W/D ratio of 2.5, we can refer to Figures 5 and 7 to obtain the larger values each for K_{tn} and K_{br}, for a conservative estimation. Looking at the two graphs gives values of K_{tn}=1.124 and K_{br}=0.899. Tensile loading during the fatigue test was performed at Pa 148.9 GPa (21.6 ksi) net section stress (the same gross stress at which the finite element models were evaluated) and R=0.05, resulting in an axial load, P, of 85.63 kN (19250 lbf). We first determine the nominal tensile and bearing areas, using Equations 2 and 3:

$$A_{tn} = (W - D)t = (75.44-30.17) \times 12.70 = 575 \text{ mm}^2$$

$$A_{br} = Dt = 30.17 \times 12.70 = 383 \text{ mm}^2$$

Now we calculate the nominal tensile and bearing stresses, f_{tn} and f_{br}, including a fitting factor, C_f (typically 1.15):

$$f_{tn} = C_f P/ A_{tn} = (1.15 \times 85.63)/575 = 0.17126 \text{ kN}/ \text{mm}^2 = 171.3 \text{ GPa}$$

$$f_{br} = C_f P/ A_{br} = (1.15 \times 85.63)/383 = 0.25711 \text{ kN}/ \text{mm}^2 = 257.8 \text{ GPa}$$

And we calculate the maximum tensile and bearing stresses, f_{tnx} and f_{brx}, using the stress correction factors:

$$f_{tnx} = K_{tn} f_{tn} = 1.124 \times 171.3 = 192.5 \text{ GPa}$$

$$f_{brx} = K_{br} f_{br} = 0.899 \times 257.8 = 231.8 \text{ GPa}$$

Now, using the larger of the two values above, f_{brx} =231.8 GPa (33.6 ksi), and referencing Mil-Hdbk-5G for fatigue data (7050-T7351 being used, for lack of 7075-T73 fatigue test data), we estimate that the ForceMate installation gives an estimated fatigue life of 740,156 cycles.

If we now consider the shrink fit bushing, the same process is used. The values of K_{tn} and K_{br} were obtained from an ABAQUS finite element model of a standard NAS shrink fit bushing, whose inside diameter is 25.41 mm (1.0005 inches), with an outside diameter of 30.22 mm (1.1898 inches). This bushing is placed into a starting hole of 30.17 mm (1.1878 inches). Using the method defined previously, the stress correction factors are calculated to be K_{tn}=2.636 and K_{br}=1.518. (Note that the stress correction factors used are for the larger values determined by Equations 2 and 3, rather than those values given by Equations 6 and 7 as was done for ForceMate). These values are calculated separately, and not referenced from the graphs, as the graphs are for

ForceMate only. The nominal tensile and bearing areas are the same as for the ForceMate bushing installation, as are the tensile and bearing stresses, f_{tnx} and f_{brx}. We can calculate new values for f_{tnx} and f_{brx}:

$$f_{tnx} = K_{tn} \, f_{tn} = 2.636 \times 171.3 = 451.5 \text{ GPa}$$

$$f_{brx} = K_{br} \, f_{br} = 1.518 \times 257.8 = 391.3 \text{ GPa}$$

Again, using the larger of the two values, $f_{tnx} = 451.5$ GPa (65.5 ksi), and referencing the applicable fatigue data for 7050-T7351 aluminum, we estimate the shrink fit installation to provide a fatigue life of 13,777 cycles.

This analysis shows that ForceMate provides an increase in fatigue life of more than 50:1. While this may seem somewhat un-conservative when determining the fatigue life improvement due to ForceMate, FTI test data [4] and numerous other tests confirm that fatigue life improvements of this magnitude are common for ForceMate bushing installations. For a similar load and geometry case, with stainless steel bushings installed in 7075-T73 parent material, at greater than the 2% applied expansion (2.7%, on average) evaluated in this exercise, the minimum life improvement was 17:1 over a shrink fit bushing installation. The fatigue life predicted for the ForceMate, 740,156 cycles, is close to the log average determined by the FTI test, 823,000 cycles. This would seem to indicate that the value determined above might be a slightly conservative value. The log average for the shrink fit installations was 26,000 cycles, indicating a very conservative shrink fit bushing fatigue life prediction.

7 CONCLUSION

This paper developed stress correction factors, using non-linear elastic-plastic finite element models, to enable any designer to obtain an estimate of the fatigue life benefits offered by ForceMate bushing installations. Various W/D ratios were investigated, at two levels of ForceMate bushing applied expansion. Based on the finite element results, stress corrections for each W/D ratio and applied expansion were determined.

Following a method for analysis set forth by Tsang [2], the stress correction factors were used to estimate the fatigue life of a typical ForceMate bushing installation. A fatigue life prediction was also performed for a similar shrink fit installation. Comparison of the two values to actual test data showed conservative estimates for predicted fatigue life, more so for shrink fit than for ForceMate. The analysis confirms the tremendous benefits in fatigue life improvement that can be realized by the simultaneous high interference fit and residual compressive stresses induced by the ForceMate process when applied to bushings installed in lugs.

1025

A possible explanation for the conservative estimates is the use of the isotropic hardening model when performing the finite element analysis. As mentioned previously, this has the effect of increasing the amount of residual stress present after the ForceMate installation, and may have increased the calculated stress correction factors accordingly.

It should be noted that the work presented here does not consider some crucial aspects of the preliminary lug design process. The effects of friction, misalignment, eccentricity, and pin bending have not been investigated here. As these aspects are by no means trivial, the reader is encouraged to reference Tsang for further information on these effects in the preliminary lug design process with respect to fatigue loading.

Continuation of this work could include the evaluation of different bushing/parent material combinations, as well as a broader range of bushing applied expansion. Additional lug geometries, as well as transverse and oblique loadings, need to be evaluated as well, to further extend the usefulness of this fatigue life estimation method. The effect of friction, while not covered in this effort, should be evaluated for its effect. Also, it is imperative that a combined hardening material model be used in any further analyses, to better predict the behavior of the materials throughout the entire ForceMate process.

8   REFERENCES

1   NIU, MICHAEL C. Y. - Airframe Structural Design, Conmilit Press Ltd., Hong Kong, 1995
2   TSANG, S. K. - Fatigue Design of Axially Loaded Semicircular Lugs, Journal of Aircraft, Vol. 32, No. 3, May-June 1995
3   LANDY, M.A. and CHAMPOUX, R. L. - Fatigue Life Enhancement and High Interference Bushing Installation Using the ForceMate Bushing Installation Technique, ASTM-STP-927, 1987
4   MIL-HDBK-5G, Metallic Materials and Elements for Aerospace Vehicle Structures, Vol.1, 1990
5   BOLSTAD, R. T. and EASTERBROOK, E. T. - Fatigue Life Variations Due to Insert Material Using the ForceMate Process, FTI Technical Report TR-9174-3, 1989
6   FATIGUE TECHNOLOGY INC. - The ForceMate System, Rev. B, FTI Engineering Handbook EH-3, 1996

Integrated non-linear FE module
for rolling bearing analysis

Hermann Golbach, INA Wälzlager Schaeffler oHG

Summary

Rolling bearings are commonly used machine elements in which the engineer encounters in the design of all kinds of machine systems. Due to their properties as non-linear, statically indeterminate systems, rolling bearings place high demands on calculation methods. A model for representing the non-linear mechanical behaviour of rolling elements and rolling bearings has been developed for static finite element analyses and converted into the form of a user-defined element for use in the ABAQUS/Standard system. This user element determines, as a kind of structural element, the non-linear contact stiffness in the Hertzian contact area between the rolling element and raceway on the basis of analytical geometrical and elastic considerations. Unlike the representation of the rolling element using conventional continuum elements and the resulting unavoidable fine mesh, this user element manages with the minimum degrees of freedom. At the same time, the accuracy achieved is very satisfactory, as has been indicated by a comparison with the results of a continuum finite element model of the particular rolling element type. In practical application, only the essential geometric parameters of the rolling element or rolling bearing are required for element definition. The user element, integrated as a type of module into the FE structure of a machine system, enables realistic analyses both of the load distribution of the rolling bearing under the influence of the elastic environment and of the stress and deformation of the components adjacent to the bearing.

1. Introduction

INA Wälzlager Schaeffler oHG is an international manufacturer of rolling bearings and a supplier to the automotive industry, employing over 20,000 people. The central calculation department at the company headquarters in Herzogenaurach/Germany develops calculation methods and tools for various products, particularly rolling bearings.

Rolling bearings are some of the most important basic components in machine, vehicle and plant construction. They are used wherever machine elements moving in relation to each other require mutual support. As their name suggests, the rolling elements roll between the raceways of the components moving in relation to each other. Among other things, rolling bearings are differentiated according to the type of movement relations in rotating and linear motion bearings and the shape of the rolling element in ball and cylindrical roller bearings. Figure 1 shows two examples from INA's extensive range of products: a rotating bearing with needle rollers as the rolling elements and a linear motion bearing with balls as the rolling elements.

Figure 1   Needle roller bearing and linear guidance system

The external load is transmitted from one raceway to the other, distributed over several rolling elements. There is a very high local stress created at the contact point between the rolling elements and the raceways, which occurs cyclically when the bearing is in motion. In order to ensure that this extreme stress can be endured reliably and for the required fatigue life, a rolling bearing made from of excellent quality, designed under consideration of various environmental contitions, is essential. Analysis of the load distribution of the bearing is a key task, and is the first step in determining the fatigue life of the bearing.

2. Calculation of the load distribution in rolling bearings

Even the best rolling bearing will fail if its system behaviour is not adapted to the machine into which it is integrated. It is therefore necessary to analyse and determine in advance the intensive interactions between bearings and adjacent components. For example, in a system comprising of bearings, shaft and housing, the bending of the shaft and the deformation of the housing influence the bearing reactions and therefore the load distribution within the bearing. In turn the stiffness of the bearing influences the elastic line of the shaft, see Figure 2.

In mechanical terms, the system components, bearing, shaft and housing, represent spring elements which form a statically indeterminate spring system. The bearing, which generally has several load-transmitting rolling elements, is itself also a high-grade statically indeterminate spring system, which is also characterised by the strongly non-linear spring behaviour of the rolling elements:
- rolling elements behave in a unidirectional fashion, i.e. they only transmit compressive and not tensile forces
- the compressive load-deflection relationship within the rolling contact is non-linear

- the behaviour of ball bearings (compared to roller bearings) is more complicated due to the fact the point-shaped contact areas may shift under load and the direction of force transmission also changes under load

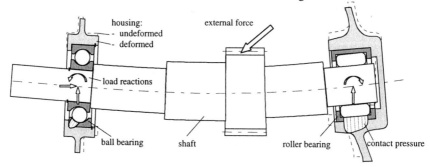

Figure 2   Deformed system comprising of bearings, shaft and housing

In the past, the calculation methods for these demanding non-linear, statically indeterminate mechanisms were developed and improved in stages. They were usually based on analytical foundations [1], [2]. At INA, these approaches have been converted into a sophisticated calculation program for rolling bearing design known as **Bearinx**®. In the case of complex machine systems, **Bearinx**® determines the equilibrium of the statically indeterminate spring systems in a closed calculation (up to the equilibrium of a single rolling element), taking precisely into account the interaction between the individual components. However, if the geometrical structure of the bearing support is complicated, **Bearinx**® must assume that the bearing rings are rigidly supported for simplification purposes. In the vast majority of cases, this assumption is sufficient, but due to the increasing trend towards lightweight designs (in the automotive industry in particular), it is becoming more necessary to take the deflection of the bearing rings into account from the beginning. In such cases, the load distribution of the bearing may only be calculated accurately on the basis of numerical approaches, such as the finite element method (FEM).

In the finite element analysis (FEA) of rolling bearings, the modelling of the rolling contact is a core problem. A model using conventional continuum elements very quickly reaches the limits of the computing power of today's computers. In order to represent the curvature of the rolling element, the non-linearity of the Hertzian contact with the changing extent of the contact surface and the local stresses and deformations, very fine meshing is required, as can be seen in Figure 3 with the typical stress and deformation state in the linear contact with a cylindrical roller.

The FEA of one or more rolling bearings integrated into a complex machine system using continuum elements for rolling element representation can therefore no longer be realised with justifiable expenditure. It is necessary to find a model of the rolling element to represent behaviour with drastically reduced degrees of freedom.

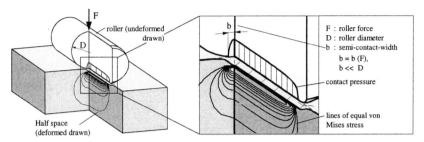

Figure 3   Stress and deformation state in linear contact

3. Model for representing rolling elements and rolling bearings

When developing a mathematical model for the description of the behaviour of the rolling element, the mechanical characteristics of the rolling element has certain advantages: it only transmits normal compressive forces, and these are distributed over a very small area in comparison with the usual geometric dimensions. One-dimensional spring stiffnesses can therefore be used to describe the local non-linear contact deflection. However, existing spring elements are not sufficiently suited, as will be shown later. The user elements provided for users by ABAQUS [3] offer a more suitable option for describing rolling element behaviour. As the user elements also include the proportion of raceways at the contact deflection, the local deformation is independent from the overall raceway deflection and the bearing rings can be modelled using conventional elements.

3.1 Fundamentals of the user-defined element (ABAQUS)

The mechanical behaviour of the ABAQUS user elements, which is generally non-linear, must be defined in a user subroutine. When calling the subroutine, ABAQUS provides the solution dependent nodal variables u^M, the nodal coordinates and the element properties to be defined in the input file as the essential information. Using this information, the contribution of the element at the residual vector of the structure F^N is defined in the subroutine:

$$F^N = F^N \left(u^M, \text{Coord.}, \text{Pr op.} \right)$$ (1)

and the contribution of the element at the Jacobian matrix K^{NM}:

$$K^{NM} = -\frac{\partial F^N}{\partial u^M}$$ (2)

1030

Once the mechanical behaviour has been coded in the user subroutine, only the nodes and the element properties need to be defined in the practical application of the user element as for conventional elements.

3.2 Model of the ball

3.2.1 Geometric relations

Although ball bearings appear to be simple machine elements, their internal geometry and therefore their motion and stress relations are quite complex. The most important geometric parameters are shown in Figure 4, with a profile representation of an angular-contact ball bearing (under thrust load).

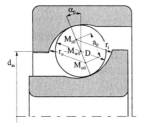 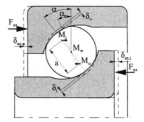

Figure 4   Geometric parameters of the unloaded and loaded ball

Two of these parameters are the ball diameter D and the pitch diameter d_m, on which the balls rotate around the bearing axis. The raceway groove curvature radii of the inner and outer ring, r_i and r_o, are slightly larger than the radius of the ball. The difference in curvature, which has been greatly magnified in Figure 4 for illustration purposes, is characterised by the osculation κ:

$$\kappa_i = r_i / D \ (\ > 0.5) \quad \text{or} \quad \kappa_o = r_o / D \ (\ > 0.5) \tag{3}$$

In the zero clearance and unloaded state, the ball makes contact with the inner and outer raceway at a single point. Under loading, the contact point increases to an elliptical contact surface. Under static loading (without considering centrifugal forces), the connection line of the contact points or the centres of the contact surfaces passes through the ball centre M_w and the centres of curvature of the raceway grooves of the inner and outer ring, M_i and M_o. The external load is transmitted from one bearing ring to the other along this line in the form of a compressive force. The angle between this line and the radial plane is called the contact angle α of the bearing. The distance a between the curvature centres is the criterion for whether the ball is loaded or unloaded. In the zero clearance and unloaded state, it is exactly a_0:

$$a_0 = r_i + r_o - D = (\kappa_i + \kappa_o - 1) \cdot D \tag{4}$$

In the loaded state, an elastic deflection δ_i or δ_o arises in the contact points from the ball to the raceway. The distance of the curvature centres from the raceway radii a is now greater than in the unloaded state. The relation between the sum of the two deflections and the distances a and a_0 is as follows:

$$\delta = \delta_i + \delta_o = a - a_0 \qquad (5)$$

3.2.2 Elasticity relations

From a purely geometric point of view, the direction of the ball force can be derived from the positions of the bearing rings in relation to each other. The theory of elasticity is required to determine the extent of the force. The classical solution for the local stress and deformation of two non-conforming elastic bodies which are pressed together was established by Hertz [5]. For steel bodies, the mutual approach of both bodies δ (in mm) at the contact point is dependent on the force F (in N):

$$\delta = 2.79 \cdot 10^{-4} \cdot \delta * \cdot \sqrt[3]{\Sigma\rho \cdot F^2} \qquad (6)$$

with: $\Sigma\rho$: sum of curvatures in the principal curvature planes

$\delta *$: Hertzian parameter (function of curvatures to be taken from table work)

3.2.3 User element ball

The position of the curvature centres of the raceway grooves and therefore the ball force F and the contact angle α can be determined from the position of two points located on the raceway grooves, provided that curvature is constant. To represent the ball behaviour, a 4-node element with two nodes on each of the raceway grooves of the inner or outer ring is defined, see Figure 5. Each of the nodes has three translatory

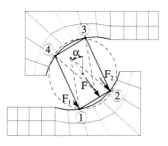

Figure 5 User Element Ball

degrees of freedom (in the global coordinate system). In addition to the geometric parameters described above, the user must also define the bearing axis as an element property. The ball force is divided into two statically-equivalent proportions on the two nodes of each raceway groove.

3.2.4 Verification

For verification of the user element ball, a very finely-meshed continuum FE model of a sector of a ball bearing was generated. For loading, the inner and outer rings of the bearing are displaced axially in relation to each other. The ball force and the contact angle are indicated by the axial displacement. A unique matching of the characteristic curves of the user element and continuum FE model can be recognised, see Figure 6.

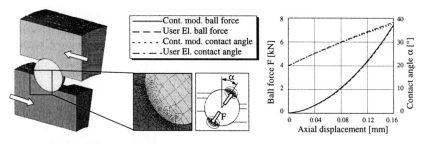

Figure 6   Continuum FE model of the ball and comparison of results

3.2.5 Limits of the model

Application of the user element is only permitted if the requirement of a constant raceway groove curvature is met. This requirement is not met in the case of thin-walled, insufficiently supported bearing rings.

3.3 Model of the roller

The following explanations are limited to the special case of cylindrical rolling elements that are loaded in the transverse direction only, and which are not subject to axial forces.

3.3.1 Geometric relations

Due the cylindrical shape of the rollers, the contact angle of radial bearings remains constant at 0°. This considerably simplifies the geometric description of the roller bearing in comparison to the ball bearing, particularly under load. Figure 7 illustrates the most important roller bearing parameters for determining the load distribution.

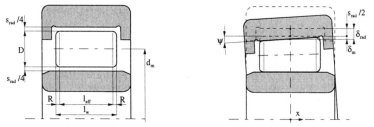

Figure 7   Geometric parameters of the unloaded and loaded roller

The effective roller length l_{eff} is equal to the total length l_w minus the two edge radii R. Although the rollers appear to be perfectly cylindrical macroscopically (except for the edge radii), the contour is profiled slightly (on the order of a few μms) to avoid edge contact pressure on the roller which would considerably reduce the fatigue life of the bearing. The profile of real rollers can usually be approximated by special functions dependent on a small number of parameters. Using the example of an untilted loaded roller, Figure

1033

8 illustrates the effect profiling has on the contact pressure near the roller edge.

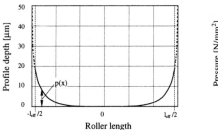 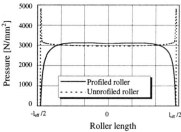

Figure 8   Roller profile and influence on load distribution

From the radial displacement δ_{rad} and the tilting ψ of the bearing rings towards each other, it is possible to specify the deflection of the roller δ depending on the axial coordinate x as follows, taking into account the bearing clearance s_{rad} and the profiling p(x):

$$\delta(x) = \delta_{rad} - s_{rad}/2 - \psi \cdot x - 2 \cdot p(x) \tag{7}$$

3.3.2 Elasticity relations

The literature provides different load-deflection relationships for the mutual approach δ (in mm) of a cylindrical roller of the finite length l (in mm) that is pressed against the infinite half space with the force F (in N). Some of these are specified on an analytical basis, and others are specified on an empirical basis. In an extensive parameter study using the FE model of an (unprofiled) roller of different lengths and diameters introduced in Section 3.3.4, the load-deflection relation specified in [2] proved to be the most suitable:

$$\delta = 3.84 \cdot 10^{-5} \cdot \frac{F^{0.9}}{l^{0.8}} \tag{8}$$

3.3.3 User element roller

The roller is discretized into a limited number of n laminae parallel to the radial plane of the bearing, see Figure 9. Each of these has 1/n of the contact stiffness of the whole roller and is connected with one node to the raceway of the inner and outer ring. The total number of nodes on the user element roller is therefore $(2 \cdot n)$. The deflection of the laminae can be determined from the radial displacements of the related nodes (taking into account radial clearance and profiling). Each of the laminae can only transmit normal forces in the direction of their original connection line, i.e. in the local coordinate system of the roller, the nodes in the directions x and z are free of forces. The load-transmitting length l that influences the deflection relation (8) is determined from the number of loaded laminae.

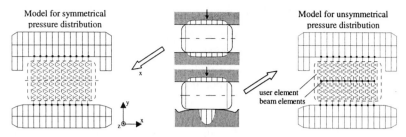

Figure 9   User element roller (two variants)

In cases where the roller shows an unsymmetrical pressure distribution of the two line contacts with regard to its axis (due, for example, to an additional profiling of one of the two raceways), the model must be modified slightly: the laminae which are independent of each other must be interlinked on the middle line by means of beam elements in order to represent both contacts of a lamina separately. Because of the highly increased computational effort use of the second model is only recommended when absolutely necessary.

3.3.4 Verification

To verify the user element of the roller, a half-symmetrical continuum FE model of the roller and of the infinite half space was created and loaded in a tilted way. The load was deliberately selected to be so extremely high that the profile was no longer sufficient to avoid edge contact pressure. With the exception of the edge contact pressure, the pressures are well matched, see Figure 10.

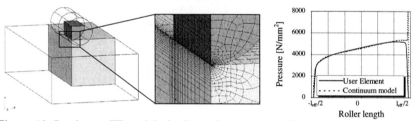

Figure 10 Continuum FE model of roller and comparison of results

3.3.5 Limits of the model

The above comparison calculation shows the limits of the laminae model: any edge contact pressures that occur cannot be determined directly. However, as these are limited to a very small area, the effects on the global load distribution within the bearing remain insignificant. Together with the roller force and moment which were sufficiently precisely determined in the analysis of the whole bearing, the continuum FE model of a single rolling element can then be used to examine rolling elements which could be critically loaded in relation to edge contact pressure.

3.4 Entire ball and roller bearing

In the analysis of shaft systems with multiple bearings, there may be only one bearing which is of particular interest. Also, due to the thick-walled design of the bearing's support structure, it may be possible to neglect local deformations of that structure. For the efficient bearing modelling of such cases, two user elements were developed to represent the behaviour of an *entire* ball or roller bearing, assuming that the bearing rings are rigid. To describe the positions of both bearing rings, a middle node with 3 translatorial and 3 rotational degrees of freedom were introduced for each ring. In the FE model, these are then coupled to the bearing support in the housing and the shaft using enforced constraints. After carrying out a few coordinate transformations, it is possible to determine the load distribution in the entire bearing from the displacements and rotations of both nodes of the bearing element using the relations (3) to (8) outlined above. In order to describe the element properties, specifications concerning the number of rows i, number of rolling elements per row z and the pitch distance between the rows t are required in addition to the geometric parameters of a single rolling element.

3.5 Overview of elements generated

Bearing support	Elastic		Rigid	
User element	Ball	Roller	Ball bearing	Roller bearing
Number of nodes	4	2·n	2	2
Degrees of freedom per node	u_x, u_y, u_z	u_x, u_y, u_z	$u_x, u_y, u_z,$ $\varphi_x, \varphi_y, \varphi_z$	$u_x, u_y, u_z,$ $\varphi_x, \varphi_y, \varphi_z$
Number of degrees of freedom of the element	12	6·n	12	12
Element properties	- x_j, j=1,6 - D - d_m - κ_i - κ_o - s - α_0	- x_j, j=1,6 - D - d_m - l_w - l_{eff} - a_j, j=1,3 - s	- x_j, j=1,6 - i - z - t - D - d_m - κ_i - κ_o - s - α_0	- x_j, j=1,6 - i - z - t - D - d_m - l_w - l_{eff} - a_j, j=1,3 - s
Output variables	- F - α	- F_j, j=1,n	- F_j, j=1,(i·z) - α_j, j=1,(i·z)	- F_j, j=1,(i·z·n)

Table 1   Overview of the four elements

Common symbols	Ball / ball bearing	Roller / roller bearing
x_j : Axis parameters i : No. of rows z : No. of rolling elements t : pitch distance D : Ball/roller diameter d_m : Pitch diameter s : Bearing clearance F : Ball/lamina force	κ_i : Inner osculation κ_o : Outer osculation α_0 : Nominal contact angle α : Operational contact angle	n : No. of laminae l_w : Overall roller length l_{eff} : Effective roller length a_j : Profile parameters

Table 2   Symbols

Table 1 provides an overview of the four user elements developed, while Table 2 summarises the element property symbols used.

4. Evaluation of the load distribution: the bearing fatigue life

Out of the load distribution of the relevant bearing, analytical relations can be used in a subsequent calculation to work out the distribution of Hertzian pressure on the contact surface and the subsurface stresses [5]. In a moving bearing, this loading of the rolling element and raceway occurs cyclically. The resulting fatigue life, which is generally limited, can also be obtained by evaluating the bearing load distribution. The new fatigue life theory according to Ioannides and Harris [2], [4] as an expansion of the classical Lundberg-Palmgren-theory is the latest state of research in this field. To carry out a fatigue life calculation, the lubrication and contamination state, as well as the material properties of the bearing, are required in addition to the bearing load distribution.

5. Application example

The application of the user element "roller" and the knowledge gained by its use with the influence of the elastic structural support on the bearing load distribution are to be illustrated using the example of a shaft bearing of a private car gear. The FE model of the gearbox, which is made of a light metal alloy, was provided by the customer and expanded to include the bearing, a drawn cup roller bearing, see Figure 11.

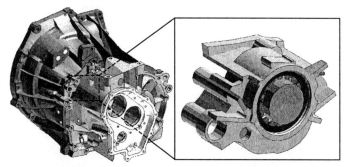

Figure 11   FE model of the gear and the bearing support (mesh unposted)

Two essential deformation proportions of the bearing supporting structure arise under load: an ovalisation in the circumferential direction and a conicity in the axial direction, see Figure 12.

In comparison to rigid support, the deformation has the following effects on load distribution: due to the ovalisation of the hub, the load zone stretches over a greater circumferential angle area, the distribution of the roller forces is more balanced overall and the maximum roller force is reduced by 23%, see

1037

Figure 13. This positive influence of deformation in the circumferential direction is opposed by the negative influence of deformation in the axial direction: the lack of support in the frontal area of the bearing support causes the rollers to be loaded in a more tilted way. These two deformation influences almost compensate for each other with regard to the maximum occurring Hertzian pressure: in comparison to the calculation with rigid support, this is reduced only slightly by 5%, while fatigue life extends by 20%.

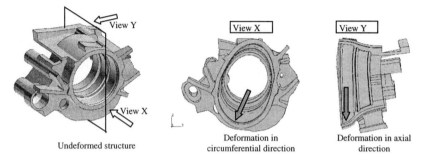

Figure 12 Deformations of the bearing supporting structure under load

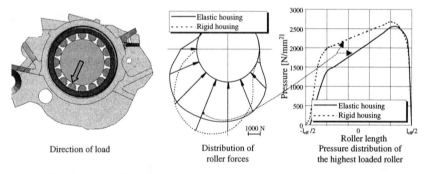

Figure 13 Load distribution of the bearing

Using this calculation, it was possible to show firstly that the hub deformation regarded as a whole in this actual application example did not cause any essential changes with reference to the conventional analysis assuming rigidly supported bearing rings, and secondly, that there is still potential for optimisation (better support over the axial length).

Generally speaking, it is not possible to say in advance with certainty whether the elastic deformation of the bearing support structure has a significant influence on the bearing load distribution, and therefore on the fatigue life of the bearing, and whether this influence is positive or negative overall. Conclusions can only be drawn in individual cases using an FE analysis with representation of the bearing supporting structure.

6. Conclusion

The user element of the rolling element displayed represents the mechanical behaviour of the contact between rolling element and raceway with great accuracy, and simultaneously with the minimum degrees of freedom. In comparison to a model with conventional continuum elements, a huge amount of calculation time is saved, and analyses of bearing load distribution in the elastic environment are enabled using current computer power. A more reliable design and fatigue life prediction of the bearing where there is significant local deformation of the bearing support structure can be carried out using a FE analysis with the user element. The element description requires little modelling expense in comparison to an explicit definition of the characteristic stiffness curves of the rolling element, as can be carried out using only a few geometric parameters.

This user-friendly type of element definition in the practical application also allows calculation engineers from non-bearing industries to integrate the user element of the rolling bearing into their models as a kind of black box, without having to be familiar with the fundamentals of rolling element behaviour. This allows them to make a more realistic analysis of the stress and deformation of their products adjacent to the bearing.

The method outlined above is a consistent further development in the course of integral solutions in the CAE process and the necessity of representing the system behaviour of products as early as possible in the development stages (keyword: virtual product). In the next step, it is planned to support the user element in the pre-processor environment also, in order to improve handling and achieve complete integration into the CAE process.

The advantages of the rolling element model outlined can be summarised as follows:
- improved quality of calculation and design
- saving of both manpower and calculation time.

7. References

1. ESCHMANN, HASBARGEN, WEIGAND - Die Wälzlagerpraxis, Oldenbourg Verlag GmbH, 1978
2. HARRIS, T. A. - Rolling Bearing Analysis, John Wiley & Sons, Inc., 1991
3. HIBBIT, KARLSSON & SORENSEN - ABAQUS Version 5.5, Hibbit, Karlsson & Sorensen Ltd., 1995
4. IOANNIDES, E. / BESWICK, J. M. - Moderne Wälzlagertechnik, Vogel Buchverlag, 1991
5. JOHNSON, K. L. - Contact Mechanics, Cambridge University Press, 1985

REVIEW OF METHODS FOR GENERATING ELEMENTS FOR STRESS SINGULARITY PROBLEMS INVOLVING ELASTO-PLASTIC MATERIALS

A. D. Nurse[1]

1 SUMMARY

The paper will provide a brief review of the techniques developed over more than twenty years for analysing singular problems through optimal positioning of mid-side and/or corner nodes in an eight-noded isoparametric element. Emphasis will be made on applications involving elasto-plastic non-linear material behaviour where nine-noded Lagrangian elements are often preferred. A new approach is then described that attempts to rectify some of the problems that exist, and is applied to a situation involving singular stress fields in a bimaterial joint. Numerical solutions compare favourably with data collected experimentally.

2 INTRODUCTION

Problems involving a singular stress field include the analysis of a crack tip, point contact between two bodies, and material interfaces in bonded joints. In each of these cases the order of the singularity varies as $r^{\lambda-1}$ where λ lies in the range $0 \leq \lambda < 1$. The finite element method, using displacement-based elements, is an established tool for analyses of all these problems. Since assumptions are made based on the resulting displacements, defined in terms of polynomial functions over elements of finite size, it is not possible to obtain exact representations of the behaviour in the region of a singularity. To overcome this problem it is usual practice to employ substantial mesh refinement at the point of the singularity though this is expensive in terms of computer power and mesh preparation time.

---

[1] Lecturer, Department of Mechanical Engineering, Loughborough University, Loughborough, LE11 3TU, U.K. a.d.nurse@lboro.ac.uk

Alternative strategies to mesh refinement are: (i) user-defined elements that model the constitutive behaviour of the problem under consideration, or (ii) selective re-positioning of mid-side nodes to generate singular stresses at a corner node. For the first strategy one needs the generalised closed-form solution of the problem under analysis, whilst, for the second one only needs the order of the singularity. The advantage of the first approach is that the angular distribution of stress about the singularity is modelled as well as the radial component. In (ii) only the radial variation in stress is modelled thereby the accuracy of the angular distribution is dependent on the number of elements employed in the immediate vicinity of the singular point. One could argue that (i) is the preferred approach, though much effort is required to model the element properties and the particular element can only be used for the specific problem in mind. Also, if elasto-plastic properties of engineering materials are to be included in the analysis then the determination of the generalised closed-form solution is extremely difficult and has only been successfully applied to a very limited number of cases. Selective re-positioning of mid-side nodes, by contrast, is a technique that is comparatively simple and readily transportable between the different types of singularity problem. The loss of accuracy with respect to the angular distribution seems a small sacrifice compared to the benefits from the simplicity of the approach.

The paper will describe some of the methods for analysing stress singularities that can be effected using displacement-based isoparametric elements as well as introducing a new approach. Consideration will be given to analyses of materials that undergo power-law hardening. A case study is described involving a bimaterial joint, comprising halves of epoxy resin and aluminium alloy, in a three-point-bend test that undergoes yielding at the interface. This configuration has singularities at the free edge of a material interface and at the point load applied to the interface edge.

3 HISTORICAL BACKGROUND

Original attempts to analyse the stress fields at singularities began with fracture mechanics' studies involving substantial mesh refinement (e.g. Ref.1). To overcome the expense in data preparation effort and computational time required, investigators then sought to include the singularity in new elements based on the exact solution to the crack tip stress field (e.g. Ref.2). The earliest approach to generating a singularity at a corner by node repositioning involved a four-noded isoparametric element with one of the sides reduced to zero length to give a triangle [3]. This gives the required singularity of the form $r^{-0.5}$ in the $[B]$ (or strain-displacement) matrix.

It was first shown by Henshell & Shaw [4], and later by Barsoum [5] that crack tip problems can be studied using the so-called "quarter-point" element. The singularity $(r^{-0.5})$ can be modelled exactly by repositioning the mid–side nodes of a quadratic isoparametric element to be at the quarter point on the side of the element. This approach is described in some detail in the next section.

Pian et al [6] showed that stress intensity factors can be determined accurately using a small number of degrees of freedom if the assumed stresses in the crack tip elements are of the order of $r^{-0.5}$. Tong and Pian [7] have shown that the convergence rates of finite element methods for singular problems are dominated by the nature of the singularity in the assumed functions. So-called "high-accuracy" elements with higher-order polynomial interpolation functions do not improve the rates of convergence.

For fracture problems involving yielding in the vicinity of the crack tip a r^{-1} singularity is predicted for strains in elastic perfectly–plastic material [8,9]. Thus, a hybrid finite-element model that exhibits this singularity would be a possible choice for the analysis of yielding at a crack tip. Levy et al [10] used a special element derived for elastic perfectly-plastic material exhibiting the strain singularity of order r^{-1}. A hybrid displacement model for finite element analyses has been developed for the analysis of both elastic and elasto-plastic stress fields around cracks (e.g. Ref.11).

Recent work has seen the development of elements for the study of singularities at the interface of two materials where the value of λ is dependent on the elastic properties of the materials involved. In this case it has been shown that they at best modelled *approximately* by selective repositioning of the mid-side nodes involved [12-21]. The techniques effectively employ a curve-fitting routine that tries to replicate the slope of the strain near the singularity but except for the quarter-point element do not actually produce an infinite strain at the corner node. Though the case of the bimaterial interface joint has recently has been improved somewhat by this author [21] through collapsing a rectangular quadratic element into a triangular element. This gives the requirement for singular stresses at one corner of the element. It only, however, gives the correct singularity *order* for some of the terms in the [B] matrix.

Concluding, the review reveals that the simplicity of the approach in moving mid-side and/or corner nodes to optimal positions has gained widespread appeal. However, the lack of a satisfactory approach to the generation of singular elements, other than for elastic or elasto-plastic fracture mechanics, gives scope for further development. The area of elastic and/or elasto-plastic analysis of bimaterial interface joints is one that requires further attention and is the subject of study at the end of this paper.

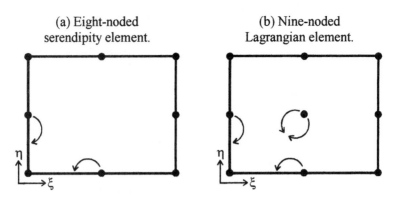

(a) Eight-noded serendipity element.

(b) Nine-noded Lagrangian element.

Figure 1: Movement of nodes in isoparametric elements.

4 GENERAL METHODOLOGY

The basic process for generating a crack tip element is now described. The eight-noded isoparametric 2-D element, originating in xy-space, is transformed into a square, in $\xi\eta$-space, with vertices at $(\pm 1, \pm 1)$. The mid-side nodes on two sides are moved away from their usual positions as shown in Fig.1(a). The central node of the Lagrangian element is also moved as in Fig.1(b). To simplify the mathematics consideration will be made to one of the sides denoted by $\eta = -1$. The shape functions for nodes along this edge take the form:

$$N_1 = -0.5\xi(1-\xi), \quad N_2 = (1-\xi^2), \quad N_3 = 0.5\xi(1+\xi) \qquad (1)$$

The x co-ordinate may be represented by the following:

$$x = \sum_{i=1}^{3} N_i x_i = N_1 x_1 + N_2 x_2 + N_3 x_3 \qquad (2)$$

Denoting the physical length of the element edge as L then the nodal positions are denoted by:

$$x_1 = 0, \quad x_2 = \alpha L, \quad x_3 = L \qquad (3)$$

where α normally has the value 0.5 for the standard element. Solving for x in terms of ξ gives:

$$x = (1-\xi^2)\alpha L + 0.5\xi(1+\xi)L \qquad (4)$$

Similarly, the displacement u in the x direction is given by:

$$u = \sum_{i=1}^{3} N_i u_i = -0.5\xi(1-\xi)u_1 + (1-\xi^2)u_2 + 0.5\xi(1+\xi)u_3 \qquad (5)$$

Differentiating equation (4) gives:

$$d\xi / dx = 1/(1 + 2(1-\alpha)\xi) \qquad (6)$$

and is singular under the conditions $\xi=-1$, $\alpha=0.25$. In other words, the singularity is created at the $x=0$ end of the element by moving the mid-side node to the quarter-point position. This yields the following expression for displacement:

$$u = u_1\left(1 - 3\sqrt{x/L} + 2x/L\right) + u_2\left(4\sqrt{x/L} - 4x/L\right) + u_3\left(-\sqrt{x/L} + 2x/L\right) \qquad (7)$$

The strain is seen to be of the form $\varepsilon = \varepsilon(1/\sqrt{x})$ giving the required order of the singularity for the crack.

Henshell and Shaw [4] show that the traction-free conditions for the face of a crack may not be satisfied if quarter-point elements are used. However, the effects of this inaccuracy quickly disappear after one element spacing or the "Saint-Venant" distance.

5 ELASTO-PLASTIC ISSUES

For fracture mechanics' problems involving yielding in the vicinity of the crack tip the elasto-plastic state surrounding the crack tip must be known. Unfortunately, other than for Mode III crack tip deformation, there does not exist closed-form solutions for the deformation field in the presence of plastic yielding. Consequently, finite element approaches to the analysis of elasto-plastic problems have grown in importance in recent years.

The well-known works of Rice and Rosengren [8], and Hutchinson [9] predict the order of the singularity for strain that could be used to develop singular elements. If the material obeys the power-law hardening rule:

$$\bar{\varepsilon} = \bar{\sigma}/E + (H\bar{\sigma})^{1/n} \qquad (8)$$

where:

$$\bar{\sigma}, \bar{\varepsilon} = \text{equivalent stress, strain}$$
$$E, H = \text{material constants}$$
$$n = \text{hardening parameter}$$

then the strain singularity is given by:

1045

$$\varepsilon \to r^{-n/(1+n)} E(\theta) \qquad (9)$$

The stresses have a singularity order given by:

$$\sigma \to r^{-1/(1+n)} \Sigma(\theta) \qquad (10)$$

noting that the singularity for stress disappears for perfectly-plastic material (i.e. $n \to \infty$). In equations (9) and (10) $E(\theta)$ and $\Sigma(\theta)$ are functions of the angle θ that need not be known. One could therefore formulate the element's displacements to have the appropriate strain singularity in one corner of the element according to equation (9). The stresses are governed by user input material data.

Bimaterial joints exhibit a singularity at the ends of the interface given by $r^{\lambda-1}$ where λ is found from the elastic constants of the constituent materials and the joint geometry [26]. The strain singularity can be fashioned in a similar way to the fracture mechanics' problem. Being consistent with the analyses by Rosengren and Rice [8], and Hutchinson [9], a good approximation to the strain singularity would be given by:

$$\varepsilon \to r^{-2n(\lambda-1)/(1+n)} E(\theta) \qquad (11)$$

where n is now the hardening parameter of the constituent material.

6 QUADRATURE RULES

For numerically-integrated elements, the quadrature rule sufficient to provide exact integrals of all terms in the stiffness matrix is known as "full integration" providing the element is undistorted. The same rule will not integrate all terms of the stiffness matrix exactly if sides of the element are curved or if side nodes are offset from the mid-points. For these cases, terms in the Jacobian matrix $[J]$ are not constant. Even if mid-side nodes are not centered, two-point quadrature for quadratic elements is still considered "full integration".

There is some disagreement as to what is the best rule for quadrature in determining the stiffness matrix in singular elements. A lower-order quadrature rule, called "reduced integration", may be desirable for the number of sampling points is reduced and it tends to soften the element. Softening comes about because certain higher-order polynomial terms do not contribute to the strain energy, i.e. some of the more complicated displacement modes offer less resistance to deformation. Use of a higher-order quadrature rule, or "fine integration", may produce exact results but tends to stiffen the element.

In elasto-plastic analyses, the Lagrangian nine-noded element is often preferred as its use can avoid certain numerical problems [22, 23]. The standard displacement-based serendipity and Lagrangian isoparametric quadrilateral elements loose performance as the material enters the plastic (incompressible) range [24]. These problems can be alleviated by reduced integration and one approach to this is the so-called "B-bar" method of Hughes [25]. The deviatoric components of the strain-displacement matrix $[B]$ are solved at regular quadrature points (3×3) whilst the volumetric components are solved at reduced quadrature points (2×2). The two components are then brought together at regular quadrature points to give the desired matrix. It should be noted that a spurious pressure mode arises in the four-noded quadrilateral element if this selective integration scheme is used. The recommended nine-noded element approach alleviates such problems.

7 A NEW APPROACH

In an attempt to rectify the issue that only crack tip singularities can be modelled correctly by nodal repositioning, i.e. the element has a singularity at a corner *and* is of the correct order, a new approach is postulated. The approach conveniently produces a "super" element with similar properties to that of the nine-noded Lagrangian element that as mentioned earlier is useful for elasto-plastic analyses.

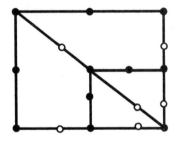
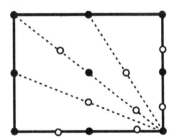

(a) Joining of four elements with movement of mid-side nodes to produce the desired singularity.

(b) Constraining of mid-side nodes to condense number of degrees of freedom.

Figure 2: Development of a Lagrangian-type "super" element from four isoparametric serendipity elements to give approximately the correct singularity in one corner.

Two eight-noded and two six-noded elements are joined as shown in Fig.2(a) and the nodes highlighted are moved to optimal positions. The triangular elements used can be quarter-point elements as they are required to have a singularity in the strain terms for one corner. Some of the mid-side nodes in the quadrilateral elements are moved to positions that give the desired singularity *order* using one of the published methods. Therefore, the actual strain singularity occurs in the triangles whilst the correct strain gradient occurs in the quadrilaterals. Intuitively, the triangles are made as small as possible, though it has yet to be decided on an appropriate sizing factor. In the study that follows their size was made to be approximately one quarter that of the quadrilaterals.

The nodes highlighted in Fig.2(b) are then constrained to have displacements that are functions, according to the desired singularity order, of the nodal displacements at the nodes on either side on a radial line from the singular point. The condensing of the degrees of freedom gives the element system the same number of degrees of freedom as a nine-node Lagrangian element.

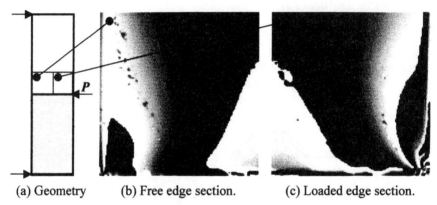

(a) Geometry   (b) Free edge section.   (c) Loaded edge section.

Figure 3: Automated photoelastic analyses of a bimaterial
epoxy resin/aluminium alloy three-point-bend specimen showing wrapped
contours of principal stress difference (*P*=350N, black=0MPa, white=5MPa).

8 SAMPLE ANALYSIS

Finally, by way of example, an elasto-plastic analysis was performed on a bimaterial three-point-bend specimen where a singularity exists at both ends of the interface. One end of the interface is subject to a point load. Under appreciable load an elasto-plastic field will exist at the singular points. For comparison, the specimen has been the subject of an experimental study using automated photoelasticity to collect data [27]. An epoxy resin half was bonded to an aluminium alloy half to form the bimaterial specimen of dimensions 200mm by 50mm by 6.35mm, Fig.3(a). Stresses in the epoxy half can be determined using the automated photoelasticity. The loads applied were 150N and then 350N. The resulting photoelastic phase maps

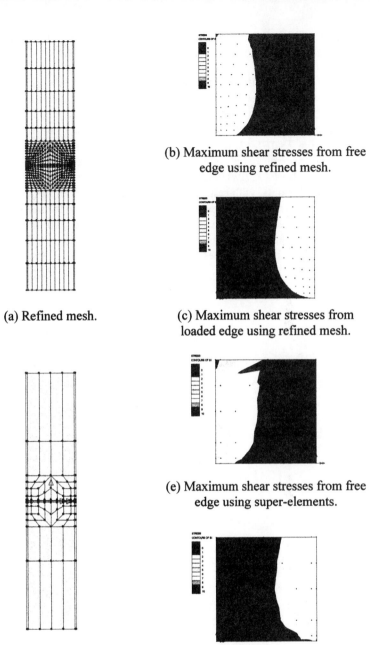

(a) Refined mesh.

(b) Maximum shear stresses from free edge using refined mesh.

(c) Maximum shear stresses from loaded edge using refined mesh.

(d) Meshing using super-elements.

(e) Maximum shear stresses from free edge using super-elements.

(f) Maximum shear stresses from free edge using super-elements.

Figure 4: Comparison of elasto-plastic finite element analyses of a bimaterial epoxy/aluminium alloy three-point-bend specimen (*P*=350N).

for the principal stress difference are shown in Figs.3(b) and (c) representing two different sections of the epoxy resin half.

Two forms of finite element analysis were performed. The first used a refined mesh with about 500 standard eight-noded elements. The other used the super-element sized at about 2mm, positioned at both ends of the interface, surrounded by about 100 standard elements. The strain singular is found using $\lambda=0.75$ [26], in equation (11). All finite element calculations were performed using the proprietary software package "LUSAS" (distributed by *FEA Ltd*). Some comparative results can be seen in Fig.4. Normalised results, with respect to load, are presented for both load cases in Fig.5 that are graphs of experimental data and nodal stresses using super-elements plotted for a radial line from each of the two singular points. There is agreement between the results thereby justifying the use of the super-element, but also, the effects of non-proportional loading can be seen.

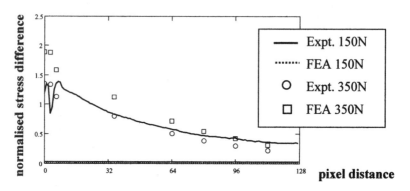

(a) Results on a radial line from the free edge.

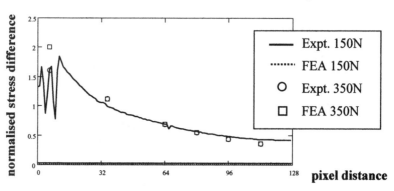

(b) Results on a radial line from the loaded edge.

Figure 5: Finite element and photoelastic results from analyses of a bimaterial epoxy/aluminium alloy three-point-bend specimen.

8 REFERENCES

1   CHAN, S.K., TUBA, I.S., WILSON, W.K., - On finite element method in linear fracture mechanics. Engng. Fract. Mechs., Vol.2, pp1-17,1970.

2   TRACEY, D.M., - Finite elements for determination of crack tip elastic stress intensity factors. Engng. Fract. Mechs., Vol.3, pp255-266, 1971.

3   LEVY, N., MARCAL, P.V., OSTERGREN, W.J, RICE, J.R., - Small scale yielding near a crack tip in plane strain. A Finite Element Analysis. Brown Univ. Tech Report for NASA, NGL-40-002-080/1.

4   HENSHELL, R.D., SHAW, K.G., - Crack tip finite elements are unnecessary. Int. J. Numer. Meths. Engng. Vol.19, pp495-507, 1975.

5   BARSOUM, R.S., - On the use of isoparametric elements in linear fracture mechanics. Int. J. Numer. Meths. Engng., Vol.10, pp25-37, 1976.

6   PIAN, T.H.H., TONG, P., LUK, C.H., - Elastic crack analysis by a finite element hybrid method. Proc. 3rd Int. Conf. Matrix Methods in Structural Mechs., Wright-Patterson Air Force Base, Ohio, 1971.

7   TONG, P., PIAN, T.H.H., - On the convergence of the finite element method for problems with singularity. Int. J. Solids Structures, Vol.9, pp.313-321, 1973.

8   RICE, J.R., ROSENGREN, G.R., - Plane Strain Deformation Near Crack Tip in a Power Law Hardening Material. J. Mech. Phys. Solids, Vol.16, pp1-12, 1968.

9   HUTCHINSON, J.W., - Singular Behaviour at the End of Tensile Crack in a Hardening Material. J. Mech. Phys. Solids, Vol.16, pp13-31, 1968

10 LEVY, N., MARCAL, P.V., - Three-dimensional elastic plastic stress analysis for fracture mechanics. Heavy Section Steel Technology Program, Fifth Annual Meeting, Oak Ridge National Laboratory, 1971.

11 ATLURI, S.N., KOBAYASHI, A.S., NAKAGAKI, M., - An assumed displacement hybrid finite element model for linear fracture mechanics. Int. J. Fract., Vol.11, pp257-271, 1975.

12 AKIN, J.E., - The generation of elements with singularities. Int. J. Numer. Meths. Engng., Vol.10, pp1249-1259, 1976.

13 TRACEY, D.M., COOK, T.S., - Analysis of power type singularities using finite elements. Int. J. Numer. Meths. Engng., Vol.11, pp1225-1233, 1977.

14 STERN, M., - Families of consistent conforming elements with singular derivative fields. Int. J. Numer. Meths. Engng., Vol.14, pp409-421, 1979.

15 HUGHES, T.J.R., AKIN, J.E., - Techniques for developing special finite shape functions with particular reference to singularities. Int. J. Numer. Meths. Engng., Vol.15, pp733-751, 1980.

16 STAAB, G.H., - A variable power singular element for analysis of fracture mechanics problems. Comput. Structures, Vol.17, pp449-457, 1983.

17 ABDI, R.E., VALENTIN, G.V., - Isoparametric elements for a crack normal to the interface between two bonded layers. Comput. Structures, Vol.33, pp241-248, 1989.

18 LIM, W.-K., KIM, S.-C.,- Further study to obtain a variable power singularity using quadratic isoparametric elements. Engng. Fract. Mechs., Vol.47, pp223-228, 1994.

19 LIM, W.-K., LEE, C.-S., - Evaluation of stress intensity factors for a crack normal to bimaterial interface using isoparametric finite elements. Engng. Fract. Mechs. Vol.52, pp65-70, 1995.

20 LEE, J., GAO, H., -A hybrid finite element analysis of interface cracks. Int. J. Numer. Meths. Engng., Vol.38, pp.2465-2482, 1995.

21 EKMAN, M.J., NURSE, A.D., - Combined experimental/numerical elasto-plastic analysis of a composite joint. Proc. 5th ICCE Int. Conf. Composites Engng., , Ed. HUI, D., Las Vegas, USA, July 5-11, pp675-676, 1998.

22 MALKUS, D.S., HUGHES, T.J.R., - Mixed finite element methods – Reduced and selective integration techniques: A unification of concepts. Comp. Meths. Appl. Mechs. Engng., Vol.15, pp63-81, 1978

23 SHIH, C.F., ASARO, R.J., - Elastic-plastic analysis of cracks on bimaterial interfaces: Part I – Small scale yielding. ASME J. Appl. Mechs., Vol.55, pp299-316, 1988.

24 NAGTEGAAL, J.C., PARKS, D.M., RICE, J.R., - On numerically accurate finite element solutions in the fully plastic range. Comp. Meths. Appl. Mechs. Engng., Vol.4, pp153-178, 1974.

25 HUGHES, T.J.R., - Generalization of selective integration procedures to anisotropic and nonlinear materials. Int. J. Numer. Meths. Engng., Vol.15, pp1413-1418, 1980.

26 BOGY, D.B., - Two edge-bonded elastic wedges of different materials and wedge angles under surface tractions. J. Appl. Mechs. Vol.38, pp377-386, 1971.

27 EKMAN, M.J., NURSE, A.D., - Absolute determination of the isochromatic parameter by load-stepping photoelasticity. Expt. Mechs., Vol.38, pp189-195, 1998.

FINITE ELEMENT STRESS DISTRIBUTIONS AT
CRACK TIPS
Colin Dickie

Associate of the PTP in conjunction with Strathclyde
University and the National Engineering Laboratory

ABSTRACT

This paper presents a method that uses finite element models to
calculate stress distributions associated with large strains when materials
exceed their yield points. The stress patterns which were being investigated
were those local to the tips of cracks. At these regions stresses are known to
be high and generally exceed the material's yield point. It is common practice
in crack problems to obtain results by using fracture mechanics techniques.
However, such techniques developed from fracture mechanics theory are often
used to analyse crack problems on the basis of linear elastic assumptions. The
method presented in this paper is not restricted to this range. This paper
compares results obtained from two classic fracture mechanics problems using
two types of finite element. A model for each problem was generated using
the two element types; eight-node isoparametric elements (PLANE 82) and
four-node structural elements (PLANE 42). Good agreement is found between
these elements for both models when the linear elastic fracture mechanics
approach is assumed. However, as stresses at crack tips become very high the
yield point of the material could easily be surpassed and linear elastic theory is
no longer valid for calculating the stress pattern. To illustrate this the two
problems with the two different types of finite element models are subjected to
conditions that ensure that yield has been surpassed Comparisons were then
made of their crack tip stress and strain distributions. ANSYS 5.4 was used
throughout this work.

1 INTRODUCTION

1053

Westergaard [1], Williams [2] and Irwin [3] are responsible for the traditional stress approach to fracture mechanics. They used stress functions as solutions to a certain class of plane stress or plane strain problems These equations are the first term of an infinite series and represent the linear elastic stress distribution in the neighbourhood of a crack tip.

Linear Elastic Fracture Mechanics (LEFM) developed from these equations and has found applications in many types of engineering problems that involve cracked components under service loads. It has even found its way into finite element analysis by way of special element types. A report by NAFEMS [4] utilises linear elastic theory to provide guidelines on obtaining important fracture parameters. This paper examines stress patterns at crack tips using finite element analysis and compares results between different theories and different finite elements.

The stresses predicted by LEFM tend towards infinity as the crack tip is approached and result in a singularity at the tip. However, infinite stresses are not produced in nature because the behaviour of most (metallic) materials changes after the stress reaches a certain value, named the yield stress. Until this point is reached the material has behaved in an elastic manner and has largely been dominated by its stress value which rises rapidly with strain. After yield the behaviour changes to one dominated by strain and the relationship with stress becomes non linear. The lack of any definite constitutive laws renders the calculation of stress from strain a major difficulty.

The constitutive relationship for homogeneous isotropic materials is experimentally derived and determines quantities that are used to calculate stress from strain. These quantities are constants in the linear elastic range and as such are readily available from textbooks for a wide range of materials. If only these constants are used in a finite element analysis then no matter what strain is encountered the stresses will be calculated linearly proportional to them.

A more representative approach is highlighted in this paper whereby an experimental stress-strain curve is first obtained for the material that is to be modelled. This curve should be the actual (or true) stress-strain curve where the change in cross sectional area is taken into account. The data obtained from this can then be used in the finite element model so allowing stresses beyond yield to be calculated.

2 THE APPROXIMATION OF STRESS DISTRIBUTIONS AT CRACK TIPS USING FINITE ELEMENT ANALYSIS

One of the important features of finite element analysis is that it allows engineers to obtain approximate distributions of stress throughout, or just at particular locations, in a problem. The tip of a crack is just such an area where finding the stress distribution is an important part of assessing the overall integrity of a component. The theoretical description of elastic stress distribution in the *neighbourhood* of a crack tip is given by the following equations

$$\sigma_{xx} = \frac{K_I}{\sqrt{2\pi r}} \cos\frac{\theta}{2}\left(1 - \sin\frac{\theta}{2}\sin\frac{3\theta}{2}\right)$$

$$\sigma_{yy} = \frac{K_I}{\sqrt{2\pi r}} \cos\frac{\theta}{2}\left(1 + \sin\frac{\theta}{2}\sin\frac{3\theta}{2}\right) \quad [1]$$

$$\tau_{xy} = \frac{K_I}{\sqrt{2\pi r}} \sin\frac{\theta}{2}\cos\frac{\theta}{2}\cos\frac{3\theta}{2}$$

r is the distance from the crack tip. x and y are Cartesian axes centred at the crack tip with x directed away from it. θ is the angle measured anticlockwise from the x-axis.

These linear elastic equations have been derived for a problem where the crack is being opened by symmetrically applied forces. This is called the opening mode or mode one in fracture mechanics terminology. Similar equations have been derived for when the crack is opened by opposing shearing forces in the plane of the crack and this is called mode two. Mode three is when the crack has applied out of plane forces trying to tear it open.

The important term in these equations is the stress intensity factor which is named after the mode with which it is associated, i.e. K_I for mode one, K_{II} for mode two and K_{III} for mode three. If two different cracks have the same value of stress intensity factor then they will have identical crack tip stress distributions according to this theory. Essentially the other terms in the equations describe the angular distribution of the stress and the level of stress as the crack tip is approached.

1055

When using a finite element model to simulate a cracked structure under load, care must be taken when modelling the tip of the crack. To account for the rapid rise in stresses in this region special crack tip elements have been designed and are now available for use in most commercial finite element packages. These elements were made so that they emulated the expected linear elastic response of a homogeneous isotropic material and are all designed to be used at the tip of the crack. These elements have been used successfully in Linear Elastic Fracture Mechanics, especially for calculating stress intensity factors. This in turn can then be applied to the LEFM equations of fracture mechanics to obtain the stress distribution around the tip of a crack.

These crack tip elements first described by Tracey [5] rapidly improved [6],[7] and eventually became part of a "family" of elements [8] derived from standard finite elements already in much use. These elements were all designed to combat the problem of the singularity at the crack tip predicted by the equations of linear elastic fracture mechanics, i.e. where the distance from the crack tip, r, tends to zero in equation [1]. Stress intensity factors can be expressed in terms of displacements [9] which, in finite element work, is the nodal displacements of elements. In ANSYS a linear extrapolation of the displacements nearest to the crack tip takes place after the crack tip node value has been normalised to zero, the result of which is inserted in the displacement equations and produces the desired stress intensity factor.

3 CRACK TIP MODELLING

What actually does happen at the tip of a cracked component that is under load? The prediction of stresses becoming unbounded at the tip of the crack is only a mathematical theory derived from linear elastic relations. To model it with the above mentioned crack tip elements would also restrict the analysis to this range. A more straightforward method would be to use simple structural elements at a crack tip. Although individually they are not designed to model a crack tip singularity, they should still produce accurate results if used in sufficient numbers. The use of many elements is no longer a problem with the computational power now available and any analyst should not hesitate to use as many elements as is necessary to produce detailed results. Accordingly, all the models that used these elements had 5000 of them in an area 3mm by 1.5mm around the tip of the crack.

Two classic fracture mechanics models were chosen to study; the single edge cracked plate in tension (SECT) and the centre cracked plate in tension (CCP). Diagrams showing their dimensions together with material properties, number and type of elements used are given in Appendices I and II.

The linear elastic results are compared with some of the finite element test cases by Pang and Leggatt [4] which in turn uses the theoretical results of Rooke and Cartwright [10] for comparison. The ANSYS commands KSCON and KCALC are used at the crack tip to produce the stress intensity factors for the two generated models.

The models using the 4-node 2D structural elements use the J integral [11] to calculate the stress intensity factor. In LEFM J and K_I are related through the equation $J = (1 - \upsilon^2) K_I^2 / E$. Many elements were used so that the energy in the cracked body is more accurately approximated. The J integral was calculated by writing a macro in the Ansys Parametric Design Language to approximate the famous integral given by J. Rice in [11].

4.1 Centre Cracked Plate, Linear Elastic Results

The results are presented in two tables, TABLE 1 comparing the stress intensity factor results obtained from the J integral and TABLE 2 comparing the stress intensity factor results obtained from the displacement substitution method employed by reference[4] with the displacement extrapolation method employed by the KCALC command of ANSYS for the PLANE 82 elements. In these tables the term K_R refers to the ratio K_I / K_O.

	REFERENCES		ELEMENT TYPES	
	10	4	PLANE 42	PLANE 82
J	2.4247	2.4334	2.4559	2.4571
K_I	742.66	744.00	747.43	747.61
K_R	1.325	1.327	1.333	1.337

TABLE 1 J Integral Results (CCP)

	DSM - [4]	KCALC
K_I	745.0	749.326
K_R	1.329	1.337

TABLE 2 K_I Obtained From Nodal Displacement Methods (CCP)

4.2   Single Edged Cracked Plate, Linear Elastic Results

In this section the results are again in two tables and they have the same headings as the previous section.

	REFERENCES		ELEMENT TYPES	
	10	4	PLANE 42	PLANE 82
J	12.4298	11.71	12.476	12.4349
K_I	1681.5	1632.0	1684.62	1681.84
K_R	3.000	2.91	3.005	3.001

TABLE 3 J Integral Results (SECT)

	DSM - [4]	KCALC
K_I	1647.4	1688.8
K_R	2.939	3.013

TABLE 4 K_I Obtained From Nodal Displacement Methods (SECT)

If the stress distributions from these results are obtained by substituting the mode one stress intensity factor from the above tables into equations [1] then the results are all going to be close to each other. However, with finite element models stress distributions can be obtained using post-processing facilities present in most commercial FE packages.

Plots of Von Mises stress a short distance (1.5mm) along the x axis from the crack tip ($x=0$) are now shown for the two models each with the two different types of elements compared in the one graph.

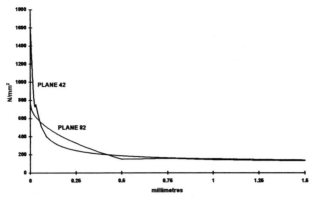

FIGURE 1   CCP   Von Mises stress   42 vs 82

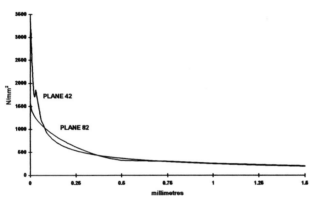

FIGURE 2   SECT Von Mises stress   42 vs 82

4.3   Comments

Although the stress intensity factor results given in Tables 1-4 are very close to each other, the Von Mises stresses near the crack tip are not close in the finite element results. In both cases the PLANE 82 element models make an abrupt change in stress value at the distance of 0.5mm from the crack tip. This is the length of the collapsed 8-node isoparametric element used to model the crack tip. Also the PLANE 42 element models exhibit an unusual glitch in stress value very close to the crack tip. This corresponds to the structural 4-node elements at the crack tip being deformed beyond their designed capacity.

These test cases are based upon Linear Elastic Fracture Mechanics models. The stress distributions near the crack tips are shown to be extremely high, especially in the SECT model

4.4 Elastic - Plastic Results

The same finite element models as before are used but non linear analysis is applied to the finite element solution phase. Oil Rig Quality steel was chosen for this analysis because it is a load bearing steel much used by industry and is susceptible to failure by cracking because of its working environment. Experimental data [12] obtained from this steel was used to constrain the calculation of stress to a particular relationship with the strain. Essentially data points are input to the FE model and the FE package then uses an iterative procedure to approximate a curve fit between the data points. The data points that were used in the finite element analysis are shown in Figure 3.

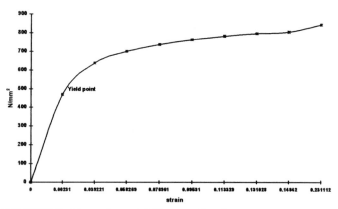

FIGURE 3   Experimental Stress - Strain curve

4.5 Stress Plots

The results presented show the stress calculated using the Von Mises equivalent stress (SEQV) and the equivalent plastic (non linear) stress (NLSEPL). They are presented separately for each problem and element type.

1060

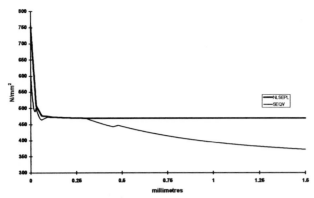

FIGURE 4   Equivalent Elastic and Plastic Stresses PLANE 42 (SECT)

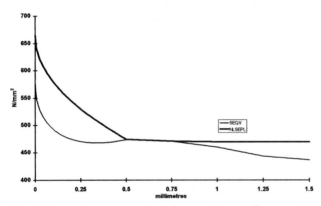

FIGURE 5   Equivalent Elastic and Plastic Stresses PLANE 82 (SECT)

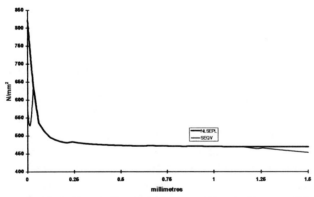

FIGURE 6   Equivalent Elastic and Plastic Stresses PLANE 42 (CCP)

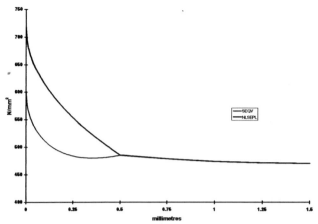

FIGURE 7   Equivalent Elastic and Plastic Stresses PLANE 82 (CCP)

<u>4.6   Comments</u>

The yield stress of this material is 469.5 N/mm^2 and stresses are shown to go beyond this, see Figures 4,5,6 and 7. Both models with the PLANE 42 elements do so at a very small distance from the crack tip whereas the models with the PLANE 82 elements do so at the start of the special crack tip elements. The stresses in all the plots are notably lower than the linear elastic ones but again the models with PLANE 42 elements show sharp increases at distances very close to the tip.

5   CONCLUSIONS

Different finite element stress distributions were obtained when different element types were used. The stresses in the non-linear elastic-plastic results are notably lower than the linear elastic ones. This is because the true stress-strain curve was used and large strains were dealt with accordingly. The use of PLANE 42 elements for LEFM models produces accurate results. The stress distribution for the PLANE 82 element models show a similar pattern in both elastic and plastic models. The calculation of strains and stresses at crack tips which become singularities will never be achieved with numerical approximation, so results are always going to be suspect at the regions closest to this phomena and care must be taken to determine what data is useful and what is not.

REFERENCES

1. WESTERGAARD, HM - Bearing Pressures and Cracks. Journal of Applied Mechanics, Transactions of ASME 61, pp. A49-53, 1961.

2. WILLIAMS, ML - On the stress distribution at the base of a stationary crack. Journal of Applied Mechanics, Transactions of ASME 24, pp.109-114 1957.

3. IRWIN, GR - Analysis of stresses and strains near the end of a crack traversing a plate. Journal of Applied Mechanics, Transactions of ASME 24, pp. 361-364 1957.

4. PANG, HLJ and LEGGAT, RH - 2D test cases in Linear Elastic Fracture Mechanics. National Agency for Finite Element Methods and Standards (NAFEMS), NAFEMS Report, 1990.

5. TRACY, DM - Finite Elements for Determination of Crack Tip Stress Intensity Factors. Engineering Fracture Mechanics, 3, pp. 255-265, 1971.

6. HENSHELL, RD and SHAW, KG - Crack tip finite elements are unnecessary. International Journal for Numerical Methods in Engineering, 9, pp. 495-507 1975.

7. BARSOUM, RS - On the use of isoparametric finite elements in linear fracture mechanics. International Journal for Numerical Methods in Engineering, 10, pp. 25-37 1976.

8. AKIN, JE - The Generation of Elements with Singularities. International Journal for Numerical Methods in Engineering, 10, pp. 1249-1259, 1976.

9. PARIS, PC and SIH, GC - Stress Analysis of Cracks. Fracture Toughness Testing and its Applications, American Society for Testing and Materials, Philadelphia, STP 381, pp. 30-83, 1965.

10. ROOKE, DP and CARTWRIGHT, DJ - Compendium of Stress Intensity Factors, HMSO, London, 1976

11. RICE, JR - A path independent integral and the approximate analysis of strain concentration by notches and cracks. Journal of Applied Mechanics, pp. 379-386, 1968.

APPENDIX I   SINGLE EDGE CRACKED PLATE
uniform tensile stress

GEOMETRY
a/b = 0.5 ; h/b = 0.5 ; b = 20mm

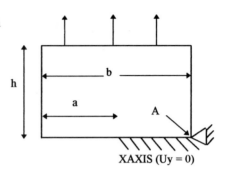

SOLUTION
bending unrestrained,

$K_I / K_O = 3.000$, $K_O = \sigma \sqrt{\pi a}$

F.E. MODEL
One half of the test geometry
using symmetry along the plate

MATERIAL PROPERTIES
E = 207000 N/mm²
υ = 0.3

BOUNDARY CONDITIONS
Uy = 0 on the X axis
Ux = 0 at point A

LOAD : uniform stress, σ = 100 N/mm²

ELEMENT ATTRIBUTES
8-node isoparametric quadrilateral elements
plus collapsed 8-node crack tip elements
total = 232, nodes = 755
ALSO
4-node structural element - total = 17173,
nodes = 17434

APPENDIX II   CENTRE CRACKED PLATE IN TENSION

GEOMETRY
a/b = 0.5 ; h/b = 1.0 ; b = 20mm

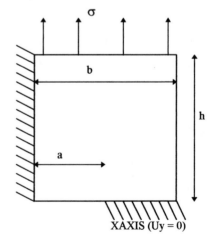

SOLUTION
$K_I / K_O = 1.325$, $K_O = \sigma \sqrt{\pi a}$

F.E. MODEL
One quarter of the test geometry
using symmetry along plate axes

MATERIAL PROPERTIES
E = 207000 N/mm²
υ = 0.3

BOUNDARY CONDITIONS
Uy = 0 on the X axis
Ux = 0 on the Y axis

LOAD : uniform stress, σ = 100 N/mm^2

ELEMENT ATTRIBUTES
8-node isoparametric quadrilateral elements plus collapsed 8-node crack tip
elements - total = 232, nodes = 755
ALSO
4-node structural elements - total = 20298, nodes = 20584

OPTIMUM BRIDGE MONITORING STRATEGY USING FINITE ELEMENT ANALYSIS

W. T. Yeung [*] and J. W. Smith [**]

* Research Student, University of Bristol, UK
** Senior Lecturer, University of Bristol, UK

SUMMARY

Much previous research on structural damage detection has focused on vibration frequencies of simple laboratory or analytical models. However, frequency is an insensitive parameter and an effective technique for large scale structures is not yet proven. In this paper a new method of damage detection is presented using a finite element model of an iron suspension bridge of significant span and considerable age. Loosening of riveted joints in the main girders has occurred in recent years. This was simulated in the model by introducing pinned joints in the girders.

An efficient vibration monitoring strategy was devised using modal analysis of the damage scenarios. To simulate realistic bridge monitoring, a transient analysis was carried out under the effect of a moving vehicle. The responses at the chosen locations were analysed in the frequency domain by Fast Fourier Transform (FFT). The results indicated that the new method is sufficiently sensitive to detect low levels of damage. Furthermore, it was shown that it could be implemented in conjunction with a pattern recognition algorithm suitable for continuous monitoring.

1 INTRODUCTION

The ability to detect structural damage of a bridge well before it endangers the structure has been of interest to engineers for many years. It has been estimated that in the United States alone as many as 40% of bridges are structurally deficient, with repairs costing in the order of billions of dollars [1]. Currently, bridge condition assessment is largely limited to visual inspection taking place at intervals typically of one to five years. For old bridges this is of some concern since significant deterioration could have accumulated in the intervening period. Continuous performance monitoring is therefore highly desirable, especially for old, decaying structures and for critical highway links.

Monitoring the behaviour of structures by observing their vibration properties is attractive because of the availability of sensitive transducers and electronic instrumentation. The basic principle is that vibration frequencies, mode shapes and damping of a structure will change when it has been damaged. Changes arise from loss of stiffness due to the damage and increase in damping due to friction at new interfaces at crack locations. This method was used centuries ago for checking the quality of clay pots and glass bottles by a tap test whereby defects were revealed by a change in the ringing sound. Early studies in the modern era were carried out by Cawley and Adams [2] who used frequency changes in modes of vibration to locate damage in a vibrating plate.

Unfortunately, damage detection by vibration methods has been less successful than expected. This has particularly been because change of natural frequency is not very sensitive to minor damage. For example Stubbs and Osegueda [3] found that the detection limit was represented by a 50% reduction in the stiffness of a major structural member. Similar conclusions were drawn by Farrar and Doebling [4] who tested full size bridge girders in the USA and found that frequency reduction only became significant when the level of damage was disastrous.

The most successful improvements in dynamic damage detection have been achieved by observation of changes in the mode shapes. A well known method is the Modal Assurance Criterion (MAC) suggested by Allemang and Brown [5]. The MAC provides an index between 0 and 1 that is a direct comparison between mode shapes before and after damage. It reduces from 1 (undamaged) down to 0 when the damage is so great that there is no longer any correlation between the mode shapes. Biswas et al [6] demonstrated the potential of the MAC by comparing the performance of an actual highway bridge which was damaged synthetically by removing splice plates in the girders. Further refinements have been suggested by Pandey et al [7] and Nalitolela [8] who both used the curvature of mode shapes which are proportional to the strain energy. In theoretical studies this approach was promising but experimental mode shapes are usually very noisy or even incomplete.

An alternative development was investigated by Wolff and Richardson [9] who suggested that the locations of stationary points of mode shapes were even more sensitive. The study was carried out on a steel plate, reinforced with a rib stiffeners and bolted along its centre line. The location of the node line was sensitive to damage in the form of removal of bolts.

There is still limited evidence that dynamic methods are useful for detecting the effects of damage in large scale structures. Moreover, there is a need to explore an effective operational procedure for continuous monitoring that will set warnings when the structural performance is not as expected. The main objective of this paper is therefore to present a systematic study on the effects of realistic damage in an actual structure, using a detailed finite element model, and to develop an efficient monitoring strategy.

2 FINITE ELEMENT MODEL

The Clifton Suspension Bridge is a wrought iron eye-bar suspension bridge of approximately 214 metres span, with iron longitudinal and cross girders supporting an asphalt paved timber deck. The finite element model of the 135 year old bridge is shown in Fig 1.

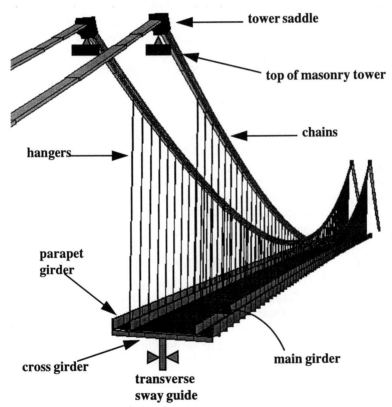

Fig. 1 Perspective view of the finite element model

There are three chains on each side of the roadway, arranged one above the other. They are made up of 175mm x 25mm x 7300mm long wrought iron bars with special eye joints forged to each end. Each link is made up of ten or eleven bars arranged side by side, interleaved with the bars of the next link, and connected with a pin through the eye joint. Successive hanger rods are attached to each of the three chains in turn. The suspended structure consists of longitudinal riveted plate girders, under each set of chains, lattice cross girders and lattice parapet girders.

A two-dimensional finite element model of the bridge was developed by Cullimore et al [10] with the purpose of investigating the structural importance of the parapet girders for purposes of fatigue analysis. However, for dynamic analysis this model was unsatisfactory because of the importance of three-dimensional behaviour and, in particular, torsion of the deck.

A three-dimensional model of the bridge was developed using the ANSYS finite element program, as shown in Fig 1. Beam elements were adopted for the chains and hangers. Beam elements were also used for the longitudinal girders, cross girders and parapet girders. The deck, which is very stiff in plan, was modelled using flat shell elements. The stone towers were not modelled, the tower saddles being free to roll longitudinally on the rigid tower tops. Another important feature is that the deck is considerably shorter than the span between the tower saddles. Thus the deck is free to move vertically at the abutments (a short articulated span is provided for traffic). Lateral sway is prevented by means of a sway guide as shown.

3 VERIFICATION OF THE ANALYTICAL MODEL

Static deflections and bending moments calculated using the numerical model were compared with the results of full scale tests carried out by Pugsley and Flint [11] and Cullimore et al [10]. These are shown in Tables 1 and 2 respectively.

Load	Position	Pugsley and Flint	3D model
10.5 ton	1/4 span	84	78
symmetric	1/2 span	66	61
	End of span	36-46	46

Table 1 Static deflections in mm

Load	Position	Main girder	Parapet girder	Total
8 ton	1/8 span	-2.1 (-5.4)	-7.0 (-5.4)	-9.1 (-10.8)
symmetric	1/4 span	-89.1 (-92.3)	-64.1 (-87.5)	-153.2(-179.8)
	3/8 span	3.3 (3.5)	5.3 (3.5)	8.6 (7.0)
	1/2 span	18.2 (16.8)	13.1 (22.9)	31.3 (39.7)

Table 2 Bending moments at 1/4 span in kNm (analysis in parentheses)

The calculated deflections are in good agreement with the tests. The bending moment results are generally satisfactory, the differences being influenced by the deck which acts compositely with the girders. The true in-plane stiffness of the deck was difficult to establish being composed of timber baulks and asphalt surfacing.

The dynamic behaviour of the model was also compared with tests. Instrumentation was set up on the bridge to monitor the dynamic displacements at the end of the span. Measured displacement time histories were analysed for their spectral contents. The spectral peaks were compared with frequencies obtained from modal analysis. The geometric stiffness of the model was included in the modal analysis utilising the ANSYS pre-stressed procedure. The results are given in Table 3.

	Mode 1	Mode 2	Mode 3	Mode 4	Mode 5
3D Model	0.278	0.356	0.358	0.401	0.482
Observed	0.290	0.350	(coupled)	0.410	0.470

Table 3 Calculated and experimental mode frequencies in Hz

Again the calculated frequencies were in good agreement with the measured ones. It should be noted that the calculated frequencies of the second and the third modes were very close together, the second being torsional and the third being a sway mode. It was not possible to distinguish these modes from the experimental observations since they were effectively coupled under transient loads.

4 DAMAGE SIMULATION

A significant form of deterioration has been the progressive loosening of riveted joints in the longitudinal girders. This is shown in Fig 2 where it should be noted that the latticework of the parapet girder is very flexible allowing considerable movement at the damaged splice. The damage was simulated by introducing pinned joints in the girders at the damaged locations. The pinned joint was modelled by including an extra coincident node, the two nodes then being coupled except for rotation about the z axis.

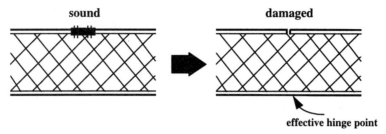

sound   **damaged**

effective hinge point

Fig. 2 Modelling of damage in longitudinal girders

A data base of different configurations of damaged locations was generated for the investigation. In total 23 different damage scenarios were considered, consisting of 8 cases of single damaged joints (randomly distributed amongst the four girders), 8 cases of two damaged joints and finally 7 cases of three damaged joints.

5 SENSITIVITY OF VIBRATION RESPONSE TO DAMAGE

In the first instance, modal analysis was carried out to determine the first five frequencies and mode shapes. An example of the way in which the frequencies change with increased damage is shown in Table 4. It will be noted that the frequency of the first mode reduces because of the reduced stiffness arising from the damage. However, the reduction is very small and not sufficient as a damage detection criterion, as noted previously.

Damage scenario	**Mode 1**	**Mode 2**	**Mode 3**	**Mode 4**	**Mode 5**
undamaged	0.278	0.356	0.358	0.401	0.482
1 joint (a)	0.278	0.355	0.358	0.400	0.482
(b)	0.278	0.358	0.401	0.461	0.484
2 joints (a)	0.278	0.357	0.400	0.454	0.506
(b)	0.277	0.358	0.380	0.400	0.482
3 joints (a)	0.276	0.352	0.400	0.478	0.583
(b)	0.275	0.346	0.400	0.479	0.588

Table 4 Sensitivity of mode frequencies to damage in Hz

The changes in frequency of the higher modes are greater. Moreover, an interesting mode switching phenomenon was noted which provides a more sensitive measure of damage. This is illustrated in Fig 3. The vertical displacements of the downstream main girder in the second mode are shown in Fig 3a. In its undamaged state this is a pure torsional mode, the displacements of the upstream girder being a mirror image of those shown. When the girders are damaged the structure becomes unsymmetrical and mode coupling occurs. In particular, the sway mode (mode 3 in undamaged state) combines with the torsion to produce a coupled sway and torsion mode as shown in Fig 3b. A consequence of this is that the vertical displacements of the girder are much reduced as can be seen in Fig 3a.

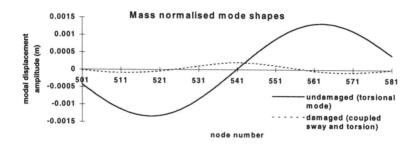

Fig 3(a) Vertical displacement of downstream girder in mode 2

Fig 3(b) Torsion in mode 2 switching to sway and torsion

Fig 3 Mode switching between mode 2 and mode 3

The greater sensitivity of mode shapes to damage, and in particular due to mode switching, has been noted. This is the main reason for interest in the Modal Assurance Criterion (MAC). For practical implementation a large number of transducers would be required to extract accurate mode shapes from response under ambient transient loading. However, for damage detection purposes, it may be sufficient to monitor only the locations which are sensitive to damage. In this study, it was proposed to monitor modal response close to the stationary points of each mode. This is different to node line monitoring as studied by Wolff and Richardson [9], because it works by measuring the response changes at the original position of the stationary nodes, as opposed to evaluating the position of the node when the structure is damaged. A new criterion called Stationary Node Response Assurance Criterion (SNRAC) was devised, similar in principle to the MAC. The difference is that the mode shape is monitored at pairs of nodes and antinodes. This will be more sensitive to change in motion at the nodes as a result of damage and will also provide correlations similar to the MAC in the case of mode switching. The SNRAC takes the value of 1.0 to 0, with 1.0 denoting no change in response and 0 denoting completely different mode shape. The new criterion was applied to the mode shapes obtained from damage simulation. The lowest SNRAC values for each of the different numbers of damaged joints is given in Table 5.

Number of damaged joints	Mode 1	Mode 2	Mode 3	Mode 4	Mode 5
1	0.999	0.000	0.092	0.000	0.952
2	0.999	0.000	0.098	0.000	0.390
3	0.999	0.000	0.100	0.000	0.150

Table 5 Sensitivity of SNRAC to damage

As expected, the sensitivity of mode 1 to damage was very low. From mode 2 onwards it will be noted that the new criterion is highly sensitive to damage. The low values of SNRAC in modes 2, 3 and 4 are all due to mode switching.

However, for practical bridge monitoring it is difficult to extract modes from vibration under normal loading. Therefore, in the second stage of the analysis, transient vibration under a moving 4 ton vehicle was conducted to simulate realistic bridge monitoring at the optimum transducer locations. It should be noted that 4 tons is the current weight limit on the bridge. The damping ratio used in the transient analysis was derived from the time series obtained in the tests. The transient motion of the vehicle was modelled as series of force impulses acting along the structure at successive time intervals. An example of one of the displacement time histories obtained from the model is shown in Fig 4. The time histories from various stationary points, comparing the undamaged and other damaged states, were analysed for their frequency contents by Fast Fourier Transform (FFT).

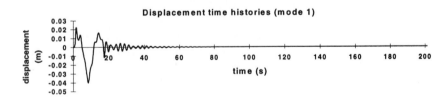

Fig 4 An example of the time histories obtained from the transient analysis

1074

It was found that there were significant changes in the characteristics of the spectra when the structure is damaged. An example of this is given in Fig 5.

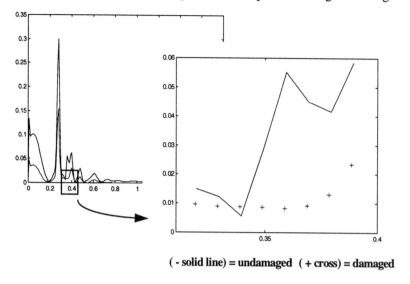

(- solid line) = undamaged   (+ cross) = damaged

Fig 5 Frequency spectrum derived from the time histories

6 CONTINUOUS MONITORING USING NEURAL NETWORK

The paper has so far demonstrated a method of damage detection with good sensitivity. However, there is a need to integrate it into an overall procedure for continuous condition monitoring. In order to achieve this some means are required for recognising that a pattern of measured behaviour is changing significantly. The approach suggested here is to make use of the pattern recognition potential of a neural network algorithm.

A neural network is a computational technique that mimics the processes in the human brain. In simple terms it consists of a network of weighted connections between a set of input nodes and a smaller set of output nodes. The weights are adjusted by computation so that the outputs can be related to the inputs. The main advantage of a neural network is that the computation for adjusting the weights is a "learning" process that has the ability to generalised the input data, providing that there are enough input vectors. In general, there are two classes of neural network, the supervised and unsupervised networks. Supervised networks require prior knowledge. The different classes of the input patterns would have to be determined prior to the learning phase, whereas unsupervised networks have the ability to establish the existence of classes within the data set. In this study, an

1075

unsupervised neural network called the Probabilistic Resource Allocating Network (PRAN) has being investigated for bridge monitoring [12]. The initial architecture of the PRAN is shown in Fig 6.

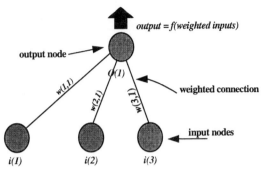

Fig 6 Initial architecture of the Probabilistic Resource Allocating Network

It consists of a fixed number of inputs nodes, all connected to a Gaussian processing unit or kernel, where the input patterns are characterised by its mean and convariance matrices. Learning takes place by updating the mean and convariance matrices of all kernels in the network for each subsequent presentation of a new input vector to the network. A new kernel is added to the network if the smallest distance between the new input and each kernel is greater than a certain threshold.

The training inputs to the network in this investigation were made up of frequency spectra derived from time histories at nodes and antinodes of the first five modes of vibration, in the undamaged state. A frequency band close to the natural frequency of each of the mode was selected and the response from the corresponding antinode was appended, to form the input vector. In total, 600 waveforms were used in training, generated by adding Gaussian noise sequences to the original uncorrupted time histories.

After training, input waveforms obtained from the damaged state of the structure were tested using the network. Waveforms that had not been "seen" by the network i.e. where they were too far away from any of the Gasssian kernels were considered to be novel implying that the structure may have been damaged. The results for 6 different damage scenarios are given in Table 6.

Damage scenario	Modes indicating novel behaviour	Damage detected
1 joint (a)	4	×
(b)	4 and 5	√
2 joints (a)	4 and 5	√
(b)	4 and 5	√
3 joints (a)	4 and 5	√
(b)	3, 4 and 5	√

Table 6 Damage identification using the PRAN

It is clear that the PRAN was effective in detecting novel waveforms from the different damaged states of the structure. For all of the six damage scenarios considered, at least one waveform from the first 5 modes has been detected as novel. However it was decided that if there were 2 or more waveforms classified as novel, this would indicate presence of damage. For medium to high levels of damage i.e. where there were 2 or 3 pinned joints, the network detected damage in all cases. For low levels of damage, the system was still capable of detecting damage for one of the two cases used in this investigation, despite the usage of a more strict criterion.

7 CONCLUSION

(a)   Mode shapes of bridges, and in particular the behaviour of stationary points, are more sensitive to damage than natural frequencies

(b)   study of the FE model of a bridge of significant span showed that a new damage criterion could minimise the number of vibration measurement points and was sufficiently sensitive for full scale monitoring purposes

(c)   the new method was suitable for integration into a pattern recognition procedure which was able to detect most damage scenarios from traffic generated vibration.

REFERENCES

[1] HUSTON, D, FUHR, P, ADAM, C, WEEDON, W, and MASER, K - Bridge Deck Evaluation with Ground Penetrating Radar. Structural Health Monitoring, Current Status and Perspectives, pp. 91-102, 1997.

[2] CAWLEY, P and ADAMS, R D - The Location of Defects in Structures From Measurements of Natural Frequencies. Journal of Strain Analysis, Vol. 14, No. 2, 1979.

[3] STUBBS, N and OSEGUEDA, R - Global Non-destructive Evaluation of Offshore Platforms Using Modal Analysis. Proceedings of the 6th International Offshore Mechanics and Arctic Engineering symposium, Vol. 2, pp.517 -524, 1987.

[4] FARRAR, C R, and DOEBLING, S W - Lessons Learned from Applications of Vibration Based Identification Methods to a Large Bridge Structure, Structural Health Monitoring, Current Status and Perspectives, pp. 351-370, 1997.

[5] ALLEMANG, R J and BROWN, D L - A Correlation of Coefficient for Modal Vector Analysis. Proceeding of the First International Modal Analysis Conference, 1982.

[6] BISWAS, M, PANDEY, A K and SAMMAN, M M - Diagnostic Experimental Spectral / Modal Analysis of a Highway Bridge. International Journal of Experimental and Modal Analysis, Vol. 5, No. 1, pp. 32-42, 1989.

[7] PANDEY, A K, BISWAS, M and SAMMAN, M M - Damage Detection from Changes in Curvature Mode Shapes. Journal of Sound and Vibration, Vol. 145, No. 2, pp321-333, 1991.

[8] NALITOLELA, N G - Localisation Of Damage in Structures By Analytical Model Improvement And Strain Energy Balance. 13th International Modal Conference, pp1400-1406, 1995.

[9] WOLFF, T and RICHARDSON, M - Fault Detection in Structures from Changes in their Modal Parameters. Proceedings of the 7th International Modal Analysis Conference, 1989.

[10] CULLIMORE, M S G, MASON, P J and SMITH, J W - Analytical Modelling for Fatigue Assessment of the Clifton Suspension Bridge. IABSE Colloquium, Copenhagen, Vol. 67, pp197-205, 1993.

[11] PUGSLEY, A G and FLINT, A R - Report on tests carried out on the Clifton Suspension Bridge. Report of Dept of Civil Engineering, University of Bristol, 1954.

[12] ROBERTS, S and TARRASENKO, L - A Probabilistic Resource Allocating Network for Novelty Detection. Neural Computation, Vol. 6, pp270-284, 1994.

Title : Integrity assessment of a combustion chamber cooling ring

Author : K.V. Delaney

Mechanical Technology Specialist, Combustion Systems-Engineering, Rolls Royce Aerospace, Filton, Bristol UK

1 SUMMARY

Large thermal gradients in aero engine combustion chambers frequently cause the initiation and growth of cracks, by a thermal - mechanical fatigue mechanism.

Such cracks can be benign, ceasing to propagate once they grow out of the immediate thermal stress field. However, in some cases they continue to grow to a point where they can damage the integrity of the chamber.

This paper discusses the finite element investigation of an observed cooling ring circumferential unzip, caused by an axial crack growing into the ring from a hot spot. The investigation successfully identifies the failure mechanism, showing that the method can be used to assess proposed redesigns.

Definition of the thermal field for the finite element analysis used the results of a combusting CFD analysis validated by engine thermal paint test results and thermocouple measurements. The subsequent strain analysis included the effects of cold plasticity and creep deformation, through a typical flight.

The axial crack initiation mechanism is discussed. Fatigue lives are assessed by comparison with Thermo-Mechanical fatigue test results. The mechanism for initiation of the cooling ring circumferential unzip is also discussed together with the assessment method using strain-life data.

Typically, gas turbine, aero engine, combustion chambers are large flexible structures, fabricated from sheet and forged or cast materials. Functionally they have to contain and control the combustion process.

The structure usually has a large number of penetrations, some of which are to direct air and fuel into the correct region of the chamber, to maximise efficiency and minimise emissions. Other holes form part of a cooling system to protect the structure from the high flame temperatures of the combustion process.

The high operating temperatures necessitate that combustion chambers should be lightly constrained to allow free expansion of the structure, so that thermal stresses are minimised.

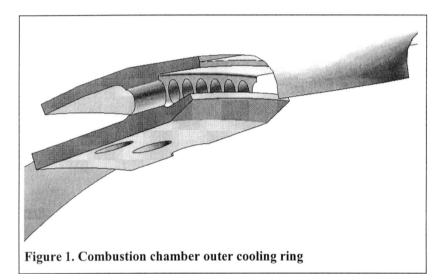

Figure 1. Combustion chamber outer cooling ring

In this style of combustion chamber the structure is protected by air flowing through the cooling ring forming a film of cooler air, which shields the downstream wall from the high temperature combustion process. This particular cooling ring, shown in figure 1, has some unusual features. Immediately upstream of the ring is a series of blown ring holes, which direct air and fuel off the chamber wall, into the combustion zone. The scoop on the outside ensures the capture of sufficient air for both the blown ring and the cooling ring holes. The hot lip is rolled to reduce the film flow area and improve cooling performance.

The complexities of the combustion process and the flow in the chamber, almost inevitably, generate large thermal gradients within the structure. The resulting thermal stresses can cause localised creep, which may lead to thermal fatigue cracking. Such cracks are fairly common in combustion chambers, the large majority cease to propagate once they grow out of the local thermal - stress field. However, some cracks can grow to an extent where they can threaten the integrity of the chamber.

Axial thermal fatigue cracks can initiate at the tip of the hot lip, such cracks can usually be tolerated, but can propagate forward into the cooling ring itself. Occasionally the presence of an axial crack in the ring can initiate a circumferential unzipping of the ring.
The objective of these analyses was to understand the causes and sensitivities of the both the crack initiation mechanism and the unzip event and to assess proposed modifications to cooling ring design.

3 AIRFLOW AND THERMAL DEFINITION

Stresses in the combustion chamber are mainly generated by short range, temperature variations, therefore an accurate thermal definition is essential for a meaningful stress analysis. In practice, knowledge of the thermal field is always incomplete and limited experimental data must be extrapolated.

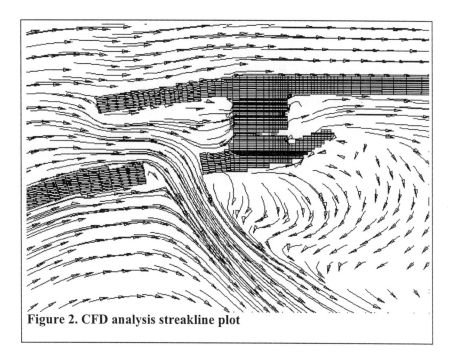

Figure 2. CFD analysis streakline plot

For this combustion chamber some full field temperature information was available from engine tests on combustion chambers coated with temperature sensitive paint. Paint measurements are of limited use as only the maximum test condition reached is recorded, and then not in fine detail necessary, because the paint colour changes occur at large intervals (typically 100°C). Thermocouple measurements were also available, which gave information about part load conditions, with generally better accuracy, but only at a limited number of point locations. Metallurgical examination of combustion chambers also gave some indication of the maximum temperatures achieved.

To fill in the gaps in the thermal picture a full combusting CFD analysis was carried out. Near wall gas conditions were used to define three dimensional gas temperatures and heat transfer coefficients for the finite element thermal analysis.

The results of the CFD analysis show, figure 2, that as well as directing fuel rich air off the chamber wall, airflow through the blown ring holes causes a recirculation downstream. The analysis also suggests that the concentrated mixture achieves stochiometric conditions close to the end of the lip, so that local gas temperatures would be extremely high.

4 FINITE ELEMENT MODEL

A symmetrical sector model of the complete axial extent of the combustion chamber was created, figure 3. Shell elements were used for the general structure, with a detailed solid model of the cooling ring embedded using multi point constraints. This allowed the model to capture the gross effects of creep on the complete structure, as well as providing a framework for future analyses of other cooling rings.

Several versions of the model have been analysed, to explore the effects of geometric and thermal variations. Typically, an analysis model contained 450 shell and 6500 solid, second order elements, with 111000 degrees of freedom.

The finite element model used for the thermal analysis included a representation of the ceramic thermal barrier coating. The coating was omitted from the structural analysis, as it was considered to have an insignificant effect on the stiffness of the structure.

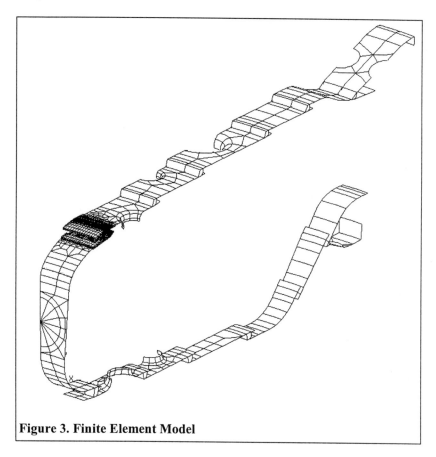

Figure 3. Finite Element Model

5 THERMAL ANALYSIS RESULTS

Thermal analyses were carried out at two operating conditions; a take off condition and a high speed flight condition. Some adjustments to the CFD derived boundary conditions were necessary to match to the thermal paint and thermocouple measurements, otherwise the thermal results successfully captured the 3D nature of the temperatures. A severe hot spot, of approximately 900°C was identified at the hot lip, which agreed with the position of cracks observed in used combustion chambers.

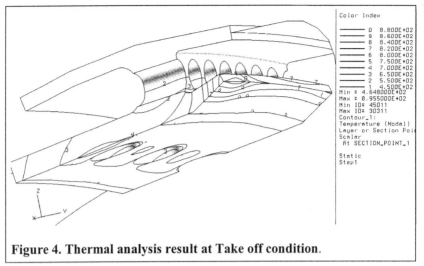

Figure 4. Thermal analysis result at Take off condition.

Thermal results at the other flight condition showed a similar distribution, but with overall temperatures generally raised, due to the increased engine inlet temperature.

6 STRUCTURAL ANALYSIS

The temperature of the structure is high enough to cause the material to creep, so that the analysis needed to account for time at condition. Time and resource limitations would not allow the generation of thermal definitions at all the varying conditions seen throughout a typical flight, so a simplified

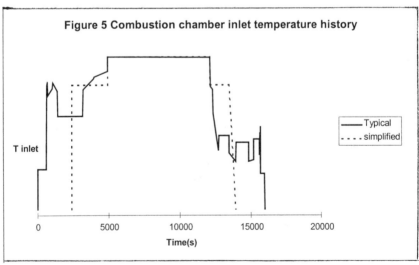

duty cycle was derived, based on two conditions, figure 5. As the analysis was to be largely comparative, examining design modifications and sensitivities, this was considered an acceptable approximation.

Including the creep calculation restricted the material plasticity model to an isotropic assumption, due to limitations in the analysis code. The creep strain calculation used a surface fit of strain with stress and time, derived from material testing.

Figure 6, shows the stress - strain history for a point in the lip hot spot. Compressive stress on the first start is limited to the yield, until the temperature becomes high enough to cause creep relaxation. During the hold at high temperature the stress relaxes, almost to zero. On cooling the stress reverses and plastically yields in tension. Repeated cycling gives a very wide hysteresis loop.

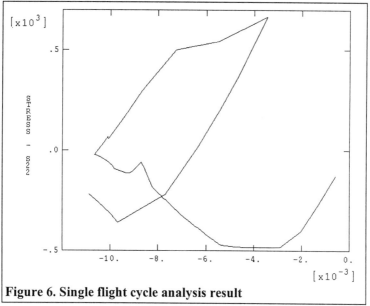

Figure 6. Single flight cycle analysis result

To reduce the computing cost and duration the model was converted to first order elements for an examination of the effect of repeated cycling. The analysis results, in figure 7, show little sign of stabilising after five cycles.

Elimination of the initial axial crack initiation would improve the integrity of the cooling ring (no circumferential unzips have been observed without an accompanying axial crack). The temperature of the lip could be reduced by increasing the thickness of the thermal barrier coating. If the temperature were reduced enough cracking should be eliminated. However, any increase in coating thickness would have to be cautious, since experience shows that thermal barrier coating can spall if it is too thick.

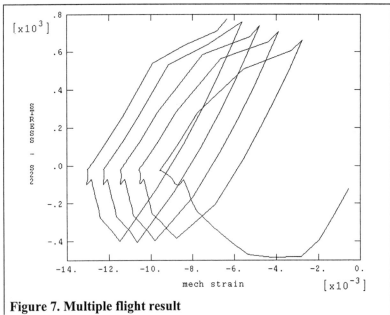

Figure 7. Multiple flight result

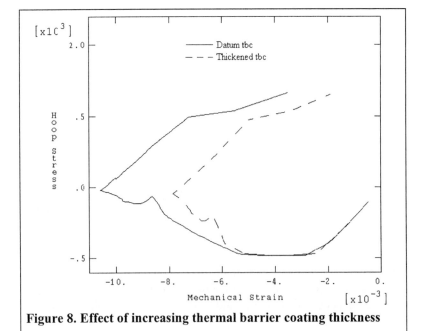

Figure 8. Effect of increasing thermal barrier coating thickness

Repeating the analyses indicated that doubling the coating thickness gave a reduction in peak temperature of approximately 50°C. The effect on the hot lip initiation site is a substantial reduction in strain, as the comparison in figure 8 illustrates.

To evaluate the results in terms of cyclic life, a series of thermo-mechanical fatigue (TMF) tests were carried out. The tests were defined, based on the results of the finite element analysis, so that a direct conversion could be made between the analysis and test results.

Although the TMF tests showed that the crack initiation life could be expected to double due to the temperature reduction, the initiation life was still expected to be only a few hundred cycles. Although worthwhile this was still considered to be an insufficient improvement, since cracking could still be expected within the service lifetime of the combustion chamber.

7 CRACKED LIP UNZIP ANALYSIS

Having concluded that reducing hot lip temperatures would only delay the onset of cracking, rather than eliminate it, the strength of the cooling ring and its resistance to circumferential unzip, in the presence of an axial crack, were examined.

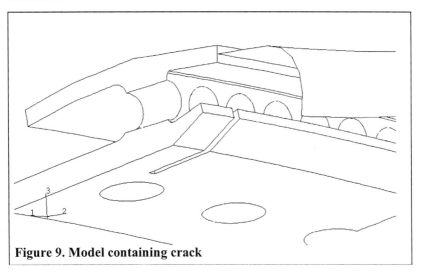

Figure 9. Model containing crack

Under operating conditions the inner part, being hotter, is in hoop compression and balanced by the outer being in hoop tension, sandwiched between the two the cooling ring webs are compressed. Introducing a crack through the inside of the ring destroys its hoop continuity, so that loads

1087

caused by the thermal fight between the inner and outer parts can not be carried by hoop stresses.

The inelastic strains in the inner part result in it shrinking when the load is removed, placing it tension. The crack tends to open and loads that were, in an uncracked ring, balanced by compressive hoop stresses in the outer part are now transferred by a shear mechanism through the cooling ring webs

At the end of the period at the maximum temperature condition, an axial crack was introduced into the hot lip, from the hottest part through to the forward face of the cooling ring. This was achieved by the rather crude, but effective, method of removing elements from the model, see figure 9. The initiation of the circumferential unzip is not affected by the stress state at the axial crack tip. The compressive creep strain in the hot lip tends to open the crack as the engine shuts down, so the load path in the ring is modelled correctly.

On removal of the loading, severe tensile stresses and strains are generated in the webs between cooling ring holes. Figure 10 shows the maximum principal strains.

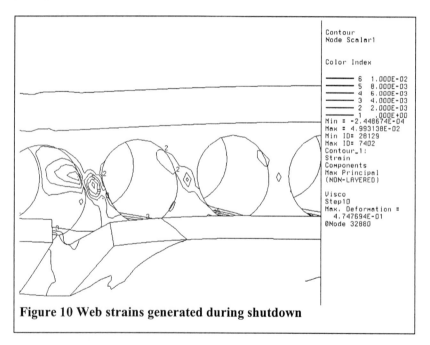

Figure 10 Web strains generated during shutdown

Comparison with strain-life data suggested a web crack initiation life, in the web of approximately 500 cycles. Sensitivity analyses have shown that modifying thermal boundary assumptions could easily reduce this estimate by a factor of two. Geometrical variations due to tolerances are also expected to affect life.

8 CONCLUSIONS

Computational fluid dynamics and finite element analysis methods have successfully been used to investigate the initiation of cracking in a combustion chamber cooling ring. The method has been shown to be an appropriate tool for the evaluation of design changes.

Improvement of initial crack initiation life in the hot lip, by increasing thermal barrier coating protection, was found to be insufficient to meet the engine life requirements.

THERMO-STRUCTURAL ANALYSIS OF A 313-PIN PLASTIC BALL GRID ARRAY (PBGA)

J.W. Zwick
Sr. Engineering Specialist, Raytheon Missile Systems Company

C.S. Desai
Regents Professor, Department of Civil Engineering and Engineering Mechanics, The University of Arizona

R.D. Ferdie
Structural Analyst Under Contract to Raytheon Missile Systems Company

1 SUMMARY

The application of a new non-linear finite element procedure for predicting solder joint reliability under cyclic thermomechanical loading is presented in this paper. The finite element procedure is based on the unified and powerful constitutive modeling approach called the Disturbed State Concept (DSC). The DSC allows, in an integrated framework, prediction of elastic, plastic, and creep strains, micro-cracking and fracture leading to degradation and fatigue failure. The DSC and finite element procedure have provided successful prediction of the measured behavior of chip-substrate systems such as Surface Mount Technology (SMT) and Ball Grid Array (BGA) packages. The parameters in the DSC are determined and calibrated based on available laboratory stress-strain tests for solders under thermomechanical loading [3, 11].

The computer method is used to back-predict fatigue life of solder joints in a group of 313 Plastic Ball Grid Array (PBGA) devices that failed after exposure to 1800 thermal cycles from -55 to +125°C. A linear elastic quarter symmetry macro-model of the package is generated using MSC NASTRAN™ to predict the displacement load conditions of each solder joint in the package. The most severe differential displacement of all solder joints in the PBGA is then imposed on a detailed 2-D plane stress structural model of a solder joint. The detailed solder joint model uses the Disturbed State Concept (DSC) to predict fatigue life under the actual test conditions. Model predictions are compared with actual test data as a validation of the proposed DSC modeling procedure.

The PBGA module selected for evaluation has a substrate made of Bismaleimide Triazine (BT) epoxy glass laminate, mold compound, silicon chip, bond wires, and micro solder balls. The solder balls are reflowed onto solder pads using a conventional forced convection nitrogen reflow oven. The PBGA is mounted to a printed wiring board (PWB) composed of FR-4 glass/epoxy laminated with copper traces or planes.

The PBGA 313 module analyzed here is 35.00 mm square and 2.33 mm high. The micro solder ball pattern is a staggered matrix composed of alternating rows (or columns) of thirteen (outer periphery) and twelve solder balls on 1.27 mm centers.

2 METHOD OF ANALYSIS

A number of experimental, analytical and finite element analysis (FEA) methods have been developed for the analysis of ceramic and plastic BGA's and other surface mounted electronic devices [4, 8 and 10]. Creep and relaxation material processes are expected to dominate since solder material is above half of its melting point even at room temperature, with other material factors such as microcracking complicating thermal fatigue studies.

Computer codes such as ANSYS™ and ABACUS™ are particularly used for FEA [4, 12] since these codes have more extensive non-linear material capabilities. However, these tools are based on conventional material constitutive relations. Although the importance of constitutive behavior is recognized, little work has been done (in the past) toward basic and unified models that can allow incorporation of factors such as elastic, plastic and creep strains, microcracking, damage and fracture, relative motions (slippage, separation and warping) at joints and cyclic fatigue failure. The DSC method described in this paper attempts to address such a unified theory.

The analytical process used here is based on a combination of classical hand calculations, and two different finite element models; a macro model of the PBGA, PWB and solder joints, and a detailed micro model of one solder joint. MSC NASTRAN™ [9] was used for the macro model of the PBGA, PWB and solder joints, with computed displacements from this model as loading in a detailed solder joint model that was simulated using a finite element code called DSC-SST2D [5].

The macro models of the PBGA/Solder Joints/PWB were run on Hewlett-Packard C180 and C200 workstations using MSC PATRAN and NASTRAN Version 70 at Raytheon Missile Systems Company in Tucson, AZ. The DSC-SST2D finite element code was run on an SGI Origin 2000 at The University of Arizona Computing Center in Tucson.

3 NASTRAN MACRO MODEL OF PBGA & PWB

3.1 Geometry

A 3-D model of the PBGA and PWB was created using NASTRAN. This is a linear elastic model that simulates the PBGA, PWB and the solder joints that mechanically join them. Drawings of the 313-pin PBGA from Amkor-Anam were used to develop the geometry for this model. The model takes advantage of symmetry by simulating only one-quarter of the PBGA/PWB. The neutral point of the package (geometric center) lies at the intersection of the two symmetry planes and forms one corner of the model. Figure 1 is a drawing of the 313 PBGA part showing the origins of the NASTRAN macro-model. A rendering of the 1/4 symmetry NASTRAN model is shown in Figure 2 .

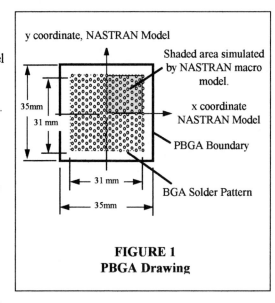

FIGURE 1
PBGA Drawing

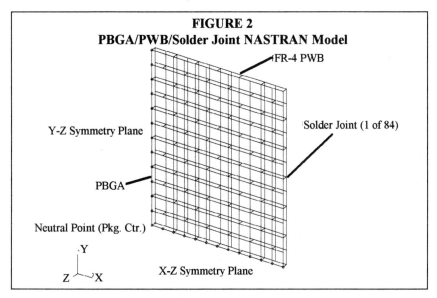

FIGURE 2
PBGA/PWB/Solder Joint NASTRAN Model

3.2 Elements

As shown in Figure 2, the PBGA and PWB are divided into 144 quadrilateral plate elements each, and connected by 84 bar elements that represent the solder joints. Note that not all 313 solder joints are modeled; only those lying within the shaded quarter shown in Figure 1. The PBGA and PWB are simulated by NASTRAN 4-node CQUAD4 elements which are plane strain quadrilateral plate elements. The thickness used in these elements is derived from photomicrographs of test hardware sections; 1.73 mm for the PBGA and 1.57 mm for the PWB.

The beams used to simulate the solder joints are NASTRAN CBAR elements 0.6 mm long and 0.4224 mm diameter. These are linear elements capable of sustaining axial loads, torque, bending moment, and shear loads. The length and bending moment of inertia properties used in the beam elements responded the same as a separate NASTRAN detailed model of the solder joint. Material properties used in this model are shown in Table I below.

TABLE I

MATERIAL PROPERTIES USED IN MSC NASTRAN MODEL

Material	Young's Modulus E, GPa	Poisson's Ratio ν	Coefficient of Thermal Expansion α, $°C^{-1}$
Molding Compound - Amoco Plaskon SMT-B-1RC	10.7	0.25	3.50E-05
PBGA Substrate Material - CCL-H832	12	0.28	1.50E-05
Solder - 63Sn/37Pb	15.7	0.4	2.45E-05
Printed Wiring Board - FR-4	12	0.28	1.60E-05

3.3 Boundary Conditions

The PBGA/PWB quarter symmetry model is constrained from motion at the neutral point in all six degrees of freedom - x,y,z translation, and rotation about each of these axes. The symmetry planes were constrained to allow motion only in the plane of symmetry; i.e. the y-axis symmetry plane was allowed to translate in the y and z directions, but rotation about these axes and translation in the x-direction was constrained. Similarly, motion in the x-axis symmetry plane was allowed in the x and z directions, but rotation in these axes and translation in the y-direction were constrained.

3.4 Load Cases

The only loading considered in this model is that induced by differential thermal expansion of the different materials over temperature, and the solder balls (simulated by the bar elements) connecting the two major components.

The reference temperature (zero stress state) is taken at 183°C, which is the 63/37 Pb/Sn solder solidification temperature. Three isothermal load cases were considered, representing the extremes of exposure and ambient conditions; 125°C, -55°C and 25°C.

3.5 Results

Analysis results for the three load cases include displacement for each node point and stresses in each element. The greatest displacement occurs in solder joints nearest the silicon die and farthest from the neutral point (package center). By evaluating the three components of relative displacement at the ends of each solder joint, the total relative displacement at the ends of each solder joint was calculated. Displacement components used in the detailed solder joint model as a function of temperature are shown below in Figure 3.

FIGURE 3

PBGA Test Exposure Profile
Induced Displacements at Critical Solder Joint

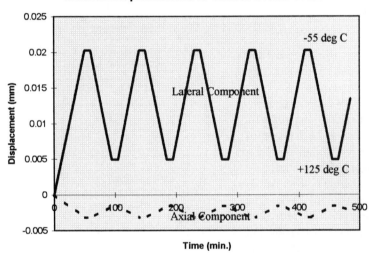

4 DSC DETAILED SOLDER JOINT MODEL

4.1 Governing Theory

The Disturbed State Concept (DSC) [1, 2, and 6] is a relatively new approach that allows the simulation of microcracking leading to damage accumulation and material degradation as a result of yielding, micro-cracking, creep, and cyclic loading. The finite element code in which it has been implemented allows the user to selectively define which aspects of material

1095

behavior will be simulated, from simple elastic analysis to elasto-visco-plastic with disturbance and temperature effects.

The concept is based on the assumption that a material element behaves as though comprising two characterizations or states, called 1.) Relative Intact (RI) and 2.) Fully Adjusted (FA). The RI state characterizes virgin, undamaged material, and can be simulated by simple elastic or elasto-plastic models of various forms. The FA state is used to characterize material that has been damaged or cracked and can no longer sustain tensile or shear loads. The finite element implementation simulates material degradation by tracking plastic strains, and correlates this with damage observed in laboratory tests on solder samples, gradually increasing the FA (or damaged) component until some predetermined limit is reached, signifying element failure. The observed response is expressed in terms of the RI and FA responses through disturbance as an interpolation and coupling mechanism. The Disturbed State Concept is implemented in a Finite Element (FE) code called DSC-SST2D [5]. The two states and their implementation in the FE code are described in detail in [13].

4.2 Model Development

A drawing of a typical solder joint was used to develop this model. The typical solder joint height is 0.600 mm with a diameter of 0.38 mm at the top and bottom, and a diameter of 0.58 mm in the middle. The top flat area interfaces with the PBGA, and the bottom flat area interfaces with the PWB. The solder joint was discretized into the 32 element mesh shown in Figure 4. Each element is an 8-node quadrilateral. The current finite element code is restricted to two-dimensional problems, so either plane strain, plane stress or axisymmetric cases can be considered.

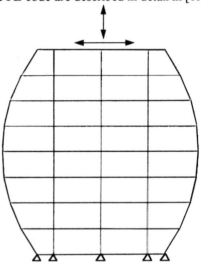

FIGURE 4
Solder Ball Finite Element Model

Initially a plane strain condition was simulated, but comparisons between these results and predictions from a 3-D NASTRAN linear elastic model showed significant disagreement. In the end, a plane stress model with a thickness equal to 0.48 mm, which is the average of the major diameter (0.58 mm) and minor diameter (0.38 mm), was used.

4.3 Boundary Conditions and Load Cases

As shown in Figure 4, the base of the solder joint is fixed, and the rest of the model is free to deform in the x and y directions.

Figure 4 also illustrates the application of displacement conditions on the top surface of the solder joint that are derived from the NASTRAN model of the PBGA/PWB. Displacement conditions from the PBGA/PWB model are imposed on the detailed solder joint model. Two displacement components are applied; a lateral component and an axial component. These displacement conditions fluctuate with temperature cycling as shown in Figure 3.

4.4 Results

The model was exercised for a full 4000 thermal cycles to simulate behavior observed in the thermal cycling tests on the 313 PBGA components at Boeing. Results from the first 20 cycle simulation are plotted in Figure 5. These extreme values occur in the upper left corner of the solder joint (Figure 3) and correspond to exposure at the two temperature levels. For example, at -55°C, σ_x reaches a maximum of about 20 MPa, σ_y peaks at 38 MPa, and τ_{xy} peaks at 2 MPa. At 125°C, σ_x reaches a minimum of -15 MPa (tension), σ_y reaches -32 MPa and τ_{xy} reaches 1 MPa. There is a slight reduction in stress levels as the cycles go on and the solder weakens under the applied cyclic displacement. This trend continues throughout the simulation.

FIGURE 5

Maximum Solder Ball Stress vs. Thermal Cycle

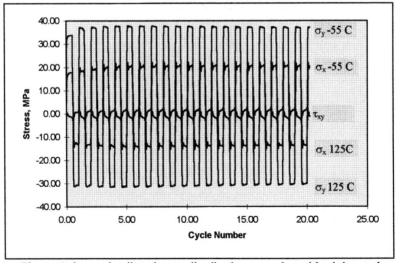

Figure 6 shows the disturbance distribution over the solder joint at the end of 2500 and 2750 thermal cycles. Generally the highest disturbance values are observed in the corner areas of the joint. It should be noted that this model does not consider geometric nonlinearities that may result from gross solder ball deformation and the resultant redistribution of load, however this simplification is probably conservative.

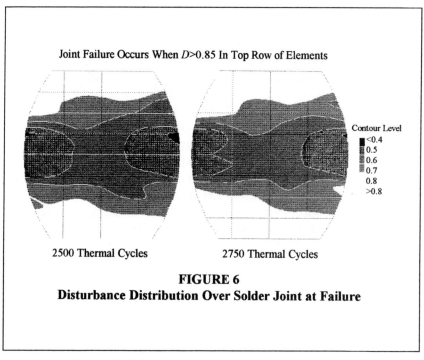

Joint Failure Occurs When $D>0.85$ In Top Row of Elements

Contour Level
■ <0.4
▦ 0.5
▦ 0.6
▦ 0.7
0.8
>0.8

2500 Thermal Cycles   2750 Thermal Cycles

FIGURE 6
Disturbance Distribution Over Solder Joint at Failure

4.5 Assumptions and Limitations

In the course of this analysis effort, a number of simplifying assumptions and realizations of this idealized approach were made. Some of these have little or no effect on the results, but others may affect the outcome, and were made in the interest of practicality. All are described in detail in [13].

5 TEST DATA

5.1 Background

In February of 1996, a number of PBGA and Ceramic Ball Grid Array (CBGA) components of different sizes and configurations were soldered to two different circuit card substrate materials and subjected to numerous thermal cycles in an effort to assess critical production parameters associated with these parts. A total of 150 test articles were fabricated and subjected to thermal cycling as part of an Interconnection Technology Research Institute (ITRI) study which was conducted among a consortium of organizations including then-GM-Hughes Electronics Missile Systems Company (now Raytheon),

Boeing, JPL, AMKOR/ANAM, Electronics Manufacturing Productivity Facility (EMPF), American Micro Devices (AMD), and Rochester Institute of Technology (RIT).

5.2 Test Conditions and Setup

The test articles were sent to Boeing, EMPF and JPL for thermal cycle testing. It is important to note that the only environmental condition studied here was thermal cycling due to external temperature excursions; no internal heating or dynamic mechanical loading conditions were imposed. The Boeing test articles were cycled from -55 to +125°C until failure. In these tests, four test articles of the type analyzed in this work (313 PBGA's on an FR-4 substrate) were thermally cycled.

Conductor paths within the test article substrates were designed to allow continuous monitoring of continuity and thus provide insight into solder joint integrity. Pads were provided to allow individual continuity checks when daisy chain discontinuity was observed.

5.3 Test Results

Results of the tests at Boeing indicated that the first PBGA package failure was observed after 1852 thermal cycles between -55 and 125°C, and the fourth PBGA failed at 3624 cycles. The first solder balls to fail were located in the rows nearest the die, at the center of the package. Failure in these tests was estimated by considering the frequency of open circuits indicated during thermal cycling. Failure was assumed when an open circuit indication was measured at least 70 times within a thermal cycle.

5.4 Comparison of Model Predictions with Test Data

These data can be compared with model predictions from the detailed solder joint model as shown in Figure 7. On this figure, failure data vs. thermal cycle number for the four 313 PBGA's tested [7] are compared with model predictions from the present work.

Predictions from DSC-SST2D are in good agreement; the solder joint at the interfaces has reached a disturbance value of 0.8 to 0.85, indicating failure between 2500 and 2750 cycles, which is between the failure of the second and third packages.

The measure of solder joint integrity, the disturbance function, D, is used to determine at what point degradation and failure occur. Solder usually begins to show signs of microcracking at $D=0.5$. As D approaches 0.85, element failure occurs. These criteria are used to evaluate the disturbance predictions as thermal cycling progresses. Figure 7 shows the maximum value of D as thermal cycling progresses.

All elements have a disturbance value less than 0.5 at 500 thermal cycles which is the critical value for the onset of microcracking. At 1000 thermal cycles, elements near the top and bottom interfaces have disturbance values of about 0.5 which means that microcracking has begun in these areas.

As the number of thermal cycles continues to climb to 1500, Figure 7 shows that the maximum disturbance has climbed above 0.8, considered the critical level for element strength. The region of microcracking has expanded considerably, to 50% of the solder joint volume. At 2000 cycles significant

Figure 7
Comparison of DSC Model Predictions and 313 PBGA Thermal Cycle Failure Data

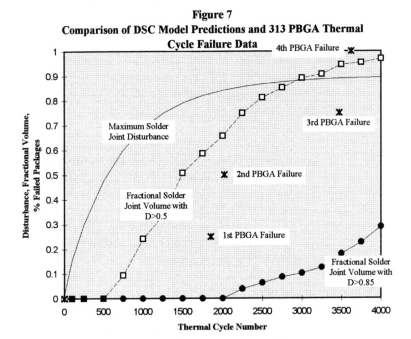

changes occur as both interface regions reach D=0.85, indicating the onset of solder failure in these areas. Also the area of microcracking, indicated by D=0.5, has spread to include nearly the entire solder joint. Based on these predictions, solder joint failure can be expected somewhere between 2500 and 2750 thermal cycles that range from -55°C to +125°C. It is interesting to note that the disturbance distribution appears to be slightly higher at the top interface with the PBGA (Figure 6), which would indicate that failure would occur here first. In fact, this is where solder joints typically fail during thermal cycling [7].

6 SUMMARY AND CONCLUSIONS

The goal of this task was to evaluate the predictive capability of a new finite element concept currently under development by Dr. C.S. Desai. During the course of this work, the code compared favorably with both hand calculations and MSC/NASTRAN finite element software, a recognized industry standard. Also, model predictions using the elasto-plastic and temperature dependent options appear to give reasonable results.

This report presents the development of a PBGA/PWB macro model using NASTRAN. Included in this model are bar elements that simulate the behavior of a detailed solder joint model, also created in NASTRAN. Results from this model were used as boundary conditions for the detailed solder joint model created in DSC-SST2D.

The development of four different detailed solder joint models using DSC-SST2D is documented in [13]. Features of the software that enable simulation of elasticity, plasticity, disturbance and temperature effects all are used. Multiple mesh densities were explored, and results between these models, NASTRAN elastic models and simple hand calculations were compared.

The ultimate test of this computer code is in the comparison with thermal cycling test data. Test data used in this work came from thermal cycle tests of four 313 PBGA packages. The models and boundary conditions developed in this work simulate the conditions of these thermal cycling tests. Results show good agreement between model predictions and test data.

Future work will focus on improvements to this DSC FE code, including an improved user interface and the development of additional DSC material parameters for an expanded list of joining materials including lead-free solders. This development work is partially funded by Raytheon Missile Systems Company and will continue at The University of Arizona under a National Science Foundation Grant under the direction of Dr. C.S. Desai.

7 ACKNOWLEDGEMENTS

The authors acknowledge the assistance of Mr. E.J. Warkomski, Mr. D.W. Campbell (Raytheon Missile Systems Company) and Mr. Zhichao Wang (The University of Arizona) who provided technical assistance and contributed to the development of this software during the course of this work.

REFERENCES

1. BASARAN, C., DESAI, C.S., and KUNDU, T.- "Thermomechanical Finite Element Analysis of Problems in Electronic Packaging Using the Disturbed State Concept; Parts I and II", Journal of Electronic Packaging, ASME, Vol. 120, No. 1, pp. 41-47 and 48-53, 1998.

2. CHIA, J., and DESAI, C.S. - Constitutive Modeling of Thermomechanical Response of Material in Semiconductor Devices Using the Disturbed State Concept, Report to the National Science Foundation, Department of Civil Engineering and Engineering Mechanics, University of Arizona, Tucson, Arizona, 1994.

3. COLE, M., CAULFIELD, T., BANKS, D.,WINTON, M., WALSH, A., GONYA, S. - "Constant Strain Rate Tensile Properties of Various Lead Based Solder Alloys at 0, 50, and 100°C", *Materials Developments in Electronics Packaging Conference Proceedings*, 1991.

4. DARVEAUX, R BANERJI, K., MAWER, A.,DODY, G. - "Reliability of Plastic Ball Grid Array Assembly", Chapter 13 of Ball Grid Array Technology edited by J. H. Lau, McGraw-Hill, Inc., pp. 223 –265, 1995.

5. DESAI, C.S. - "DSC-SST2D: Computer Code for Static, Dynamic, Creep and Thermal Analysis: Solid, Structure, and Soil-Structure Problems" and User's Manual, Tucson, Arizona, 1998.

6. DESAI, C.S. - Mechanics of Materials and Interfaces: The Disturbed State Concept, Under Preparation, 1999.

7. GHAFFARIAN, R. - "Ball Grid Array Packaging RTOP, Interim Environmental Test Results", Jet Propulsion Laboratory, California Institute of Technology, 1997.

8. JU, T., LIN, W., LEE, Y. C., LIU, J. J. - "Effects of Ceramic Ball-Grid-Array Packages Manufacturing Variations on Solder Joint Reliability", Paper No. 93-WA/EEP-2, ASME Winter Annual Meeting, New Orleans LA, November 28 – December 2, 1993.

9. MACNEAL-SCHWENDLER CORPORATION- NASTRAN Structural Analysis Code, V70, 1997.

10. MAWER, A., DARVEAUX, R., PETRUCCI, A. M.- "Calculation of Thermal Cycling and Application Fatigue Life of the Plastic Ball Grid Array (BGA) Package", Proceedings 1993 International Electronics Packaging Conference, pp. 379 – 442, 1993.

11. SKIPOR, A., HARREN, S., and BOTSIS, J.- "Constitutive Characterization of 63/37 Sn/Pb Eutectic Solder Using the Bodner-Partom Unified Creep-Plasticity Model", *Advances in Electronic Packaging*, ASME, 1992, pp. 661-672, 1992.

12. WONG, T. E., COHEN, H. M., JUE, T. Y., TESHIBA, K. T. - "Ceramic Ball Grid Array Solder Joint Thermal Fatigue Life Prediction Model", Raytheon Systems Co., Sensors and Electronic Division, 1998.

13. ZWICK, J.W. -"Thermo-Structural Analysis of a 313-Pin Plastic Ball Grid Array (PBGA)", Masters Report, The University of Arizona, Department of Civil Engineering and Engineering Mechanics, 1998.

Recovery of notch stress and stress intensity factors in finite element modeling of spot welds

Shicheng Zhang
Research and Technology, FT1/FB, E222
Daimler-Chrysler AG, 70546 Stuttgart, Germany
E-mail: shicheng.zhang@daimlerchrysler.com

Abstract: The fatigue strength of spot-welded structural components is often undermined by stress concentrations at spot welds. The stress concentrations are preferably described by notch stress or stress intensity factors. Two approaches will be presented in this paper where the notch stress and stress intensity factors at spot welds are determined or estimated by the structural stresses or interface forces and moments around or in the spot welds. Appropriate modelings of spot welds are discussed which are capable of offering the applicable structural stresses or interface forces and moments. The stress intensity factors and notch stress predicted by the two approaches are compared for a tensile-shear specimen with the results from other sources. As example of application the fatigue strength of a spot-welded box-section member is predicted by the tensile-shear specimen in terms of equivalent stress intensity factor or notch stress.

1. Introduction

There are stress concentrations at spot welds in spot-welded structures under service loading conditions. As a result, the fatigue strength of the spot welds may be decisive for the integrity and durability of the whole structures. There have been evidences that the fatigue strength of spot welds can be better predicted based on stress intensity factors or notch stress at spot welds because the fatigue test data of spot welds collapse or much less scatter if making use of equivalent stress intensity factor or notch stress as correlating

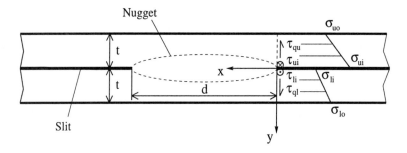

Figure 1: Structural stresses around a spot weld

parameter [1-3]. However, the information about stress intensity factors and notch stress at spot welds is normally lost in finite element analysis of spot-welded structures because the finite element mesh is rarely so fine that the stress intensity factors or notch stress can be extracted.

Two approaches will be presented in this paper which try to recover the notch stress and stress intensity factors in finite element analysis of spot-welded structures without using detailed solid elements. The approaches are based mainly on the structural stresses or the interface forces and moments around or in spot welds. In one approach, the notch stress and stress intensity factors are determined by the structural stresses around the spot welds based on some analytic solutions and in the other, they are estimated by the interface forces and moments in the spot welds according to some analytic approximations.

Finite element models for spot welds will be discussed which are capable of offering the applicable structural stresses or interface forces and moments around or in the spot welds. The stress intensity factors and notch stress predicted by the two approaches are compared with the results from other sources for a tensile-shear specimen. As example of application the fatigue strength of a spot-welded box-section member is predicted by the tensile-shear specimen in terms of equivalent stress intensity factor or notch stress at the spot welds.

2. Stress intensities converted from stresses

If the structural stresses σ_{ui}, σ_{uo}, σ_{li}, σ_{lo}, τ_{ui}, τ_{li}, τ_{qu} and τ_{ql} around a spot weld (see Fig. 1) are available, the stress intensities (notch stresses σ_k and τ_k and stress intensity factors K_I, K_{II} and K_{III}) can be determined according to

[4-10] by establishing a relationship between the structural stresses and the stress intensities. Based on [8] the following equations are derived for the notch stresses at spot welds with sheet thickness t and notch root radius ρ at the edge of weld nugget

$$\sigma_k = \sigma_n + \frac{1}{4\sqrt{3\pi}}\sqrt{\frac{t}{\rho}}\left\{\sigma_{ui} - \sigma_{uo} + \sigma_{li} - \sigma_{lo}\right.$$
$$\pm\left[2(\sigma_{ui} - \sigma_{li})^2 + (\sigma_{uo} - \sigma_{lo})^2\right.$$
$$\left.\left.+2(\sigma_{ui}^2 + \sigma_{li}^2 - \sigma_{ui}\sigma_{uo} - \sigma_{ui}\sigma_{lo} - \sigma_{uo}\sigma_{lo} - \sigma_{li}\sigma_{lo})\right]^{1/2}\right\} \quad (1)$$

$$\tau_k = \tau_n + \frac{1}{\sqrt{2\pi}}(\tau_{ui} - \tau_{li})\sqrt{\frac{t}{\rho}} \quad (2)$$

where σ_n and τ_n are nominal stresses at the spot welds. The nominal stresses at spot welds are usually difficult to determine and can normally be neglected because of small notch root radius at the spot welds. The combined sign "$\pm$" in (1) takes the plus if $\sigma_{ui} - \sigma_{uo} + \sigma_{li} - \sigma_{lo} \geq 0$ and the minus if $\sigma_{ui} - \sigma_{uo} + \sigma_{li} - \sigma_{lo} < 0$. The notch stress τ_k occurs just at the root of the notch (just on the edge of the weld nugget) and the notch stress σ_k may deviate from it as shown in Fig.2. The angle θ of deviation depends on the mixed loading condition at the root of the notch and it is given by the following equation

$$\theta = -\arctan\frac{\sqrt{3}(\sigma_{ui} - \sigma_{li})}{\sigma_{ui} - \sigma_{uo} + \sigma_{li} - \sigma_{lo}}. \quad (3)$$

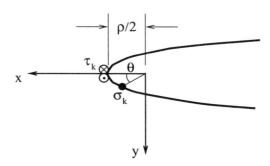

Figure 2: Notch stresses at spot welds

The stress intensity factors are determined by the structural stresses as follows

$$K_I = \frac{1}{6}\left[\frac{\sqrt{3}}{2}(\sigma_{ui} - \sigma_{uo} + \sigma_{li} - \sigma_{lo}) + 5\sqrt{2}(\tau_{qu} - \tau_{ql})\right]\sqrt{t} \quad (4)$$

1105

$$K_{\text{II}} = \left[\frac{1}{4}(\sigma_{ui} - \sigma_{li}) + \frac{2}{3\sqrt{5}}(\tau_{qu} + \tau_{ql})\right]\sqrt{t} \tag{5}$$

$$K_{\text{III}} = \frac{\sqrt{2}}{2}(\tau_{ui} - \tau_{li})\sqrt{t}. \tag{6}$$

The contributions of the structural stresses σ_{ui}, σ_{uo}, σ_{li}, σ_{lo}, τ_{ui} and τ_{li} to the stress intensity factors in the above three equations are quoted from [8] whereas the contributions of τ_{qu} and τ_{ql} are formulated according to the analytic results from [5] and to the proposals given in [11] for tackling the transverse stresses at the crack tip (*ibid.*, p. A.5). The effect of the transverse stresses at the notch root on the notch stress σ_k is neglected in (1). But the bending effect of the transverse shear stresses outside the notch root has been included in (1). The weld nugget diameter d is assumed to be significantly larger than the sheet thickness t ($d \geq 4t$). The relation between stress intensity factors and structural stresses at spot welds was first given in [12] based on boundary element analysis. Equations (4)-(6) are the analytic version of the relation. Equations (1)-(6) are valid for spot welds between sheets of identical material and identical thickness. For spot welds between sheets of different thicknesses and dissimilar materials similar but more complex equations are yet to be derived from a general solution for stress intensity factors at spot welds [9].

The structural stresses in (1)-(6) normally vary from edge point to edge point along the periphery of the weld nugget. Therefore, it may be necessary to calculate the stress intensity factors and notch stresses at a number of edge points in the periphery of the nugget in order to locate the maxima of the stress intensities around the spot weld. Attention should be paid to the coordinate system. The origin of the coordinate system is moving along the periphery of the nugget with the positive direction of the abscissa always pointing at the nugget center as shown in Fig. 1.

3. Stress intensities converted from forces

The notch stresses and stress intensity factors at a spot weld can be estimated according to [8] by the interface forces (F_x, F_y and F_z) and moments (M_x, M_y and M_z) which the spot weld transfers from one sheet to another

$$\begin{aligned}
\sigma_k &= \frac{4F}{\pi dt}\left(1 + \frac{\sqrt{3} + \sqrt{19}}{8\sqrt{\pi}}\sqrt{\frac{t}{\rho}}\right) + \frac{6M}{\pi dt^2}\left(1 + \frac{2}{\sqrt{3\pi}}\sqrt{\frac{t}{\rho}}\right) \\
&\quad + \frac{4F_z}{\pi d^2}\left(1 + \frac{5d}{3\sqrt{2\pi t}}\sqrt{\frac{t}{\rho}}\right)
\end{aligned} \tag{7}$$

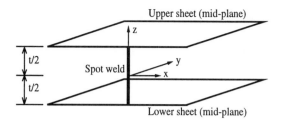

Figure 3: Coordinate system for interface forces and moments

$$\tau_k = \left(\frac{F}{\pi dt} + \frac{2M_z}{\pi d^2 t}\right)\left(1 + \sqrt{\frac{2}{\pi}}\sqrt{\frac{t}{\rho}}\right) \tag{8}$$

$$K_{\mathrm{I}} = \frac{\sqrt{3}F}{2\pi d\sqrt{t}} + \frac{2\sqrt{3}M}{\pi dt\sqrt{t}} + \frac{5\sqrt{2}F_z}{3\pi d\sqrt{t}} \tag{9}$$

$$K_{\mathrm{II}} = \frac{2F}{\pi d\sqrt{t}} \tag{10}$$

$$K_{\mathrm{III}} = \frac{\sqrt{2}F}{\pi d\sqrt{t}} + \frac{2\sqrt{2}M_z}{\pi d^2\sqrt{t}} \tag{11}$$

with resultant force $F = \sqrt{F_x^2 + F_y^2}$, resultant moment $M = \sqrt{M_x^2 + M_y^2}$ in the interface (joint face) and F_z, M_z along or about the z-axis which is perpendicular to the interface as shown in Fig. 3. The interface forces and moments are also called the joint face forces and moments or cross-sectional forces and moments. When spot welds are modeled by beam elements as is often the case, the interface forces and moments are simply obtained from the output of the beam elements. It is noted that the forces and moments are referred to the interface and not to the mid-plane of the plate elements simulating the spot welded sheets. The stress intensity factors and notch stresses given by (7)-(11) are the maxima of the stress intensities at a spot weld with K_{I}, K_{II} and σ_k at the leading vertex of the weld spot in line with the principal loading direction and K_{III} and τ_k at the side vertex which is at an angle of 90° from the principal loading direction.

Equations (7)-(11) are only estimations of the stress intensity factors and notch stresses at spot welds because only the forces and moments which are transferred by the spot welds are considered. The forces and moments running through the sheets which are not transferred by the spot welds may also cause stress intensities at the spot welds. Therefore, these equations are mainly valid for load-bearing spot welds. It is assumed in the derivation of (7)-(11) that the principal direction of the resultant force and the resultant moment in the interface coincides with each other in order to allow a simple superposition of the stress intensities caused by the force and the moment.

4. FE-models for spot welds

Reliable structural stresses or interface forces and moments around or in spot welds are preconditions for applying the two approaches given in the last two sections. These quantities are normally available from the output of finite element analysis provided spot welds are modeled appropriately. There are different finite element models for spot welds. They were reviewed in [13,14] according to the nature of the elements used to model the spot welds. These models can also be divided into force-reliable models, i.e. they can offer reliable interface forces and moments in the spot welds, stress-reliable models, i.e. they can offer reliable structural stresses around the spot welds, and force-and-stress reliable models, i.e. they can offer both reliable interface forces/moments and reliable structural stresses in and around the spot welds.

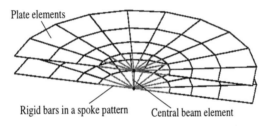

Figure 4: Finite element model for spot welds: a central beam element with diameter of nugget diameter d coupled through rigid bars in a spoke pattern (also with diameter of d) with plate elements

If spot welds are modeled by beam elements as is often the case, the interface forces and moments in the spot welds are normally applicable. If local refinements of mesh are additionally made around the beam elements, the structural stresses around the spot welds may be applicable, too. Obviously, any force-reliable model can be used in conjunction with the approach based on the interface forces and moments. Similarly, any stress-reliable model can be used in conjunction with the approach based on the structural stresses.

As an example, a finite element model with spot weld modeled by a beam element which is coupled through rigid bars in a spoke pattern with plate elements (mid-planes) as shown in Fig. 4 is used in the application examples in the next section. Mesh refinements around the beam element are made to obtain the structural stresses around the spot weld. The central beam element (in the center of the spot weld) is cylindric with diameter of the nugget diameter d and length of the sheet thickness t. The beam element is elastic and has the same material properties as the plate elements. The beam element can be set stiffer than the plate elements in view of the material hardening due to spot welding. But this has little influence on the structural stresses and the interface forces/moments.

5. Example of application and discussions

The tensile-shear specimen with single spot weld as shown in Fig. 5 is well analyzed and widely used in spot weld testings. The stress intensity factors for the specimen from different sources [4, 15-17], inclusive of detailed three-dimensional finite element results, have been available. Therefore, the two approaches given in the last two sections are first applied to the tensile-shear specimen for comparison.

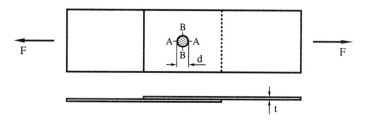

Figure 5: Tensile-shear specimen

The stress intensity factors for the tensile-shear specimen with nugget diameter of 5 mm and sheet thickness of 1 mm under tensile-shear force of 1 kN are predicted by the two approaches and compared in Table 1 with the results from other sources. The width of specimen differs to some extent among the different sources but this does not matter because it is significantly larger than the nugget diameter. The deviations in sheet thickness and nugget diameter in [16-17] are considered by the condition that $F/(d\sqrt{t})$ is a constant for the tensile-shear spot welds according to [8]. A local refinement of mesh with smallest plate elements of 0.5 mm (finer than the mesh shown in Fig. 4) is introduced around the spot weld and the structural stresses have been extrapolated to the nugget edge in the present predictions. It can be seen from the table that the stress intensity factors converted from both stresses and forces are close to the predictions from the different sources, especially for the leading stress intensify factor of K_{II}.

The notch stress for the tensile-shear specimen is so far not well analyzed like the stress intensity factors. According to a finite element analysis [18] the notch stress at the spot weld (point A in Fig. 5) in the tensile-shear specimen with nugget diameter of 6 mm, sheet thickness of 1.5 mm and notch root

Table 1: Stress intensity factors K_I and K_{II} at the leading vertex (point A in Fig. 5) and K_{III} at the side vertex (point B in Fig. 5) of the weld spot, all in N/mm$^{3/2}$, under tensile-shear force F=1 kN with nugget diameter $d = 5$ mm and sheet thickness $t = 1$ mm

Authors	K_I	K_{II}	K_{III}
Radaj [4], FEM	54.0	137.5	64.3
Smith and Cooper [15], FEM	63.2	132.4	99.0
Swellam et al. [16], FEM	64.2	134,6	-
Yuuki and Ohira [17], FEM+BEM	55.5	123.2	90.3
Zhang, converted from stresses	54.9	138.7	64.4
Zhang, converted from forces	55.1	127.3	90.0

radius of 0.3 mm is σ_k =247.6 MPa under tensile-shear force of 1 kN. The values predicted by the present approaches are 237.0 and 277.5 MPa converted from stresses and forces, respectively. The deviation of the current prediction from the finite element result seems acceptable.

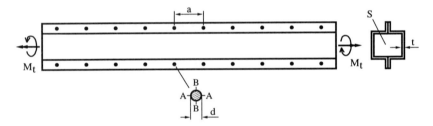

Figure 6: Spot-welded box-section member (spot pitch $a = 45$ mm, cross-sectional area $S = 50 \times 50$ mm^2)

As another example of application, the fatigue test data [19,20], in terms of force or moment range versus lifetime to failure, for a tensile-shear specimen and a spot-welded box-section member under torsion as shown in Fig. 6 are converted by the present approaches into the form of equivalent stress intensity factor range or notch stress range versus lifetime to failure (see Figs 7 and 8). The original test data in the two cases are not comparable because the load is force for the tensile-shear specimen and it is moment for the box-section member. The weld geometries in the two cases are also different. To save some calculations, especially for the box-section member, only the approach based on interface forces and moments is used in recovering the stress intensity factors and notch stress at the spot welds in this example. The finite element meshes used here are, then, much coarser than those used in the first application example.

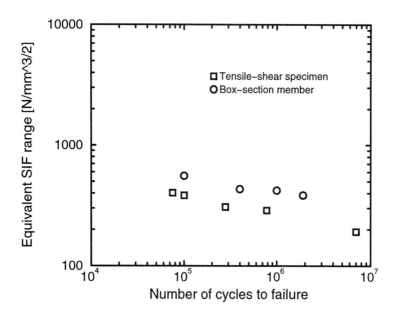

Figure 7: Equivalent stress intensity factor range versus lifetime to failure for the tensile-shear specimen and the box-section member

The following equivalent stress intensity factor according to [8] is used

$$K_{eq} = \pm\sqrt{K_{\mathrm{I}}^2 + \alpha K_{\mathrm{II}}^2 + \beta K_{\mathrm{III}}^2} \qquad (12)$$

where the square root takes the plus sign when $K_{\mathrm{I}} \geq 0$ and the minus sign when $K_{\mathrm{I}} < 0$. In this example $\alpha = \beta = 1.0$ is introduced. In the evaluation of the notch stress a fictitious rounding of $\rho = 1.0$ mm according to [4] is assumed. It can be seen from Figs 7 and 8 that the test data in terms of both stress intensity factor and notch stress have fallen into a relatively narrow band. In other words, the fatigue strength of the box-section member can be predicted by the tensile-shear specimen. This indicates the feasibility of the correlation based on the notch stress or the equivalent stress intensity factor. The leading vertex of the spot welds in the box-section member is approximately at point A as shown in Fig. 6 and the side vertex at point B.

In the application of the two approaches the main efforts more or less lie in the evaluation of the structural stresses or the interface forces and moments at spot welds. The evaluation of the structural stresses is quite straightforward given the coordinate system shown in Fig. 1. The stress intensity factors and

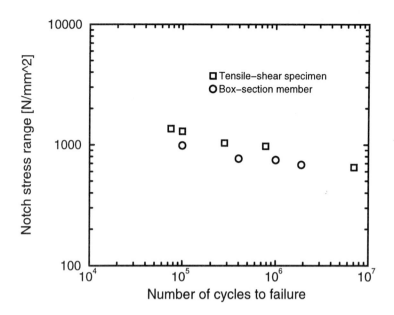

Figure 8: Notch stress range versus lifetime to failure for the tensile-shear specimen and the box-section member

notch stress derived are also referred to the same coordinate system. For the approach based on the interface forces and moments one should be cautious with the sign of the resultant forces and moments in the interface.

6. Concluding remarks

Two approaches are given for recovering the notch stress and stress intensity factors at spot welds based on appropriate finite element modelings of spot welds. The key elements of the approaches are the analytic relations between the stress intensities and the structural stresses or the interface forces/moments around or in the spot welds.

Generally speaking, any finite element model of spot welds which is capable of offering acceptable structural stresses or interface forces/moments at the spot welds can be used in conjunction with the analytic relations. The approach based on the interface forces/moments which normally demands no mesh refinement around spot welds in finite element analysis is only an estimation of the stress intensities due to the fact that the stress state at a spot weld cannot be determined by the interface forces/moments alone. This approach is, therefore, mainly suitable for load-bearing spot welds. For spot welds under general loading conditions the approach based on the structural stresses which, nevertheless, demands mesh refinements around spot welds is more reliable.

The approaches are valid only for spot welds between sheets of identical material and identical thickness and primarily for high cycle fatigue or brittle fracture of spot welds.

References

[1]   Pook, L. P., Fracture mechanics analysis of the fatigue behaviour of spot welds. *International Journal of Fracture*, Vol. 11 (1975), 173-176.

[2]   Yuuki, R., Ohira, T., Nakatsukasa H. and Yi W., Fracture mechanics analysis of the fatigue strength of various spot welded joints. in symposium on *Resistance Welding and Related Processes*, Osaka, 1986.

[3]   Lawrence, F. V., Wang, P. C. and Corten, H. T., An empirical method for estimating the fatigue resistance of tensile-shear spot-welds. SAE Paper 830035, 1983.

[4]   Radaj, D., *Design and analysis of fatigue resistant welded structures*. Abington Publishing, Cambridge, 1990.

[5]   Radaj, D. and Zhang, S., Stress intensity factors for spot welds between plates of unequal thickness. *Engineering Fracture Mechanics*, Vol. 39 (1991), 391-413.

[6]   Radaj, D. and Zhang, S., Simplified formulae for stress intensity factors of spot welds. *Engineering Fracture Mechanics*, Vol. 40 (1991), 233-236.

[7]   Radaj, D. and Zhang, S., Stress intensity factors for spot welds between plates of dissimilar materials. *Engineering Fracture Mechanics*, Vol. 42 (1992), 407-426.

[8]   Zhang, S., Stress intensities at spot welds. *International Journal of Fracture*, Vol. 88 (1997), 167-185.

[9]   Zhang, S., Stress intensities derived from stresses around a spot weld. *International Journal of Fracture*, 1998, accepted for publication.

[10] Zhang, S., Approximate stress intensity factors and notch stresses for common spot-welded specimens. *Welding Journal*, submitted for publication.

[11] Tada, H., Paris, P. C. and Irwin, G. R., *The Stress Analysis of Cracks Handbook*, Second edition, 1985.

[12] Radaj, D., Stress singularity, notch stress and structural stress at spot welded joints. *Engineering Fracture Mechanics*, Vol. 34 (1989), 495-506.

[13] Sheppard, S. D. and Strange, M., Fatigue life estimation in resistance spot welds: initiation and early growth phase. *Fatigue Fract. Engng Mater. Struct.*, Vol. 15 (1992), 531-549.

[14] Sheppard, S. D., Estimation of fatigue propagation life in resistance spot welds, ASTM STP 1211, 1993, 169-185.

[15] Smith, R. A. and Cooper, J. F., Theoretical predictions of the fatigue life of shear spot welds. *Fatigue of welded constructions*, Publ The Welding Institute, Abington, Cambridge, 1988.

[16] Swellam, M. H., Ahmad, M. F., Dodds, R. H. and Lawrence, F. V., The stress intensity factors of tensile-shear spot welds. *Computing Systems in Engineering*, Vol. 3 (1992), 487-500.

[17] Yuuki, R. and Ohira, T., Development of the method to evaluate the fatigue life of spot-welded structures by fracture mechanics. IIW Doc. III-928-89, 1989.

[18] Kuang, J-H. and Liu A-H., A study of the stress concentration factor on spot welds. *Welding Journal*, December 1990, 468s-474s.

[19] Mizui, M., Sekine, T., Tsuzimura, A., Takasshima, T. and Shimazaki, Y., An evaluation of fatigue strength for various kinds of spot-welded test specimens. SAE Paper 880375, 1988.

[20] Mabuchi, A., Niisawa, J. and Tomioka, N., Fatigue life prediction of spot-welded box-section beams under repeated torsion. SAE Paper 860603, 1986.

Application of a method of finite elements for an estimation crack-resisting of welded connections.
Nurguzhin M. R., Danenova G.T., Katsaga T.Ja.

The first pro-rector of Karaganda State Technical University, teachers

Summary

Survivability's prediction of welded connections with defects as a crack on the basis of the linear and nonlinear fracture mechanics assume use dependencies of crack-resisting's parameters from length of a crack, mechanical properties of a material and applied load. The coefficient of stress intensity (CSI) and energy J-integral are considered as parameters of the fracture mechanics in this article. The analytical expressions for the specified parameters are received on the basis of a method of finite elements, numerical experiment and methods of regression and correlation analyses. Is shown, that the most effective method of CSI estimation is a method of J-integral. The elastic - plastic approach at the solution of a nonlinear problem is realized on the basis of a method of initial stress in a combination to methods of radial return.

The concept of elastic coefficient of stress intensity (CSI) has played a fundamental role in development of the fracture mechanics and, in particular in researches of dynamic distribution of a crack. This parameter of the linear fracture mechanics is applied not only to the analysis of the reasons of destruction of already destroyed constructions or search of methods of destruction prevention. But also the parameter is applied with success for revealing connection between the stress-strain state (SSS) of a crack top and velocity of distribution of a fatigue crack. In this connection the main problem is the determination of analytical dependencies between hardness parameters and geometry of element defects. Using the program complex CRACK based on the finite element method (FEM), the method of rational design of experiments and regression analysis we obtained the formulas for calculation of coefficients of stress intensity for the typical patterns that simulate the behavior of welded connections with crack (figure 1).

Two factors were considered: l/b and P/P$_l$. Here l – a crack length; b – a pattern thickness; P - external load; $P_l = \sigma_t(b-l)$- load causing stress in netto-section of samples. We used the thickness of a welded slice δ instead of b for T-joint joint pattern.

We determined the parameter P_l for a pattern-bend and a –pattern three-point bend from the condition of causing of stress in netto-section. The stresses are equal to a yield point of material σ_T.

So, the load P_l was determined for a pattern three-point bend as

$$P_{np} = \sigma_T 4W/S, \qquad (1)$$

Where $W = t(b-l)^2/6$.

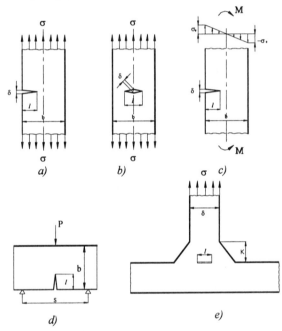

Fig.1. The settlement circuits(schemes) standard is exemplar

a - a pattern with one edge crack; b - a pattern with a central crack; c - a pattern-bend with one edge crack; d - a pattern-three point-bend; e - T- joint pattern

The ultimate load for a pattern-bend is estimated on the basic of formulas for a moment

$$M_{np} = \frac{\sigma_T \cdot t \cdot (b-l)^2}{6} \qquad (2)$$

1116

For T-joint pattern the ultimate load is determined as

$$P_{np} = \sigma_T (\delta' - 1)t, \qquad (3)$$

Where $\delta' = \delta + 1.4K$ - indicated thickness at defect with allowance for seam legs. Analytical dependencies (table 1) for the connections (figure 1) and their parameters of reliability are obtained on the basic of methods of non-traditional simulation of many dimensional associations [4]. The obtained formulas give a good hit rate for calculation of CSI with known experimental and numerical data [2, 5]. It is necessary to note that the application of the factor P/P_l allows easily to proceed to the analysis of elastic-plastic parameters of hardness such as a J-integral and a crack opening.

Table 1

The calculated formulas for coefficient of stress intensity

№	Pattern	Formula N/mm$^{3/2}$	Factor space	Parametrs of equation reliability		
				R	SD β	1-α
1	Pattern with one edge crack	$\dfrac{\sigma \cdot \sqrt{l}}{0,55 - 0,699 \cdot (l / b)}$	0,02< l/b≤ 0,5	0,999	2,8% 5	0,995
2	Pattern - three point bend	$6,67 \cdot \sigma \cdot \sqrt{l} \cdot ((l/b)^2 - 0,322 \cdot l/b + 0.282)$	0,02< l/b≤ 0,5	0,996	8,2% 5	0,995
3	Pattern - bend	$6,06 \cdot \sigma \cdot \sqrt{l} \cdot ((l/b)^2 - 0,305 \cdot l/b + 0.333)$	0,02< l/b≤ 0,5	0,999	2,1% 5	0,995
4	Pattern with a central crack	$0,86 \cdot \sigma \cdot \sqrt{l} \cdot (l/b + 1.26)$	0,02< l/b≤ 0,5	0,951	27,7% 4,5	0,995
5	T-joint pattern	$0,117 \cdot \sigma \cdot \sqrt{a} \cdot ((a/\delta)^2 x$ $x 0.524 \cdot (a/\delta) + 5.393) \cdot (K/\delta)^{-0.5}$	0,02<a/δ≤ 0,8 0,5<K/δ≤1,0	0,980	19,7% 4,5	0,995

In the considered above settlement cases the flat problem for open cracks was decided. The given situation is characteristic of majority of practical situations. At the same time, surface semielliptical cracks are frequently meeting defects in real constructions. Let's consider a procedure of CSI determination for such crack by the finite element method (program ANSYS). The beam with a rectangular cut 80x16 mm with a surface semielliptical crack (depth $l = 9.53$ mm and length $c = 2l$) was considered [6].

The three-dimensional isoparametric elements of the second order with 20 knots were used at the solution of the problem. Half of cross-section of a sample (figure 2) is divided on 44 tetragons that were the basis of prismatic elements. Total of knots was 2552. The characteristic sizes at a crack front were accepted equal (0.2... 0.25)l. The significances CSI were determined under the formula [2] in knots of a grid lying on shores of a crack

$$K(S) = \frac{EW(S)\sqrt{2\pi}}{4(1-v)\sqrt{S}}, \qquad (4)$$

Where W (S) - displacement of a knot that was on the distance S from a crack front on normal to it's shores; E and ν- modulus of elasticity and Poisson's constant.

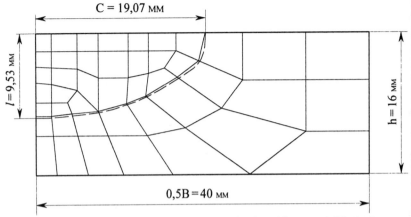

Fig.2. ½ Parts of finite-element model of the beam basis with a semielliptical crack

The significance K(S) in knots of a grid lying on large ($\varphi=\pi/4$) and small ($\varphi=0$) semiaxes of an ellipse are indicated on Figure 3.

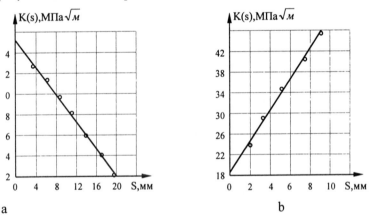

Fig.3. Linear extrapolation K (S)

a - S = 0 for $\varphi=\pi/2$; b - S = 0 for $\varphi = 0$

The significance $K_I = 18$ MPa $\sqrt{m}$ for $\varphi=0$ and the significance $K_I = 45$ MPa $\sqrt{m}$ for $\varphi=\pi/2$ were obtained by the method of extrapolation of significance K(S) to S=0. Comparison CSI obtained by different methods [6] is represented in table 2. These data testify that exact results were received in the field of an output of a crack front on a surface and less exact in the most detailed point of front of a semielliptical crack ($\varphi = 0$). At the same time is shown that using FEM it is possible to calculate significance CSI for any configuration of a crack and sample. The error of account can be reduced to minimum using a more small-sized partition.

Table 2

Method of determination	K_I, MPa $\sqrt{m}$		Δ,%	
	$\varphi = 0$	$\varphi = \pi/2$	$\varphi = 0$	$\varphi = \pi/2$
Data [8]	15.8	43.1	0	0
The analytical formulas [7]	16.8	43.7	6.3	1.4
FEM [6]	19.6	46.0	24.0	6.7
FEM in the present work	18.0	45.0	13.9	4.4

Let's consider a procedure of determination of universal dependencies for J-integral. In work [2] there is shown that the linear fracture mechanics can be applied down to netto-stress $\sigma_H = 0.8\sigma_T$ for edge cracks in massive samples by depth less than 0,25 cuts or for surface cracks by a size less than half of cut. The analysis of obtained results has shown that the limits of application of the linear approach hardly depend on a degree of a strain constraint, defect sizes and level of stress in netto-section. So for an edge crack in case of the plane stress state the magnitude of the boundary of applicability of a linear fracture mechanics is changed in limits $(0.6 \div 0.25)\sigma_T$ with an error in 15 % for a crack $l/b = 0.1 \div 0.5$. For the same cracks in conditions of a plane strain this parameter is changed from $0.85\sigma_T$ up to $0.40\sigma_T$. By analogy for samples with a central crack of length $(0.1 \div 0.6)l/b$ in case of the plane stress state the boundary of applicability of the linear approach is evaluated as $(0.75 \div 0.90)\sigma_T$, and for a plane strain - $(1.0 \div 1.1)\sigma_T$. These data essentially update earlier obtained results of other scientist [2].

The above information allows to proceed to determination of analytical dependencies J - integral.

Analytical expressions for J-integral in the function of an applied stress σ, geometry of a construction element with a crack and the material properties are offered in [9,10] as

$$J = \alpha \sigma_T \varepsilon_T \frac{l}{2}\left(1 - \frac{l}{b}\right) g_1\left(\frac{l}{b}, n\right)\left[\frac{b\sigma}{(b-l)\sigma_T}\right], \qquad (5)$$

where g_1- a function of the relation of a crack length l to a sample width b and of the parameter of material hardening n. It is supposed, that a material hardness changes under the degree function

$$\frac{\varepsilon}{\varepsilon_T} = \alpha \left(\frac{\sigma}{\sigma_0} \right)^n , \qquad (6)$$

where α - a constant of a material; $\varepsilon_T = \sigma_T/E$. The function is applied in a tabulated kind [10].

The analysis of expression (5) allows to conclude that the parameter $J' = J/(\alpha \sigma_T \varepsilon_T)$ depends only on relative length of a crack, load and a hardening parameter n. The tabulated significance of function $g(l/b, n)$ for patterns with a central crack for case of the plane-stress state are indicated in work [10]. To determination of similar expression for other settlement cases that are characteristic of welded connections with incomplete penetration on the basis of FEM and numerical experiment.

Let's consider the bilinear hardening. In this case we can exclude the factor n = 1. We established that parameter $J_p E_T / \sigma_T^2$ doesn't depend on the material at plastic deformations when stress in a netto section $\sigma_H > \sigma_T$. Here J_p - plastic J - integral that defined as

$$J_p = J_{ep} - J_e , \qquad (7)$$

Where J_e - elastic J - integral;

J_{ep} - elastic-plastic J - integral,

With allowance for the above information it is possible to note

$$J_{ep} = J_e + J_p = \frac{K_I^2}{E'} + \frac{\sigma_T^2 l(1-l/b)}{E_T} f_1(l/b, P/p_{np}), \qquad (8)$$

Where $E' = E$ in case of the plane stress state;

$E' = E/(1-v^2)$ In case of a plane strain;

$f_1(l/b, P/p_{np})$ - some function dependent on relative length of a crack and a level of an external load.

In expression (8) magnitude is

$$\frac{P}{P_{np}} = \frac{\sigma b}{\sigma_T (b-l)}$$

Using the equations for K_I (table 1) noted in a general view the expression(8) is the next

1120

$$J_{ep} = \frac{\sigma^2 l}{E'} f_0(l/b) + \frac{\sigma_T^2 l(1-l/b)}{E_T} f_1\left(l/b, P/P_{np}\right) \qquad (9)$$

The emerging of a multiplicand $(1-l/b)$ is connected to necessity of sufficing of expressions (8) and (9) to boundary conditions of a problem.

The tabulated significance for function $f_1\left(l/b, P/P_{np}\right)$ are obtained on the basis of numerical experiment. The graphic submission of a parameter $J_p E_T / \sigma_T^2$ is given in a fig. 4 and 5.

The submitted data gives good concurrence to the research of other authors and as against them are universal for considered steels and hardening dependencies of materials [11].

Thus, knowing viscosity of destruction J_{IC}, K_{IC}, properties of a material, geometry of pattern and it's SSS, the ultimate load is possible to determine. At the load a crack will begin to be distributed. We recommended using the two parametrical criteria on the basis of coefficient of stress intensity and J-integral as fracture criteria. The given approach is realized on the basis the computer system ANWELD of prediction of survival of pattern with defects as a crack [3].

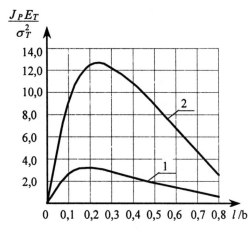

Fig.4. Dependence of a parameter $J_{ep} E_T / \sigma_T^2$ for a pattern with a central crack (100x200 mm)

1 – plane stress state; 2 - plane strain; $P/P_l = 1{,}4$

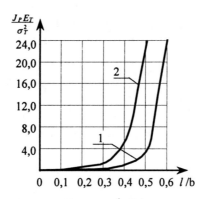

Fig.5. Dependence of a parameter $J_{ep}E_T\big/\sigma_T^2$ for a pattern with a edge crack

1 – plane stress state; 2 - plane strain; $P/P_l = 1,4$; (100x200 mm)

Reference

1. НУРГУЖИН М.Р. - Применение конечно-элементных графов в расчетах прочности металлоконструкций // Известия ВУЗов. Машиностроение. – 1988. – N 7

2. PARTON V.Z., MOROZOV E.M. - Mechanics of elastic-plastic problems. N.-Y., Hemisphere Publ., 1989. - P. 440

3. НУРГУЖИН М.Р., ДАНЕНОВА Г.Т., ВЕРШИНСКИЙ А.В.- Автоматизированный анализ разрушения сварных конструкций // Автоматическая сварка. - 1996. - N 10 - С.10-14

4. ПРОТОДЪЯКОНОВ М.М., ТЕНДЕР М.И. Методика рационального планирования эксперимента. – М.: Наука, 1970. – С. 75.

5. СИРАТОРИ М., МИЕСИ Т., МАЦУСИТА Х. - Вычислительная механика разрушения. – Москва: Мир, 1986. – 334 с.

6. ЛЕТУНОВ В.И. и др. - Закономерность развития поверхностных трещин в низколегированной стали при асимметричном циклическом изгибе. Сообщение 1 // Проблемы прочности. – 1985. – N 11. – С. 41-46

7. ВАЙНШТОК В.А. - Расчет коэффициентов интенсивности напряжений для поверхностных трещин в конструкциях // Проблемы прочности. – 1991. – N 3. – С. 29-39

8. АТЛУРИ С., КОБОЯСИ А., НАГАГАКИ М. и др.-
Вычислительные методы в механике разрушения – М.: Мир,
1990. – С. 391.

9. МАТВИЕНКО Ю.Г., ГОЛЬЦОВ В.Ю.- Некоторые аспекты
практического применения J-интеграла в расчетах на
прочность // Изв. ВУЗов. Машиностроение. – 1984. – N 10

10. GOLDMAN N.L., HUTCHINSON J.W.- Fully plastic crack
problems: The center crack strip under plane strain. Int. J. Solids
Structures. - 1975. - Vol 11, N 5.

11. NURGUZHIN M.R., DANENOVA G.T.- Application of J-integral
in the analysis of welded constructions. DESIGN, SIMULATION
and OPTIMISATION. Reliability and Applicability of
Computational Methods. // Proceeding of NAFEMS World
Congress'97. Stuttgart. Germany. Vol. 1. - p. 513-521.

CARES/*Life* SOFTWARE FOR CHARACTERIZING AND PREDICTING THE LIFETIME OF CERAMIC PARTS

Noel N. Nemeth[1]
Lesley A. Janosik[2]
Joseph L. Palko[3]

E-mail: cares@lerc.nasa.gov

ABSTRACT

The NASA Lewis Research Center has developed award-winning software that enables American industry to establish the reliability and life of brittle material (e.g., ceramic, intermetallic, graphite) structures in a wide variety of 21st century applications. The CARES (Ceramics Analysis and Reliability Evaluation of Structures) series of software is successfully used by numerous engineers in industrial, academic, and government organizations as an essential element of the structural design and material selection processes. The latest version of this software, CARES/*Life*, provides a general-purpose design tool that predicts the probability of failure of a ceramic component as a function of its time in service. In addition, the first version of the CARES/*Creep* program (for determining the creep life of monolithic ceramic components) has begun to be released to selected organizations. An overview of the development status of these programs is presented - with the emphasis placed on CARES/*Life*. Examples of components designed with CARES/*Life* are described, including a description of the reliability analysis performed for the New Millennium Mars Microprobe aeroshell .

1   INTRODUCTION

Transferring technological innovations beyond their primary aerospace applications into products with a much wider range of impact continues to be an integral part of the space program. The National Aeronautics and Space Administration (NASA) strives to ensure that its technological advances are readily accessible to industrial and academic organizations as well as to other government laboratories. At the NASA Lewis Research Center, a major effort has been dedicated to understanding and predicting the complex behavior of advanced, high-temperature ceramic materials for aerospace and terrestrial propulsion systems. Advanced ceramics offer the unique combination of being abundant materials that have a lighter weight and a greater

---

[1] Materials Research Engineer, NASA Lewis Research Center, Cleveland, Ohio
[2] Materials Research Engineer, NASA Lewis Research Center, Cleveland, Ohio
[3] Research Engineer, Connecticut Reserve Technologies L.L.C., Cleveland, Ohio

capacity to sustain load at a higher use temperature than metals. In addition, substitution of structural ceramics for traditional metals will reduce costs by increasing durability and efficiency. The increasing importance of ceramics as structural materials places high demand on assuring component integrity while simultaneously optimizing performance and cost. Successful utilization of these brittle materials as structural components requires that all potential failure modes must be identified and accounted for in a comprehensive design methodology.

In the United States, development of brittle material design technology has been actively pursued during the past two decades. In the late 1970s, NASA initiated development of the SCARE (Structural Ceramics Analysis and Reliability Evaluation) [1,2] computerized design program under the Ceramic Applications in Turbine Engines (CATE) program sponsored by the Department of Energy (DOE) and NASA. The goal of this project was to create public domain ceramics design software to support projects for the DOE, other government agencies, and industry. SCARE has since evolved into the CARES (Ceramics Analysis and Reliability Evaluation of Structures) and, most recently, CARES/*Life* (Ceramics Analysis and Reliability Evaluation of Structures Life Prediction Program) [3-6] integrated design programs. In addition, this effort has led to the newly-developed program, CARES/*Creep*, which predicts the creep life of ceramic components.

The CARES series of software is a general-purpose design tool that predicts the probability of failure of a ceramic component as a function of its time in service. Although created primarily to foster the introduction of ceramic materials in demanding space and aeronautics propulsion systems, this series of ceramic life prediction software and associated technology are being implemented for numerous other "dual-use" or "spinoff" applications. The CARES software has been used worldwide by hundreds of organizations representing industries such as aerospace, automotive, electronics, medical, and power generation.

Showing the customer that the government is a partner and is interested in a successful transfer of technology involves cultivation of interest in the technology as well as cooperation with both existing and potential customers. As further acknowledgment of their technical accomplishments, exemplary efforts, and outreach, the CARES development team has received numerous prestigious awards: 1994 NASA Software of the Year Award, 1995 NASA Engineering Excellence Award, 1995 R&D 100 Award from R&D Magazine, and the 1997 American Ceramic Society Corporate Technical Achievement Award. As improved ceramics emerge for structural applications, the need for design tools such as the CARES software will be essential to encourage commercial application of these materials.

This paper briefly describes an overview of the probabilistic ceramic component design procedure, the life prediction capabilities, and several recent new enhancements incorporated into the CARES design software. In addition, several typical "dual-use" ceramic technology spinoff applications designed using this software are presented, including a description of the reliability analysis performed for the New Millennium Mars Microprobe ceramic aeroshell .

2 PROBABILISTIC COMPONENT DESIGN

Significant improvements in aerospace and terrestrial propulsion, as well as in power generation, for the next century require revolutionary advances in high temperature materials and structural design. There is an increasing use of certain types of brittle materials (e.g., ceramics, intermetallics, and graphites) in the fabrication of lightweight components. From a design engineer's perspective, brittle materials often exhibit attractive high strength properties at service temperatures that are well beyond use temperatures of conventional ductile materials. For advanced diesel and turbine engines, ceramic components have already demonstrated functional abilities at temperatures well beyond the operational limits of most conventional metal alloys. However, a penalty is paid in that these materials typically exhibit low fracture toughness. This inherent undesirable property must be considered when designing components. Lack of ductility (i.e., lack of fracture toughness) leads to low strain tolerance and large variations in observed fracture strength. When a load is applied, the absence of significant plastic deformation or micro-cracking causes large stress concentrations to occur at microscopic flaws. These flaws are unavoidably present as a result of fabrication or in-service environmental factors. Non-destructive evaluation (NDE) inspection programs can not be successfully implemented during fabrication, because the combination of high strength and low fracture toughness leads to relatively small critical defect sizes that can not be reliably detected by current NDE methods. As a result, components with a distribution of defects (characterized by various sizes and orientations) are produced which leads to an observed scatter in component strength. Catastrophic crack growth for brittle materials occurs when the crack driving force or energy release rate reaches a critical value and the resulting component failure proceeds in a catastrophic manner. In addition, the design engineer must also be cognizant that the ability of brittle material components to sustain load degrades over time due to a variety of effects such as oxidation, creep, stress corrosion, cyclic fatigue, and elevated-temperature use. Stress corrosion cracking, cyclic fatigue, and elevated- temperature sustained load response are different aspects of the phenomenon called subcritical crack growth (SCG). SCG refers to the progressive extension of a crack over time. An existing flaw extends until it reaches a critical length, causing catastrophic crack propagation. Under the same conditions of temperature and load, ceramic components display large variations in rupture times from SCG. Unfortunately, the small critical flaw size and large number of flaws make it difficult to detect beforehand the particular flaw that will initiate component failure. Time-dependent crack growth can also be a function of chemical reaction, environment, debris wedging near the crack tip, and deterioration of bridging ligaments.

Traditional material failure analyses employing a deterministic approach, where failure is assumed to occur when some allowable stress level or equivalent stress is exceeded, are not adequate for brittle material component design. Such phenomenological failure theories are reasonably successful when applied to ductile materials such as metals. However, since analysis of failure in components fabricated from ceramics is governed by the observed scatter in strength, statistical design approaches must be used to accurately reflect the stochastic physical phenomena that determine material fracture response. Accounting for these phenomena requires a change in philosophy on the design engineer's part that leads to a reduced focus on the use of safety factors in favor of reliability analyses. If a brittle material with an obvious scatter in tensile strength is selected for its high strength attributes, or inert behavior, then components should be designed using an appropriate design methodology rooted in statistical analysis. However, the reliability approach demands that the design engineer must tolerate a finite risk of unacceptable performance. This risk of unacceptable performance is identified as a component's probability of failure (or alternatively, component reliability). The primary concern of the engineer is minimizing this risk in an economical manner.

Probabilistic component design involves predicting the probability of failure for a thermomechanically loaded component from specimen rupture data. Typically these experiments are performed using many simple geometry flexural or tensile test specimens. A static, dynamic, or cyclic load is applied to each specimen until fracture. Statistical strength and SCG (fatigue) parameters are then determined from these data. Using these statistical parameters, a time-dependent reliability model, and the results (i.e., stress and temperature distributions) obtained from a finite element analysis, the life of a component with complex geometry and loading can be predicted. This life is interpreted as a component's reliability as a function of time. When the component reliability falls below a predetermined value, the associated point in time at which this occurs is assigned the life of the component. This design methodology combines the statistical nature of strength-controlling flaws with the mechanics of crack growth to allow for multiaxial stress states, concurrent (simultaneously occurring) flaw populations, SCG, and component size/scaling effects. Additionally, although it will not be discussed in detail here, one approach to improve the confidence in component reliability predictions is to subject the component to proof testing prior to placing it in service. Ideally, the boundary conditions applied to a component under proof testing simulate those conditions the component would be subjected to in service, and the proof test loads are appropriately greater in magnitude over a fixed time interval. This form of testing eliminates the weakest components and, thus, truncates the tail of the strength distribution curve (the low probability of failure end of the distribution curve).

After proof testing, survived components can be placed in service with greater confidence in their integrity and a predictable minimum service life. With this type of integrated design tool, a design engineer can make appropriate design modifications until an acceptable probability of failure is achieved, or until the design has been optimized with respect to some variable design parameter. These issues are all addressed within the CARES/*Life* software.

3 CARES LIFE PREDICTION CAPABILITIES AND ENHANCEMENTS

Designing ceramic components to survive at higher temperatures than the capability of most metals and in severe loading environments involves the disciplines of statistics and fracture mechanics. Successful application of advanced ceramics depends on proper characterization of material properties and the use of a probabilistic brittle material design methodology. The CARES/*Life* integrated design software combines multidisciplinary research—in the areas of fracture analysis, probabilistic modeling, model validation, and brittle structure design—to determine the lifetime reliability of monolithic ceramic components. The CARES/*Life* software describes the probabilistic nature of material strength using the Weibull cumulative distribution function [7]. For uniaxially stressed components the 2-parameter Weibull distribution for surface residing flaws describes the component failure probability, P_f, as

$$P_f = 1 - \exp\left[-\frac{1}{\sigma_0^m} \int_A \sigma(x,y,z)^m \, dA \right] \quad (1)$$

where A is the area, $\sigma(x,y,z)$ is the uniaxial stress at a point location on the body surface, and m and σ_0 are shape and scale parameters of the Weibull distribution, respectively. The shape parameter is a measure of the dispersion of strength while the scale parameter is the strength of a unit area of material at 63.21% probability of failure. An analogous equation based on volume can be shown for flaws residing within the body of the component. The Weibull equation is based on the weakest-link theory (WLT). WLT assumes that the structure is analogous to a chain with many links. Each link may have a different limiting strength. When a load is applied to the structure such that the weakest link fails, then the structure fails. For CARES/*Life* the effect of multiaxial stresses on reliability is predicted by using either the principle of independent action (PIA), [8,9] the Weibull normal stress averaging method (NSA), [10] or the Batdorf theory [11,12]. For the PIA model the reliability of a component under multiaxial stresses is the product of the reliability of the individual principal stresses acting independently, which for surface distributed flaws is

$$P_f = 1 - \exp\left[-\frac{1}{\sigma_0^m} \int_A \sum_{i=1}^2 \sigma_i(x,y,z)^m \, dA \right] \quad (2)$$

where i is the individual principal stress component. The NSA method involves the integration and averaging of tensile normal stress components evaluated about all possible orientations and locations. This approach is a special case of the more general Batdorf theory and assumes the material to be shear insensitive. The Batdorf theory combines the weakest link theory and linear elastic fracture mechanics (LEFM). Conventional fracture mechanics analysis requires that both the size of the critical crack and its orientation relative to the applied loads determine the fracture stress. The Batdorf theory includes the calculation of the combined probability of the critical flaw being within a certain size range and being located and oriented so that it may cause fracture. The probability of failure for a ceramic component using the Batdorf model for surface flaws can be expressed as

$$P_f = 1 - \exp\left[-\frac{k_B}{\pi} \int_A \int_0^\pi \sigma_{Ieqc_{max},0}(\Psi)^m \, d\alpha \, dA \right]$$

(3)

where A is the surface area, ω is the arc length of an angle α projected onto a unit radius semi-circle in principal stress space containing all of the crack orientations for which the effective stress is greater than or equal to the critical crack propagation stress, k_B and m are Weibull parameters describing the material, and $\sigma_{Ieqc_{max}}(\Psi)$ is far-field equivalent maximum normal stress (the peak stress from a cyclic load). Equation (3) can be shown to reduce to a simple 2-parameter Weibull distribution.

Input for CARES/*Life* includes material data from simple experiments involving rupturing specimens, and results from a finite element analysis of a complex component. Finite-element heat transfer and linear-elastic stress analyses are used to determine the component's temperature and stress distributions. The reliability at each element (a very small subunit of the component) is calculated assuming that randomly distributed volume flaws and/or surface flaws control the failure response. The overall component reliability is the product of all the element survival probabilities. A data file containing element risk-of-rupture intensities (a local measure of reliability) for graphical rendering of the structure's critical regions can be generated. Activities regarding validation of the software include example problems obtained from the technical literature, in-house experimental work, Monte Carlo simulations (computer-generated data sets), beta-testing by users, and participation in a round robin study of probabilistic design methodology and corresponding numerical algorithms [13].

CARES/*Life* is an extension of the CARES (formerly SCARE) program, which predicted the fast-fracture (time-independent) reliability of monolithic ceramic components (see equations (1) through (3)) and estimated fast-fracture statistical material parameters. CARES/*Life* retains all of the fast-fracture capabilities of the CARES program, and additionally includes the ability to perform time-dependent reliability analysis due to subcritical crack growth (SCG). CARES/*Life* accounts for the phenomenon of subcritical crack growth by utilizing the power law, Paris law, or Walker equation. The power law expresses incremental crack growth, da where a is crack size, versus time t as

$$\frac{da(\Psi,t)}{dt} = A_1 K_{Ieq}(\Psi,t)^{N_1} + f_c A_2 K_{Ieqc_{max}}(\Psi,t)^{N_2-Q} \Delta K_{Ieq}(\Psi,t)^Q$$

$$(4)$$

and the Walker equation expresses incremental crack growth versus loading cycles n as

$$\frac{da(\Psi,t)}{dt} = A_1 K_{Ieq}(\Psi,t)^{N_1} + f_c A_2 K_{Ieqc_{max}}(\Psi,t)^{N_2-Q} \Delta K_{Ieq}(\Psi,t)^\zeta$$

$$(5)$$

where A_1, A_2, N_1, N_2, and Q are material constants which depend on temperature and environments, $K_{Ieq}(\Psi,t)$ is the mode-I equivalent stress-intensity factor at time t, $K_{Ieqc_{max}}$ is the maximum mode-I equivalent stress-intensity factor (*i.e.*, when the cyclic stress reaches maximum), and $\Delta K_{Ieq}(\Psi,t)$ is the range of mode-I equivalent stress-intensity factor. The parameter Q describes the material degradation from the effects of stress range. The Paris law is obtained from equation (5) when N_2 and Q are equal. A combined law formulation [14], superimposing equations (4) and (5), is also available for the special case of N_1 and N_2 being equal. From linear-elastic fracture mechanics, the stress-intensity factors can be expressed as

$$K_{Ieq}(\Psi,t) = \sigma_{Ieq}(\Psi) Y \sqrt{a(\Psi,t)}$$

$$(6)$$

$$K_{Ieqc_{max}}(\Psi,t) = \sigma_{Ieqc_{max}}(\Psi) Y \sqrt{a(\Psi,t)}$$

$$(7)$$

$$\Delta K_{Ieq}(\Psi,t) = \left[\sigma_{Ieqc_{max}}(\Psi) - \sigma_{Ieqc_{min}}(\Psi)\right] Y \sqrt{a(\Psi,t)}$$

$$(8)$$

where $\sigma_{Ieqc_{max}}(\Psi)$ is far-field equivalent maximum normal stress, $\sigma_{Ieqc_{min}}(\Psi)$ is far-field equivalent minimum normal stress, Y is a geometric factor, and $\sigma_{Ieq}(\Psi)$ is far-field equivalent normal stress. For constant amplitude and frequency cyclic loading, superimposing equations (4) and (5) and solving for inert strength (before SCG takes place at time $t = 0$) of a crack that causes component failure at time t_f from applied peak cyclic stress $\sigma_{Ieqc_{max}}(\Psi)$ yields

$$\sigma_{Ieqc_{max},0}(\Psi) = \left[\frac{\sigma_{Ieqc_{max}}(\Psi)^N \left\{ g(\Psi) + f_c \dfrac{A_2}{A_1}(1-R)^Q \right\} t_f}{B} + \sigma_{Ieqc_{max},f}(\Psi)^{N-2} \right]^{\frac{1}{N-2}} \tag{9}$$

where

$$B = \frac{2}{A_1 Y^2 K_{Ic}^{N-2}(N-2)} \tag{10}$$

and K_{Ic} is the critical mode I stress intensity factor, N is the fatigue exponent (for combined law), g is a constant factor for cyclic loadings for the power law (values of which range between 0 and 1), f_c is the loading frequency, and R is the R-ratio of the minimum cyclic stress to the maximum cyclic stress. The term $\sigma_{Ieqc_{max}}(\Psi)$ refers to the original strength of the crack at time $t = 0$. Equation (9) can be directly substituted into equation (2) to give the reliability of the component as a function of time, load, and loading frequency. Note that assuming either A_2 or A_1 equals to zero yields the pure power law or the pure Walker law formulation, respectively. An example of equation (9) is shown in Figure 1 for yttria-stabilized tetragonal zirconia (3Y-TZP) cyclically loaded ($f_c = 1$ Hz) uniaxial tensile specimens. The figure shows median (50%) regressed curves for various R-ratio's compared to experimental data. Note that at low R-ratio's the material strength degradation increases, while at high R-ratio's this trend is reversed.

Figure 1. Median stress-life curves for 3Y-TZP tensile specimens for various R-ratio's

Within CARES/*Life* parameter estimation routines are available for obtaining inert strength and fatigue parameters for the various crack growth laws from rupture strength data of naturally flawed specimens loaded in static, dynamic, or cyclic fatigue. In addition, CARES/*Life* can predict the effect of proof testing on component service probability of failure. Creep and material healing mechanisms are not addressed in the CARES/*Life* code. However, the newly-developed program, CARES/*Creep*, which determines the creep life of a monolithic ceramic component, addresses the creep rupture mechanism, as briefly discussed later in this section.

Because the modeling and prediction capability provided by the software is in demand by an extraordinary range of industries for a wide array of products, several new enhancements have recently been developed or are under development to increase the functionality and utilization of the CARES/*Life* software, including:

- **ANSCARES**-New interface module that supports the ANSYS finite element analysis program and includes an automatic surface recognition feature of solid element modeling for surface flaw reliability analysis.
- **WinCARES**-A graphical user interface (GUI) shell program controlling the CARES/*Life* FORTRAN-based numerical algorithms for PC computers. WinCARES also includes Graphics software for showing specimen rupture data, including Weibull plots, and static, cyclic, and dynamic fatigue plots. This feature is currently under development.
- **CARES/DB**-An interactive database program based on Microsoft Access containing representative specimen fracture data. This feature is currently under development.

Methodology to handle non-random orientation of flaws resulting from finishing operations such as grinding and the capability to characterize material data from any type of specimen geometry have also been added to the CARES/*Life* software.

Additionally, CARES/*Creep* (Ceramics Analysis and Reliability Evaluation of Structures/*Creep)* [15], the newest software in the CARES series, has been developed to predict the creep life of monolithic ceramic components. Currently, most ceramic researchers utilize deterministic approaches to predict lifetime due to creep rupture. Stochastic methodologies for predicting creep life in ceramic components have not reached a level of maturity comparable to those developed for predicting fast-fracture and SCG reliability. CARES/*Creep* utilizes the ANSYS finite element package and takes into account the time-varying creep stress distributions (stress relaxation). Asymmetric creep response from both tension and compression can also be considered. The creep life of a component is divided into short time steps, assuming a constant stress distribution for each step. The damage is calculated for each time step based on a modified Monkman-Grant creep rupture criterion. The criterion for failure to commence is that the normalized damage accumulated at a point in the component must reach or exceed unity. Rupture is assumed to occur when the damage zone is large enough so that the component cannot sustain load. The corresponding time defines the creep rupture life for the component. The current methodology incorporated into the CARES/*Creep* program is purely deterministic. Plans for further enhancements to CARES/*Creep* include developing the capability to account for stochastic creep rupture behavior.

4 CERAMIC TECHNOLOGY APPLICATIONS

In addition to their primary space and aeronautics propulsion applications, the CARES/*Life* software is used to design ceramic and glass parts for an extensive range of other demanding applications. These include hot section components for turbine and internal combustion engines, laser windows on test rigs, radomes, radiant heater tubes, cathode ray tubes (CRTs, or TV picture tubes), electronic packaging for microprocessors, heat exchangers, and prosthetic devices such as artificial hip joints, knee caps, and teeth. Engineers and material scientists also use these programs to reduce data from specimen tests to obtain statistical parameters for material characterization. Illustrated below are some typical design and analysis applications that have utilized the CARES and CARES/*Life* software and associated technology for several diverse applications.

4.1 Turbine and Internal Combustion Engine Parts

Solar Turbines Incorporated is using CARES/*Life* to design hot-section turbine parts for the CSGT development program [16] sponsored by the DOE Office of Industrial Technology. This project seeks to replace metallic hot section parts with uncooled ceramic components in an existing design for a natural-gas-fired power-generation turbine engine operating at a turbine rotor inlet temperature of 1120°C (2048°F). First stage blades and vanes, as well as the combustor liner, have been replaced with ceramic parts. Early field trials demonstrated the ability of the ceramic components to survive engine operational temperatures and loads. However, due to incidents of foreign object damage (FOD), the components are in the process of being redesigned to be more impact resistant. Ultimately, demonstration of the technology will be proved with a 4000-hr engine field test.

The CARES algorithm has been successfully used in the development of ceramic automotive turbocharger wheels at AlliedSignal's Turbocharging and Truck Brake Systems [17]. Specifically, the CARES algorithm was utilized to design the CTV7301 silicon nitride turbocharger rotor which was implemented in the Caterpillar 3406E diesel engine. The reduced rotational inertia of the silicon nitride ceramic rotor compared to a metallic rotor significantly enhanced the turbocharger transient performance and reduced emissions. Note that this was a joint effort involving AlliedSignal and Caterpillar and represents the first design and large-scale deployment of ceramic turbochargers in the United States. Over 1700 units have been supplied to Caterpillar Tractor Company for on-highway truck engines. These units together have accumulated a total of over 120 million miles of service. In Japan, ceramic automotive turbochargers have been in production since 1985.

Ceramic poppet valves for spark ignition engines have been designed by TRW's Automotive Valve Division [18], as well as by General Motors and other automotive manufactures. These parts, depicted with other engine components in Figure 2, have been field tested in passenger cars, with excellent results. Potential advantages offered by these valves include reduced seat insert and valve guide wear, improved valve train dynamics, increased engine output, and reduced friction loss using lower spring loads. Manufacturing costs and handling issues have been the largest impediments to their widespread use thus far.

Figure 2. Automotive valves and engine components.(Courtesy of TRW Automotive Valve Division.)

Figure 3. Stress plot of an evacuated 27-in. (68-cm-)-diagonal cathode ray tube (CRT). (Courtesy of Philips Display Components Company.)

4.2 Cathode Ray Tubes (CRT's)

Glass components behave in a similar manner as ceramics and must be designed using reliability evaluation techniques. Using the CARES/*Life* software, Philips Display Components Company has developed superior 27-inch (68-cm)-diagonal tubes optimized for safety, reliability, performance, and efficiency. Philips analyzed the possibility of alkali strontium silicate glass CRTs spontaneously imploding [19]. CRTs are under a constant static load due to the pressure forces placed on the outside of the evacuated tube. Tubes were analyzed both with and without an implosion protection band. The implosion protection band reduces the overall stresses in the tube and, in the event of an implosion, also contains the glass particles within the enclosure. Stress analysis (Fig. 3) showed compressive stresses on the front face and tensile stresses on the sides of the tube. The implosion band reduced the maximum principal stress by 20%. Reliability analysis with CARES/*Life* showed that the implosion protection band significantly reduced the probability of failure to about 5×10^{-5}. Philips was able to develop CRTs with a new design that reduced glass consumption, tube weight, hazardous waste, and x-ray emissions.

4.3 New Millennium Mars Microprobe

The Mars Surveyor '98 program is the next generation of spacecraft to be sent to Mars. The orbiter, which was launched in late 1998, is to achieve Mars orbit in September 1999 while the lander, which was launched in early 1999, is to touchdown near the southern polar cap in December that same year. Piggybacking on the lander are two small basketball-sized microprobes which will penetrate into the Martian polar subsurface in an attempt to detect water ice. The microprobes are attached to the Polar Lander Spacecraft's cruise stage, underneath the Lander's solar panels (Fig. 4). The microprobes are contained within protective ceramic aeroshells. Just prior to atmospheric entry, the microprobes separate from the lander, freefall through the atmosphere and impact on the Martian surface, whereupon the aeroshell shatters allowing the probe to penetrate below the surface. The aeroshells are made of silicon carbide (0.76 mm wall thickness) and coated with an ablative material (Fig. 5). While the aeroshells were designed to shatter on Mars, there was concern that stresses induced during launch could be large enough to fracture the aeroshells prematurely . Consequently, the NASA Jet Propulsion Laboratory requested Connecticut Reserve Technologies (CRT), LLC to conduct a reliability analysis of the aeroshell assembly from anticipated stresses.

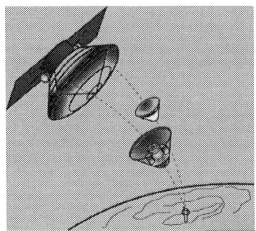

Figure 4. New Millennium Mars Microprobes separating from the Mars Polar Lander Spacecraft (Courtesy of NASA/JPL/Caltech)

Figure 5. Aeroshell. Fully assembled with ablative coating (Courtesy of NASA/JPL/Caltech)

For the analysis the assembly was assumed to be subjected to acceleration loads of 60 g's. Since the aeroshell was fabricated using a brittle silicon carbide material, the analysis needed to account for the inherent scatter in tensile strength associated with this ceramic material. Mechanical loads were analyzed using the ANSYS finite element code, and the CARES/*Life* program was used to conduct the reliability analysis of the assembly.

a) Approach

The CARES/*Life* algorithm utilizes the results from finite element analyses along with material failure data to arrive at a reliability value for a component subjected to given service load conditions. Here the material-specific information consisted of estimates of the inert strength Weibull distribution parameters obtained from 29 four-point bend bar failures. Note that slow crack growth was

1137

not considered since silicon carbide is relatively resistant to SCG. This data was generated by tests conducted at NASA JPL and provided to CRT. Estimates of Weibull parameters were determined from maximum likelihood technique (which conforms to ASTM C 1239). Nonlinear regression techniques discussed by Duffy et al. [20] were also used to estimate values for the three parameter Weibull distribution for comparison purposes. The results of the parameter estimation analysis appear in Table 1, and are presented graphically along with the failure data in Figure 6. Note that the estimates in Table 1 are at the low end of a 95% confidence interval and are therefore conservative estimates of parameters, while Figure 6 shows those values that represent the best-fit estimates relative to the data.

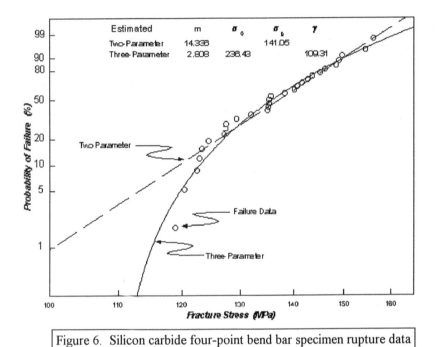

Figure 6. Silicon carbide four-point bend bar specimen rupture data

Figure 6 shows the specimen rupture data of rupture strength (maximum stress in the body at failure) versus probability of failure. This data can be characterized by a special case of equation (1) as

$$P_f = 1 - \exp\left[-\left(\frac{\sigma_f}{\sigma_\theta}\right)^m\right] \qquad (11)$$

where σ_f is the failure stress and σ_θ is the characteristic strength - the strength at which 63.21% of specimens fail.

Table 1: Weibull parameter estimates used for the CARES/*Life* Reliability analysis.

	m	σ_0 (MPa · mm$^{(2/m)}$)	σ_θ (MPa)	σ_{th} (MPa)
Two Parameter Distribution	10.103*	244.90*	137.07*	0.0
Three Parameter Distribution	2.808	236.43	N/A	109.31

*Parameters based on the low end of a 95% confidence interval for estimated parameter values.

It was also assumed (although this was not verified with fractographic analyses at the time of the analysis) that the data cited above came from failures that were generated at the surface of the bend bars. Note that the bend bars and the aeroshell have the same surface finish.

Three load cases were considered – the application of a 60 g acceleration in the global X direction, in the global Y direction, and the global Z direction. The two-parameter Weibull distribution was utilized in the reliability analyses. Since there was relatively little volume in comparison to the surface area of the aeroshell, a choice was made to utilize the surface analysis here.

b) Analysis and Results

Once the stress solution and Weibull parameters were obtained, the reliability of the aeroshell component was computed for each of the three load cases identified above. Table 2 contains the results for the reliability analyses using the PIA multaxial theory (equation (2)). The worst load case (based on the results from the reliability analysis) was generated by the 60 g acceleration in the global Z direction. A Von Mises stress plot of the aeroshell assembly for this acceleration is depicted in Figure 7. For the Z direction acceleration the reliability of the aeroshell assembly was 0.9950 (i.e., 5 failures out of 1,000). For the acceleration in the X direction, the reliability increased to 0.9984 (16 failures in 10,000) and for

1139

the acceleration in the Y direction, the component reliability was effectively 1.0 (no failure).

Table 2: Summary of component Reliability for various load conditions.

Load Case	Aeroshell Reliability	Aeroshell Probability of Failure
X-Acceleration	0.9984	0.0016
Y-Acceleration	1.000	0.000
Z-Acceleration	0.9950	0.0050

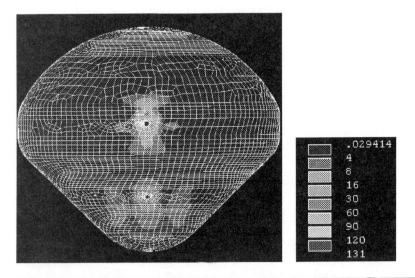

Figure 7. Von Mises plot of aeroshell model with 60g load in the Z direction (into the page). Note stress concentration about attachment. holes.

The acceleration load in the Z direction produced high stresses that were localized in the regions around the support attachments. The Z direction acceleration produced a maximum Von Mises stress of 131 MPa. This localized effect is caused by the presence of the rigid elements used to model the support attachments at the 6 holes equally spaced around the aeroshell. This high stress has an adverse effect on the reliability of the component since the probability of failure (which is equal to one minus reliability) involves raising stress to an exponent. One can see the association of high stress with the degradation in reliability in Figure 8.

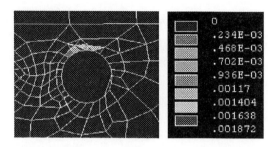

Figure 8. Risk of rupture intensity plot of attachment hole.

Here the risk of rupture intensity associated with each finite element in the vicinity of the support attachment is presented. This quantity is an area independent measure of the individual element reliability. Note that the highest values correspond to the elements where the stress is highest. The same effects are seen in the assembly for the acceleration in the X direction, only to a lesser extent.

For all three load cases the regions in the near vicinity of the support struts contributed most to the degradation in the reliability of the assembly. The fact that the strut connection was modeled in a simplistic manner with rigid elements and the use of the 2-parameter Weibull distribution for the reliability analysis indicates that the reliability values reported above are conservative. The net result of the analysis was that the aeroshell design was deemed acceptable for deployment.

5 CONCLUSION

Showing the customer that the government is a partner and is interested in a successful transfer of technology involves cultivation of interest in the technology as well as cooperation with both existing and potential customers. The technology transfer activities associated with NASA's CARES project have generated considerable good will and enthusiasm among customers. The impact of the technology transfer effort associated with this project is evident from the extensive customer base that has been developed. Numerous essential contacts with industry, government laboratories, and the university community, as well as a number of cooperative efforts, have been initiated as a result of this willingness to consult with and accommodate customers. The CARES software series has already been shown to be an essential tool in the successful utilization of advanced brittle materials for component designs encompassing a wide range of demanding applications. As improved ceramics emerge for structural applications, the importance of this work to industry, government agencies, and academia should continue to grow and gain further recognition.

ACKNOWLEDGEMENTS
The authors wish to thank all the CARES and CARES/LIFE users referenced in this paper. Further acknowledgement is given to those who have provided project descriptions and current status information.

REFERENCES

1. Gyekenyesi, J. P., "SCARE: A Postprocessor Program to MSC/NASTRAN for Reliability Analysis of Structural Ceramic Components," *Journal of Engineering for Gas Turbines and Power*, Vol. 108, 1986, pp. 540-546.
2. Gyekenyesi, J. P., and Nemeth, N. N.: "Surface Flaw Reliability Analysis of Ceramic Components With the SCARE Finite Element Postprocessor Program," *Journal of Engineering for Gas Turbines and Power*, Vol. 109, 1987, pp. 274-281.
3. Nemeth, N. N., Powers, L. M., Janosik, L. A., and Gyekenyesi, J. P.: "Time-Dependent Reliability Analysis of Monolithic Ceramic Components Using the CARES/LIFE Integrated Design Program," *Life Prediction Methodologies and Data for Ceramic Materials*, ASTM STP 1201, C. R. Brinkman, and S. F. Duffy, Eds., American Society for Testing and Materials, Philadelphia, 1993, pp. 390-408.
4. Nemeth, N. N., Powers, L. M., Janosik, L. A., and Gyekenyesi, J. P.: "Lifetime Reliability Evaluation of Structural Ceramic Parts with the CARES/LIFE Computer Program," AIAA paper 93-1497-CP, Proceedings of the 34th AIAA/ASME/ASCE/ASC Structures, Structural Dynamics, and Materials Conference, April 19-21, 1993, La Jolla, California. American Institute for Aeronautics and Astronautics, Washington, D.C., 1993, pp. 1634-1646.
5. Powers, L. M., Janosik, L. A., Nemeth, N. N., and Gyekenyesi, J. P.: "Lifetime Reliability Evaluation of Monolithic Ceramic Components Using the CARES/LIFE Integrated Design Program," Proceedings of the American Ceramic Society Meeting and Exposition, Cincinnati, Ohio, April 19-22, 1993.
6. Nemeth, N. N., Powers, L. M., Janosik, L. A., and Gyekenyesi, J. P.: "Ceramics Analysis and Reliability Evaluation of Structures Life Prediction Program (CARES/LIFE) Users and Programmers Manual," NASA TM-106316, to be published.
7. Weibull, W. A., "A Statistical Theory of the Strength of Materials," *Ingenoirs Vetenskaps Akadanien Handlinger*, 1939, No. 151.
8. Barnett, R. L.; Connors, C. L.; Hermann, P. C.; and Wingfield, J. R.: "Fracture of Brittle Materials Under Transient Mechanical and Thermal Loading," U. S. Air Force Flight Dynamics Laboratory, AFFDL-TR-66-220, 1967. (NTIS AD-649978)
9. Freudenthal, A. M., "Statistical Approach to Brittle Fracture," *Fracture, Vol. 2: An Advanced Treatise, Mathematical Fundamentals*, H. Liebowitz, ed., Academic Press, 1968, pp. 591-619.
10. Weibull, W. A., "The Phenomenon of Rupture in Solids," *Ingenoirs Vetenskaps Akadanien Handlinger*, 1939, No. 153.
11. Batdorf, S. B. and Crose, J. G., "A Statistical Theory for the Fracture of Brittle Structures Subjected to Nonuniform Polyaxial Stresses", *Journal of Applied Mechanics*, Vol. 41, No. 2, June 1974, pp. 459-464.
12. Batdorf, S. B.; and Heinisch, H. L., Jr.: Weakest Link Theory Reformulated for Arbitrary Fracture Criterion. *Journal of the American Ceramic Society*, Vol. 61, No. 7-8, July-Aug. 1978, pp. 355-358.
13. Powers, L. M., and Janosik, L. A.: "A Numerical Round Robin for the Reliability Prediction of Structural Ceramics," AIAA paper 93-1498-CP, Proceedings of the 34th AIAA/ASME/ASCE/ASC Structures, Structural Dynamics, and Materials Conference, April 19-21, La Jolla, California, 1993, pp. 1647-1658.
14. Rahman, S., Nemeth, N.N., and Gyekenyesi, J.P.: "Life Prediction and Reliability Analysis of Ceramic Structures Under Combined Static and Cyclic Fatigue," Presented at the International Gas Turbine and Aeroengine Congress and Exposition, Stockholm, Sweden, June 2-5, 1998
15. Powers, L. M., Jadaan, O. M., Gyekenyesi, J. P., 1996, "Creep Life of Ceramic Components Using a Finite Element Based Integrated Design Program (CARES/Creep)," ASME paper 96-GT-369.
16. van Roode, M., Brentnall, W.D., Norton, P.F., and Pytanowski, G.P, "Ceramic Stationary Gas Turbine Development", ASME Paper 93-GT-309, Presented at the

International Gas Turbine and Aeroengine Congress and Exposition, Cincinnati, Ohio, May 24-27, 1993

17. Baker, C. and Baker, D., "Design Practices for Structural Ceramics in Automotive Turbocharger Wheels", *Engineered Materials Handbook*, Volume 4: Ceramics and Glasses, ASM International, 1991, pp.722-727

18. Wills, R.R, and Southam, R.E., "Ceramic Engine Valves," *Journal of the American Ceramic Society*, Vol. 72, No. 7, pp. 1261-1264, 1989

19. Ghosh, A, Cha, C.Y., Bozek, W., and Vaidyanathan, S., "Structural Reliability Analysis of CRTs", *Society for Information Display International Symposium Digest of Technical Papers Volume XXIII*, Hynes Convention Center Boston, Massachusetts, May 17-22, 1992. Society of Information Display, Playa Del Ray, CA, pp. 508-510.

20. Duffy, S.F., Powers, L.M., and Starlinger, A.,"Reliability Analysis of Structural Components Fabricated from Ceramic Materials Using a Three-Parameter Weibull Distribution, *Transactions of the ASME - Journal of Engineering for Gas Turbines and Power*, Vol. 115, No. 1, pp. 109-116, January, 1993 (also published as NASA TM-105370).

DESIGNING OPTIMUM FIBRE LAYOUTS IN COMPOSITES

M.J.Platts and S.E.Jones

University of Cambridge, Manufacturing and Management Division
Mill Lane, Cambridge CB2 1RX, UK. mjp@eng.cam.ac.uk

SUMMARY

A vectorial design approach to placing fibres in the optimal directions is described, based on Michell's minimum weight theorem. Following a stress analysis, appropriate quantities of fibres are placed in the three principal stress directions. This creates a new structure which has to be re-analysed, creating an iterative process which leads to an optimum structure. This follows the natural process used in bone development. A computational model is demonstrated.

1. VECTORIAL DESIGN

Structural design in composites has always suffered from the gap between the micro-analytical level of the materials scientist, concerned with fibre-matrix adhesion etc., and the macro-analytical level of the conventional structural designer, wanting bulk material properties, which are preferably uniform throughout the component, and can thus be taken as a 'given' in the design process, just as metal properties are taken as a separately selected 'given'. This approach misses the fundamental advantage offered by composite materials, which is that at meso-level they offer strength and stiffness as a *vector* and thus ask the meso-level vectorial question "where do you want the fibres?". This vectorial question requires a vectorial answer. Once the designer faces it he recognises that the problem at meso-level is a problem of stress fields - vectorial patterns of stress - which ask for matching fibre fields to carry the stresses efficiently.

Once the switch is made to thinking about stress fields, a whole new way of thinking is opened up to the designer, drawing on the full body of work on field theory - magnetic fields, flow fields, stress fields etc. - developed a century ago. Within this body of work stands Michell's minimum weight theorem, which is a vectorial structural answer to the vectorial stress field problem[1]. In a way that is of pivotal relevance to composites designers, Michell thought not in terms of solid material but in terms of open frameworks. Taking as his starting point a notional stress field which at every point has three orthogonal principal stresses defining the stress field, he proved that the minimum weight structure to carry that stress field is not a solid structure but an open framework, a curvilinear mesh of members in the three orthogonal directions, the members being thicker or thinner, and spaced closer or wider, as appropriate to the local load (figure 1). At a stroke, Michell redefined the rules of structural analysis, in that for most components we are concerned with uniformly distributed material, within which stresses are non-uniformly distributed. Michell observed that efficiency is in fact achieved by the opposite arrangement, in which material is non-uniformly distributed in space, so that the stresses *are* uniformly distributed within the material:- this second being the defining characteristic of structural efficiency.

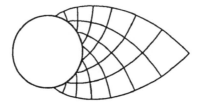

Figure 1. An analytically derived Michell truss cantilever[2].

2. MATCHING STRESS FIELDS

Civil engineers have always understood this, and when designing reinforced concrete structures will separately think of the overall concrete shape, and then put the steel reinforcing bars in as appropriate to carry the tensile stress field. Nature does it even more elegantly, bone being a continuously edited and refined 3D fibrous structure tuned to match a dynamic stress field. The challenge for composites design is to follow this same logic and achieve not only highly efficient but also highly resilient designs, as nature does. In the process, the analytical nature of the task changes. Whereas most FE analysis of composites is concerned with finding failure zones in secondary places, where despite the local stresses wishing to go in directions 1,2,3 the fibres happen to be going in directions A,B,C, the design process now is a geometric orientation problem so that there *are* no secondary stress zones. Not only are the fibres being used to their optimum advantage, the matrix is being defended to the maximum as well.

Interestingly, optimisation software based on this approach and mimicking exactly what bone does as an iterative calculation, progressively applying both quantity and orientation of material to design a fibre field to match an applied stress field, gives simple code and produces elegant and practically build-able structures, suitable for development as preforms for use in resin transfer moulding, for instance. This development for the first time enables composites to properly address the seriously 3D mechanical components or parts of components (e.g attachment points) which still tend to be left in metal. While the central algorithm produces elegant code however, it poses difficulties for the peripheral data handling which surrounds the finite element analysis, as FE packages are written assuming material properties will be specified and then applied to a whole block of elements. When, as a fundamental part of the conceptual thinking, every element will have different material properties, and these different properties involve genuinely different physical entities (different quantities of fibres, in different orientations), the representation of the material properties during the optimisation process, and the representation of material distribution after the completion of the optimisation process, requires some tuning of FE packages.

Returning to the fundamental concept of generating a curvilinear orthogonal mesh of fibres to achieve a minimum weight structure, the importance of orienting fibres in the principal stress directions is obvious. What is less obvious is how keeping the fibres orthogonal to each other protects the resin. As figure 2 shows[3], the secondary stresses in the resin matrix in a fibre reinforced composite component rise rapidly for non-orthogonal fibre layouts and are usually the source of failure, denying the component the potential strength offered by the fibres. Optimisation of meso-level fibre orientation is thus a major design tool in serious design work.

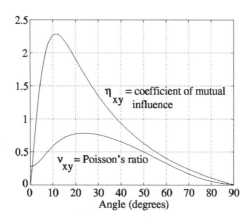

Figure 2. Secondary effects of non-orthogonal fibres in composites.

From the point of view of quality assurance in manufacture, smooth curvilinear patterns also generate fibre arrangements which are both easy to lay and easy to 'read' visually for correctness of shape. Exploration of off-geometry variants (of fibre direction or load direction), such as that of fibres round a loaded hole (figures 3,4) show that optimised layouts are also highly resilient, being very visibly distorted before any significant strength loss has occurred[4]. They are thus 'optimised' from a QA point of view as well.

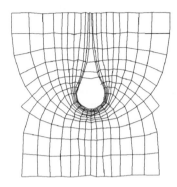

Figure 3. Fibre pattern distortion giving 30% loss of strength round a loaded hole.

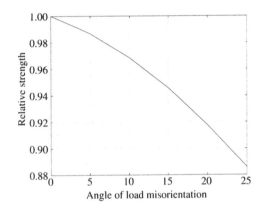

Figure 4. Variation in strength with load angle.

3. BONE - A NATURALLY OPTIMISED MATERIAL

Bone is a fibrous material. As a foetus develops, what will become bone begins as cartilage. The basic mechanism in bone design is a pair of cells which constantly feel the strains being experienced by the cartilage and respond to it. There is obviously a summation process involved and a certain speed of response, which together create the damping necessary for any control system and this damping has to be included in the FE representation of this process, to get stable, optimised solutions. One of these cells steadily sums the strain levels and if they are too high, lays down calcium to add stiffness and resist the strains. Its colleague sums the same strains but does the opposite, in that if the strains are too low it eats calcium away. It is immediately apparent that these two cells produce not only a bone creation mechanism but also a structural auditing and editing process, which continuously maintains an optimum of cancillarious material, optimised both in quantity and direction (figure 5)[5].

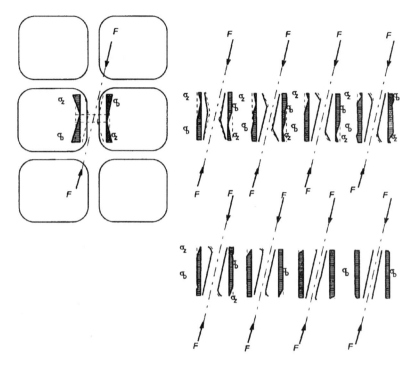

Figure 5. Bone's progressive geometric adaptation to load.

The observation that bone produces orthogonal fibrous patterns following stress trajectories was first made by Culmann and Meyer in 1867 (figure 6)[6,7,8] and the understanding of the mechanism of how it does so, and of the structural significance of that, is a fascinating story in its own right proceeding from then to the present, with a series of insights coming alternately from medicine and engineering (Culmann was an engineer and Meyer was a doctor) at the rate of about one key insight per generation. Bone's mechanism of adding or removing fibrous material both in quantity and direction to match the stress field is easy to replicate as an iterative finite element routine, progressively editing a structure until it is uniformly stressed.

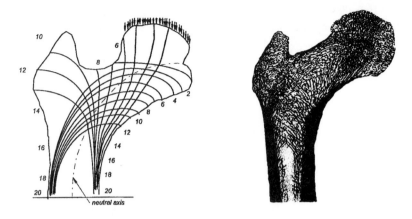

Figure 6. Lines of stress in the upper femur compared to a bone section.

4. COMPUTATIONAL APPROACHES TO OPTIMISATION

If the important detail of direction, i.e. orthogonality, is left aside, and the addition and subtraction of material is considered, this can be simulated in a FE package by the hardening and softening of a material through temperature effects. For instance this can be done in ABAQUS using UMAT - user defined material properties. Mattheck, at Karlsruhe, has produced some very elegant software using this approach, his inspiration also coming from nature, namely the shape-adaptive growth process demonstrated by trees[9]. Mattheck has more recently extended this work to allow for the orthogonality which is present in all fibrous materials, including wood[10]. In an optimisation sense the only difference between wood and bone is that wood is only ever an additive process. One of the key observations made about bone, made by Roux in 1895 was that because mammals move, minimum weight structures are of central importance to their bodily energy efficiency[5]. Thus the ability to remove unneeded bone is an important part of a mammal's strategic design, making the bone editing process a fundamentally more complete optimisation process.

A different and more abstract approach to both the addition and removal of material and the orientation of it has been developed by Kikuchi at Michigan[11,12]. Kikuchi developed a special code using elements containing a void. This void can grow and shrink, can change orientation and change its proportions and can thus in a rather abstract way represent the editing process which bone demonstrates. This route involves some rather fierce mathematics but, more importantly, results in a somewhat abstract suggestion of an optimised structure, which has to be remeshed as real material before a proper final analysis can be done (figures 7,8). Approaching the subject as a mathematician, Kikuchi observed, more or less as an interesting aside, that his analyses tended to produce Michell-like structures.

1150

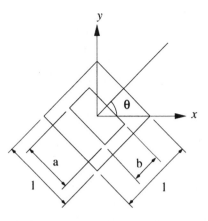

Figure 7. A volume element with a rectangular cavity whose proportions and direction can be varied.

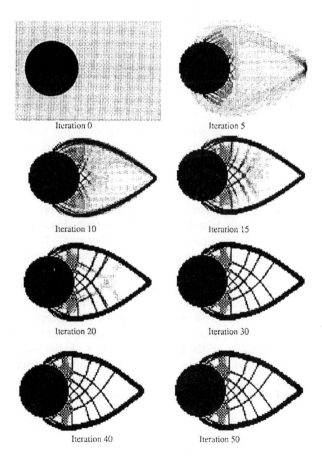

Figure 8. An indication of a cantilever shape developed using voided elements.

1151

Cambridge's approach [13,14] has been in many ways the opposite of Kikuchi's. Firstly, the addition and subtraction of the material itself has been used as the core concept, again using UMAT in ABAQUS, in a similar way to Mattheck. However, whereas Kikuchi finds similarities with Michell's structures as a conclusion, Cambridge has taken the fundamental assumption of Michell's theorem - that a minimum weight structure will always be a framework, always with three sets of fibres in orthogonal directions - as an equally important core concept. This enables the placement of fibres in composites design to be modelled directly using fairly simple assumptions, which include making simple assumptions about the non-orthogonal properties which help the speed and stability of the iterative calculations (when the off-orthogonal properties are significantly involved) and effectively disappear when the final solution is reached (when the off-orthogonal properties have lost their significance).

The Cambridge work has always had two threads to it, one seeking an absolute optimisation algorithm and the other seeking pragmatic, build-able solutions. The difference between these two is that, in the first, an ideal design route might be to define a free space and then allow the optimisation algorithm to generate an ideal minimum weight structure within it. In most cases such a structure will then have to be geometrically constrained and edited to produce a component which will fit. This will necessarily make it heavier, but there is considerable attraction in a design route which always goes through the minimum weight solution as a fundamental first step. As a minimum, what it keeps constantly visible is the extra 'cost' of a more constrained but pragmatically acceptable design.

This is now easy to calculate but, in a fully three-dimensional solid, gives problems concerning how to draw it. However, many composites components turn out to be essentially two-dimensional shell-type structures, in which the third stress direction is small compared to the other two. This is compounded by the fact that the resin in composites does have noticeable strength and when, in an optimised design, the proportion of fibres in the third direction drops to a small percentage, there comes a point where, simply from a fibre packing point of view, it becomes more efficient to leave it out altogether. This means that many analyses can be done using shell elements, giving rapid solution times. On a shell, the resulting values of 'density of fibres' in the two orthogonal directions simply translate as 'thickness of fibre layers' in the two directions and this is easily represented by colour contours on the fibre orientation plots, as in the examples of the fibre layers required for a loaded lug (figure 9) or for a formula 1 Grand Prix car body loaded in torsion (figure 10).

Tensile load case, Primary fibre directions

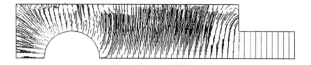

Tensile load case, Secondary fibre directions

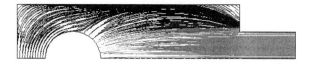

Compressive load case, Primary fibre directions

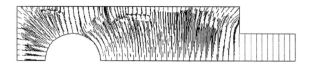

Compressive load case. Secondary fibre directions

Figure 9. Pairs of orthogonal layers of fibres for a lug, optimised for tension and compression.

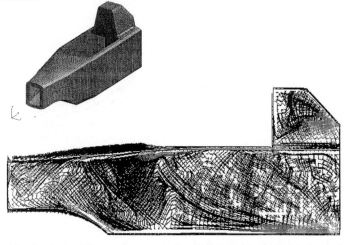

Figure 10. Optimised fibre orientations for a Formula One chassis loaded in torsion.

1153

The second thread however starts from a designer's sense of how he is going to make something, and so starts with a set of postulated shapes which might have bundles of fibres round things in particular ways which are conceptually right but which need to be optimally sized. Here the problem is one of relieving the chore of specifying material directions for every element and linking it to what must inevitably be the fibre bundle placement geometry, so that the two can be optimised together. Here, judicious use of UMATHT in ABAQUS allows fibre directions to be specified - thus specifying the material properties for each element - by imagining the fibres to be heat flow lines through that zone. This enables much more pre-specified shapes to be optimised, such as the end of a tubular tie/strut shown in figure 11, where longitudinal fibres bunch and wrap round a metal bobbin and are held in place structurally by a circumferential binding.

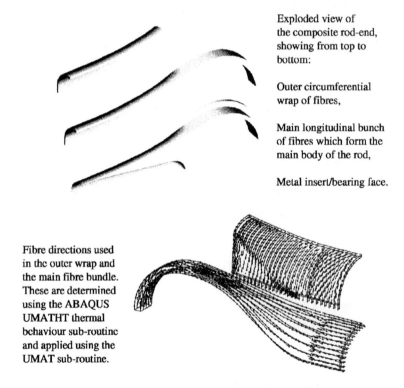

Exploded view of the composite rod-end, showing from top to bottom:

Outer circumferential wrap of fibres,

Main longitudinal bunch of fibres which form the main body of the rod,

Metal insert/bearing face.

Fibre directions used in the outer wrap and the main fibre bundle. These are determined using the ABAQUS UMATHT thermal behaviour sub-routine and applied using the UMAT sub-routine.

Figure 11. Composites eye at the end of a tubular composites strut/tie.

5. CONCLUSIONS

A central aim of the Cambridge work is to introduce a design route which automatically establishes a minimum weight design as the first step in the design process. This is a key psychological step towards always producing

1154

efficient designs. The functional introduction of optimisation as an automatic part of the design process involves three stages. The first is the development of usable algorithms which help the designer think about the principles involved in efficient design. In this respect Michell's theorem is very clear and helpful. The second stage is the development of usable computer code to apply the algorithm. What is visible here is that Michell's theorem generates elegant core code but its full utilisation depends on peripheral software improvements to the data handling aspects of FE packages, since these iterative design processes ask fundamentally different things of an FE package compared to the straightforward analysis processes for which they were originally written.

Finally however, pragmatic issues of manufacturability constrain composites design to an unreasonable degree. Whereas a designer designing a machined metal part has hundreds of thousands of different cutting tools to select from, a designer designing a composite part has only a handful of techniques available for placing fibres, all of which are highly restrictive. An efficient vectorial approach to design works. An efficient vectorial approach to fibre placement has yet to be developed.

ACKNOWLEDGEMENTS

This work is supported by EPSRC Grant RG23105.

REFERENCES

1   MICHELL, A G M - The limits of economy of material in frame structures. Phil. Mag, Vol. 8, No. 6, pp. 589, 1904
2   PRAGER, W - Optimisation in structural design, Mathematical Optimisation Techniques, Bellman, R, RAND Corp. Report P-396-PR, pp. 279-289, 1963
3   HERAKOVICH, C T - On the relationship between engineering properties and delamination of composite materials, J. Comp. Mater., Vol. 15, pp. 336-338, 1981
4   JONES, S E and PLATTS, M J - Using internal fibre geometry to improve the performance of pin-loaded holes in composite materials, Appl. Comp. Mater., Vol. 3, No. 2, pp. 117-134, 1996
5   PAUWELLS, F - Biomechanics of the locomotor apparatus, Chs 9,19, Springer-Verlag, Berlin, 1980
6   CULMANN, K - Die graphische statik, Zurik, 1866
7   MEYER, H - Die architectur der spongiosa, Archiv f. Anat. u. Physi., 1867
8   WOLFF, J - The law of bone remodelling, Ch. 2, Springer-Verlag, Berlin, 1986
9   MATTHECK, C - An intelligent CAD method based on biological growth, Fatigue Fract. Engng Mater. Struct., Vol. 13, No. 1, pp. 41-51, 1990

10   REUSCHEL, C and MATTHECK, C - Three dimensional fibre optimisation with computer aided internal optimisation (CAIO), Multidisciplinary design and optimisation, pp. 10.1-10.11, R. Ae. Soc., London, 1998

11   BENDSØE, M P and KIKUCHI, N - Generating optimal topologies in structural design using a homogenization method, Computer methods in applied mechanics and engineering, Vol. 71, pp. 197-224, 1988

12   BENDSØE, M P, DFAZ, A and KIKUCHI, N - Topology and generalised layout optimisation of elastic structures, Topology design of structures, Bendsøe, M P and Mota Soares, C A, pp. 159-205, Kluwer, Netherlands, 1993

13   MAKIYAMA, A M and PLATTS, M J - Topology design for composite components of minimum weight, Appl. Comp. Mater., Vol. 3, No. 1, pp. 29-41, 1996

14   JONES, S E and PLATTS, M J - Practical matching of principal stress field geometries in composite components, Composites Part A, Vol. 29A, pp.821-828, 1998

CERAMIC SHAPE OPTIMISATION USING A GROWTH TECHNIQUE

W.M. Payten and P. Bendeich

Australian Nuclear Science and Technology organisation, (ANSTO), Material, Division, PMB 1, Menai, NSW, 2234 Australia.

SUMMARY

Ceramic materials tend to be brittle in comparison to metallic materials. Thus shape optimisation of ceramic components requires an alternative approach to that of ductile materials. For ceramics, a suitable objective function is the failure probability. This can be calculated using the Weakest-Link-Theory of Weibull. Many techniques are available to optimise this function, examples being sequential linear programming, quadratic programming, gradient search routines or stochastic type approaches. The design variables are commonly the external shape of the component, modelled using either polynomials or B-splines with a number of control points. In order to utilise traditional optimisation techniques, it is then necessary to derive the gradient and Hessian of the design variables. For two dimensional shapes, this can be accomplished with only minor difficulties. However, for three dimensional objects this can be become intractable. Using a heuristic method based on a modified growth law, coupled with Weibull's failure theory, can overcome these difficulties and allow the use of probabilities rather than tensor stress properties. The method is applied to the optimisation of a 3-dimensional component and is shown to be successful in reducing the probability of failure.

1. INTRODUCTION

Engineering ceramics have a wide scatter in strength parameters making a deterministic approach unsuitable in designing for load bearing applications. Statistical approaches can be used that incorporate this variability, usually based on Weibull's weakest link theory with extensions to triaxial stress states based on Freudenthal Principle of Independent Action (PIA). The PIA model does not consider the mutual influence of the principal stresses on failure probability, thus failing to account for shear effects. More accurate and complex approaches are possible by extensions to linear elastic fracture mechanics based on the Batdorf model. In this paper the simpler PIA method is used.

The field of optimisation is extensive, particularly in regards to metallic and composite components, however only a limited amount of work has been done in the area of ceramic optimisation. Lahtinen and Pramila [I] optimised a ceramic piston head. This model incorporated 200 2-D axi-symmetric elements with 32 design variables. They were successful in reducing the volume, stresses and the probability of failure. Their methodology was based on the PIA Weibull method with optimisation via a quadratic programming method using a Taylor expansion of the constraint function.

Eschenauer and Vietor [II] used stochastic optimisation to solve the problem. Both sequential linearisation and generalised reduced gradient with quadratic approximation were used. The optimisation of a turbine disk was undertaken, again the example was two dimensional. The disk contour was approximated using a B-spline approach with 4 control variables.

In both the above examples the number of design variables is small and thus the recovery of the sensitivities (the gradients) of the design variables is not computationally difficult. If these techniques are extended to 3 dimensional shapes then the number of design variables and the three dimensional nature of the sensitivities greatly complicates the optimisation approach. One of the problems of these approaches, particularly the quadratic formulation, is that it is necessary to derive the gradients of the Hessian matrix $\nabla^2 f(x)$, which can be computationally expensive.

For this reason a gradient-less approach is attractive due to simplification of the search routines. A number of techniques are feasible such as genetic algorithms, and simulated annealing. One methodology that has considerable merit is the so-called biological growth or bulking swelling techniques [III],[IV]. This technique is based on the axiom of constant stress, and is easily modified to accommodate reliability rather than von Mises stresses.

2. RELIABILITY ANALYSIS

The Weibull's weakest link theory describes the probabilistic failure of ceramic components. In a one dimensional case this can be described by the following expression.

$$P_s = e^{-(1/V_o)\int((\sigma-\sigma_u)/\sigma_o)^m \, dV} \, . \tag{1}$$

The integral is performed over the entire material volume, σ_u is the threshold stress usual taken as zero, σ_o is the 63.5% characteristic strength within the volume V_o and m is the Weibull's modulus. Thus the probability of failure is

$$P_f = 1 - P_s \, . \tag{2}$$

In a three dimensional stress state the triaxial nature of the stress field must be taken into account. The PIA method is an appropriate assumption, where the reliability is assumed to be locally independent. Taking the failure probability to be

$$P_f = 1 - \exp(-\chi) \, , \tag{3}$$

with

$$\chi = (\frac{1}{m}!)^m (\frac{1}{\sigma_o})^m \frac{1}{V} \int_V (\sigma_1^m + \sigma_2^m + \sigma_3^m) dV \tag{4}$$

3. OPTIMISATION

The use of biological growth laws to optimise structures has proven to be an effective and flexible method. It is a heuristic method that is based on the idea of adding material to overstressed regions and eliminating material from under stressed regions. The driving force for the algorithm in its simplest form is the *constant surface stress axiom*. Mathematically the optimization problem can be stated as

$$\min F(x_1, x_2, ..., x_n) \tag{5}$$

The objective function F is a function of the design variables x. In this paper the objective function is chosen to be the failure probability, which has the form

$$F = 1 - \prod_{i=1}^n \exp(\frac{1}{m}!)^m (\frac{1}{\sigma_o})^m \frac{1}{V} \int_V (\sigma_1^m + \sigma_2^m + \sigma_3^m) dV \, . \tag{6}$$

The growth swelling strain can be defined at each node point as follows.

$$\varepsilon_T = (\frac{p_f - P_{ave}}{P_{ave}}) \cdot h \ , \tag{7}$$

where h is the incremental growth rate, p_f is the individual elemental reliability and P_{ave} is

$$P_{ave} = F/n \ , \tag{8}$$

where n is the number of elements. The swelling strain is transferred to a fictitious temperature field that is applied in a further finite element analysis as the only applied loading,

$$\varepsilon_T = \alpha \Delta T$$
$$T = T_{ref} + \Delta T \ . \tag{9}$$

The results of the second analysis are the displacements. These displacements are added to the coordinate of the first step to construct an updated structural shape.

$$\begin{Bmatrix} x \\ y \\ z \end{Bmatrix} = \begin{Bmatrix} x \\ y \\ z \end{Bmatrix} + d \begin{Bmatrix} u \\ v \\ w \end{Bmatrix} \tag{10}$$

Where x,y,z are the nodal spatial co-ordinates. u,v,w are the nodal displacements in the respective directions d is the scaling constant.

The ceramic reliabilities are calculated on an element by element basis. As the design variables are the nodal co-ordinates it is necessary to transfer the elemental reliabilities to nodal reliabilities. Each element reliability is first calculated by averaging the nodal stresses. The reliability is then calculated on the averaged stress. At each node a connectivity search is performed for the surrounding elements. The element reliabilities are then summed and averaged back to the node. Figure 1 shows the flowchart of the process.

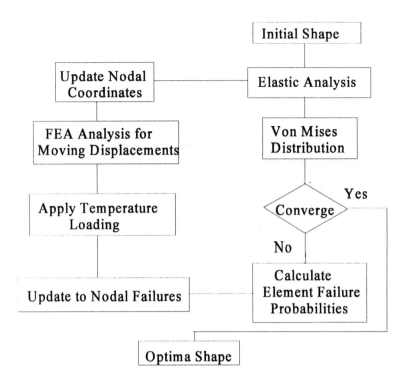

Figure 1 Flowchart of optimisation process

4. RESULTS

A three dimensional FEA model was constructed of a high temperature ceramic component that operates in a corrosive environment. The geometry of the component is detailed in figure 2. It has sides of 75 mm by 130 mm, a thickness of 10 mm with a circular hole of 50 mm diameter. It is to be manufactured from silicon nitride. The boundary conditions are fixed on the edge adjacent to the hole and loaded in the centre directly opposite the hole with a force of 36000 Newtons. The number of elements and nodes are 2240 and 2970 respectively.

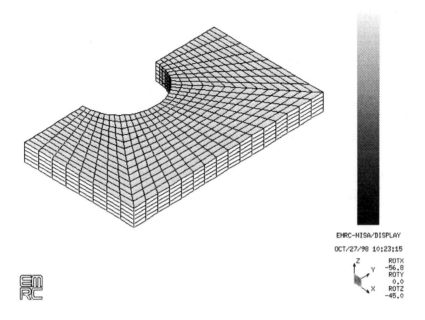

Figure 2 Finite Element Geometry

For the optimisation example here, the x, y co-ordinates are frozen with the design vector being the z axis, thus neglecting the constraint nodes, there are 2970 design variables. Three cases are examined. The first case assumes the component is to be made from metal. The objective function is then based on the von Mises stresses and the algorithms used match the original Mattech expressions [III]. The second case used the first principal stress as the objective function. This is based on the normal stress hypotheses according to Rancine and Lame for brittle materials. The third case uses equation 7 to 10 with the objective function being the probability of failure.

To compare the three cases the probability of failure is also calculated for cases 1 and 2 assuming that the final component is to be made from a ceramic. Figure 3,4,5 display the optimised shape for the metallic (case 1), SIP (case 2) and the ceramic component (case 3).

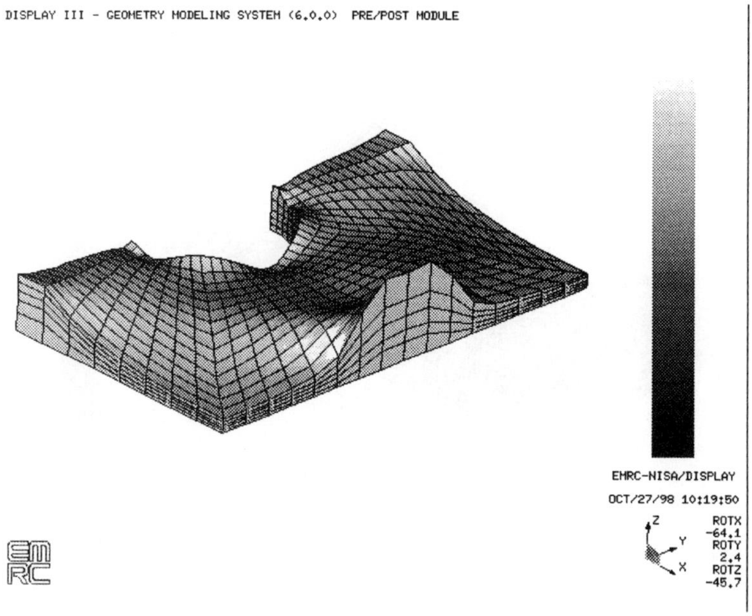

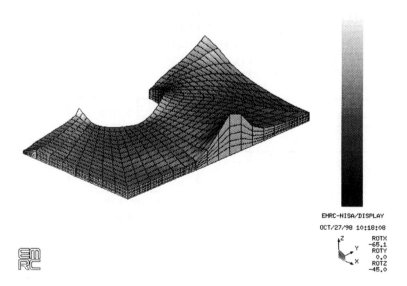

Figure 3   Optimised shape using Von Mises stresses for a metallic component

Figure 4 Optimised shape using S1P Stresses

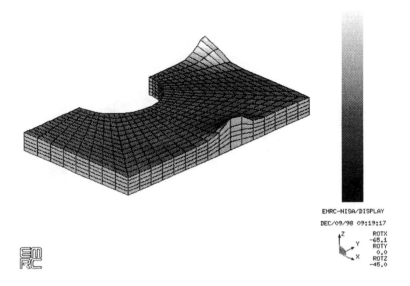

EMRC-NISA/DISPLAY

DEC/09/98 09:19:17

ROTX
-65.1
ROTY
0.0
ROTZ
-45.0

Figure 5 Optimised shape using failure probabilities for a ceramic component

Table 1 shows the changes in various parameters associated with the three cases.

Table 1 Analysis Parameters statistics

Parameter ---------------- Case	Probability of failure	von Mises (MPa)	S1P (MPa)	Volume (mm³)	Strain Energy
Initial	0.3036	354	299	87641	11671
1. Von Mises	0.1602	293	242	86605	11413
2. S1P	0.2665	356	288	83506	12069
3. Probability of failure	0.1199	390	235	86041	12712

Note: Columns represent measured parameter, rows represents the three optimisation cases. The first row represents the initial un-optimised case. S1P represents the first principal stress.

Figure 6, 7 shows the stress and volume reduction versus iteration for the various methods

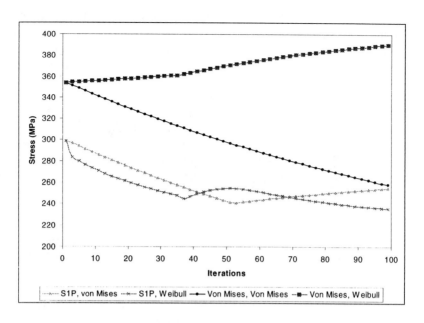

Figure 6   Stress   reduction versus number of iterations for von Mises optimisation Case 1 and Probability optimisation Case 3.

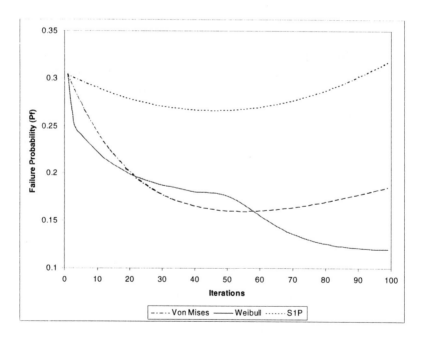

Figure 7   Volume reliability reduction versus number of iterations

1165

Optimisation based on the three different cases presented lead to distinct differences in the final shape. As expected the von Mises cases leads to symmetrical optimisation for both tensile and compressive regions. Both the Principal stress optimisation and the Probabilistic optimisation lead to asymmetrical designs. If the failure probabilities for the optimised von Mises and SIP design are calculated using the same material properties as for the ceramic, both fail to achieve the reduction in failure probability seen using the Weibull approach. These results show that the failure probability must be incorporated as part of the objective function and not as a constraint. The lowest volume and first principal stress were also seen using failure probability as an objective function. This is due to the reduction in compressive stress regions as these do not influence the probabilities.

Optimisation under the first principal stress does poorly in all areas except the volume. The probability stays high due to the influence of the second and third positive principal stresses that are effectively reduced in area causing higher stresses and a resultant higher failure probability.

Using the PIA method reduces the first principal stress to the lowest of the three methods but conversely the von Mises stress rises to the maximum value of the three methods, (Figure 6). The stiffest shape, as expected, results from using the von Mises method (case 1) as a direct relationship between the strain energy, stiffness and von Mises stress exists.

The optimised shape using reliability shows the least shape change of the three methods. Most shape change has occurred in the small highly stressed regions around the top constraint and loading point, with very little change in the rest of the model. This can be attributed to the Weibull distribution. A few of the elements contribute most to the overall failure probability, while many of the elements have extremely low probability of failure resulting in little sensitivity. The result is the majority of the elements in the low stresses regions slowly reduce in a very even fashion.

5. CONCLUSIONS

The concept of axiom of constant stress has a useful analogue in ceramic optimised designs by replacing the stress with failure probability as the objective function. The biological growth algorithm works because the algorithm strives to reduce the maximum stress and also reduces the volume of the component. For ceramics this is of benefit, firstly by reducing the peak stress and secondly by decreasing the component volume. In both instances the effect is to reduce the probability of failure (P_f). This is apparent by the following equation which shows the volume effect on failure probability for a uniform stress state.

1166

$$\frac{\sigma_{v1}}{\sigma_{v2}} = \left(\frac{v_2}{v_1}\right)^{1/m} \tag{11}$$

The modified algorithm for failure probabilities delivers a shape design improvement over the other cases. According to the results the probability of failure decreased by 60%. Using the von Mises objective function resulted in a decrease of 47%.

6. FUTURE DIRECTIONS

One problem with the probabilistic approach used here is its inability to model shear sensitivity effects. Thus although ceramics can withstand large magnitude compressive stresses, it is possible that these stresses can become sufficiently large that the component will fail in shear. The PIA method cannot account for this effect.

To account for shear, effects the Batdorf shear sensitivity model is currently being incorporated into the model. This also includes the concept of a statistical distribution of cracks. Many ceramics show R curve behaviour (sub critical crack growth) thus a fully brittle failure model is inadequate. It is possible to account for static and dynamic fatigue using extensions to the Batdorf model, this can then be integrated into an optimization routine. A further direction also includes combined shape and topology optimisation thus holes will form in areas where no material is needed.

7. REFERENCES

[I] Eschenauer H.A., Vieter T. - Application of stochastic optimization techniques at the example of ceramic turbine disks. Advances in Design automation, Vol 2, ASME, pp. 445-452, 1993

[II] Lahtinen H, and Pramila, A. -Improvement of the shape of a ceramic piston head by optimization and FEM. Int. J. Of Materials and Product Technology, Vol 6 no.4, pp. 361-370, 1991

[III] Mattheck C. and Burkhardt, S.- A new method of structural shape optimization based on biological growth. Int. J. Fatigue, Vol 12 no 3, pp. 185-190, 1990

[IV] Chen, J.L. and Tsai, W.C. - Shape optimization by using simulated biological growth approaches. AIAA, Vol 31, No. 11, pp. 2143-2147. 1993

Modeling of Bone Microcracking

Ulrich Hansen

Centre for Biomimetics & Department of Engineering,
The University of Reading, Reading RG6 6AY, United Kingdom.

SUMMARY

This study concerns the development of damage in bone. It is generally accepted that this damage consists of the initiation and evolution of microcracks and that these microcracks result in a reduction of the modulus of the material. Over the past few years several techniques modeling the development of damage in bone have been proposed. However, these models are unable to address non-uniform stress fields as well as the damage induced history dependence of these fields. In the present paper an analytical/ numerical approach capable of addressing these issues is proposed. This new formulation of bone constitutive behavior is incorporated into the finite element program MARC using the user subroutine HYPELA2. The proposed FEA model is capable of describing the development of damage resulting from static loads as well as fatigue loads.

1 INTRODUCTION

As early as 1974 Currey and Brear [1] reported a significant post-yield region before failure when testing bone specimens. These findings were soon complimented by Carter and Hayes in 1977 [2] who observed bone micro-damage prior to final failure. Since these early works there has been numerous studies confirming these experimental results and it is now generally accepted that prior to final failure considerable accumulation of internal microdamage in the form of microcracks occurs. Significantly, this microdamage is observed to result in a reduction of bone modulus.

More recently a considerable effort has been aimed at quantifying this microdamage and relating it to the monotonic stress level during static testing, the fatigue stress level and number of cycles during fatigue testing, as well as, the creep stress level and time during creep testing. There are two reasons to the great interest in understanding the damage phenomena in bone.

One is that a better understanding will allow us to predict more precisely when bone will fail. This is particularly important around orthopedic implants where the stress situation is both complex and unnatural. The other reason is the current interest in identifying the bone remodeling signal that allows bone to adapt to its mechanical environment. It may be that it is damage that stimulates remodeling.

One approach attempting to quantify micro-damage has been to compare the experimentally observed damage regions, around for example a hole, with predicted maximum principal stress, the strain energy density function or other quantities which may serve as stimuli for bone remodeling. However, these predictions have all been based on analyses that assume a linear material. This is obviously in contrast to the desire of analyzing the effect of a significant post-yield region. In particular these types of analyses are unable to take into account that because of yielding, the original elastic stress will be redistributed as damage progresses [3].

The other approach commonly adopted is the Continuum Damage Mechanics (CDM) method that has its origins in Kachanov's early work [4]. According to this approach the change in bone modulus is used as a quantitative indicator of damage. Modulus is a well-defined engineering property directly involved in the mechanics calculations (stress analysis) through the constitutive relations. This method, therefore, does take into account the effects of evolving microdamage or yielding. However, the formulation adopted in earlier works related to bone microcracking only allows the analysis or uniform stress fields. That is, the effects of microcracking in for example specimens with a hole or the effects on the stress fields around an orthopedic implant can not be directly assessed using this method in it's currently adopted form.

In this paper we also use the concept that a change in bone modulus is a quantitative indicator of damage. However, the aim of this paper is to propose a method enabling us to predict evolving damage fields in bone specimens with non-uniform stress fields. Two different formulations are outlined in the following, however, the derivation of the two formulations is similar. One formulation addresses the static case where damage increases as a function of applied stress. This approach was first suggested by Brockenbrough and Suresh [5] as a method for modeling microcracking in ceramic materials. The other formulation includes a time dependency that permits us to address fatigue and creep problems. This method was proposed by Hansen [6] for predicting damage evolution in woven polymer composites subjected to fatigue loading.

2 ANALYSIS

In the model to be described the bone material is assumed to behave as a homogeneous isotropic material. This is probably a reasonable assumption for woven bone such as the skull bones but obviously a very questionable assumption for bone in general. However, we consider the proposed method a step in the process towards a better understanding of bone mechanical behavior and the further development of the method to include anisotropic material behavior a topic for future work.

In the next sections the derivation of the static model is shown first followed by the derivation of the time dependent model.

2.1 Static model

We consider an isotropic solid in which damage initiates and evolves as a result of increased static loading. The damage is described through a damage parameter β and we assume the following relationship between mechanical properties and β:

$$E = E_0(1 - \beta) \quad \text{and} \quad \nu = \nu_0(1 - \beta) \tag{1}$$

where E_0 and ν_0 are the Young's modulus and Poisson's ratio prior to testing. Furthermore, the model assumes a damage evolution function of the form:

$$\beta = A\left(\frac{\sigma_e}{\sigma_0} - 1\right)^n \quad \text{for} \quad |\sigma_e| > |\sigma_0|$$

$$\tag{2}$$

$$\beta = 0 \quad \text{for} \quad |\sigma_e| \le |\sigma_0|$$

where 'A' and 'n' are constants and σ_0 is the effective stress at which damage is initiated. The parameters A and n can be determined from uniaxial tests of specimens with no stress raisers. An effective stress σ_e is used in Equation (2). The following form of σ_e is assumed

$$\sigma_e = \sqrt{\sigma_{ij}\sigma_{ij}} \tag{3}$$

This choice of effective stress implies isotropic damage or in the case of microcracks that the orientation of the microcracks is without preferred directions. Considering the damage to be microcracks it would perhaps seem more appropriate to select a maximum principal tensile stress criterion. Indeed, there is some evidence to suggest that cracks in bone at the microscale level develop mostly normal to the tensile field [7]. However, this evidence is not conclusive and we have chosen the above form as it facilitates the subsequent derivation. Furthermore, this choice may be entirely appropriate for the woven type of bone that, as mentioned earlier, the model in its present form is most suited for.

2.2 Numerical Formulation of the static model

For an isotropic elastic solid the stress-strain relationship can be described as

$$\varepsilon_{ij} = M_{ijkl}\sigma_{kl} \tag{4}$$

where ε_{ij} is the strain tensor, σ_{ij} the stress tensor and M_{ijlk} is the compliance tensor, which depends on the level of damage through Equation (1) and Equation (2). For the numerical calculations, it is desirable to express $\dot{\varepsilon}_{ij}$ the strain increment in terms of an increment in stress. Differentiating Equation (4) with respect to stress we obtain

$$\dot{\varepsilon}_{ij} = M_{ijkl}\dot{\sigma}_{kl} + \dot{M}_{ijkl}\sigma_{kl} = M_{ijkl}\dot{\sigma}_{kl} + \frac{\partial M_{ijkl}}{\partial\beta}\dot{\beta}\sigma_{kl} \tag{5}$$

Using Equation (2) and Equation(3) the following expression for $\dot{\beta}$ is derived

$$\dot{\beta} = nA\left(\frac{\sigma_e}{\sigma_0} - 1\right)^{n-1} \bullet \frac{\sigma_{mn}}{\sigma_0\sigma_e}\dot{\sigma}_{mn} \tag{6}$$

The standard expression of the compliance tensor M_{ijlk} can be written as

$$M_{ijkl} = \frac{1+\nu}{E}\left(\delta_{ik}\delta_{jl} + \delta_{il}\delta_{jk}\right) - \frac{\nu}{E}\delta_{ij}\delta_{kl} \tag{7}$$

Through Equation (1) M_{ijkl} depends on β. Differentiating Equation (7) with respect to β we get

$$\frac{\partial M_{ijkl}}{\partial\beta} = \frac{E_0}{E^2}\left(\delta_{ik}\delta_{jl} + \delta_{il}\delta_{jk}\right) \tag{8}$$

Inserting Equation (6) and Equation (8) into Equation (5) we then get

$$\dot{\varepsilon}_{ij} = M_{ijkl}\dot{\sigma}_{kl} + \frac{nAE_0}{E^2}\left(\frac{\sigma_e}{\sigma_0} - 1\right)^{n-1}\frac{\sigma_{ij}\sigma_{kl}}{\sigma_0\sigma_e}\dot{\sigma}_{kl} \tag{9}$$

1172

where the last term on the right-hand side is the non-linear contribution to the strain increment. Equation (9) can be expressed in a similar manner to Equation (4) as

$$\dot{\varepsilon}_{ij} = M_{ijkl}\dot{\sigma}_{kl} \tag{10}$$

where M_{ijkl} is the incremental compliance tensor. For numerical calculations the inverse expression of Equation (10) is often more convenient. To obtain the inverse expression we assume that the strain increment can be separated into an elastic and a non-elastic component as follows

$$\dot{\varepsilon}_{ij} = \dot{\varepsilon}_{ij}^{ela} + \dot{\varepsilon}_{ij}^{ne} \tag{11}$$

Then using the relationship between the stress increment and the elastic strain increment

$$\dot{\sigma}_{ij} = l_{ijkl}\dot{\varepsilon}_{kl}^{ela} \tag{12}$$

where l_{ijkl} is the standard elastic stiffness tensor, which depends on the level of damage through Equation (1) and Equation (2). Combining Equation (11) and Equation (12) we get

$$\dot{\sigma}_{ij} = l_{ijkl}\left(\dot{\varepsilon}_{kl} - \dot{\varepsilon}_{kl}^{ne}\right) \tag{13}$$

Finally, inserting in this expression the last term on the right hand side of Equation (9) followed by some manipulation of the expressions the incremental relationship between stress and strain is obtained

$$\dot{\sigma}_{ij} = L_{ijkl}\dot{\varepsilon}_{kl} \tag{14}$$

where L_{ijkl}, the incremental stiffness tensor is determined as

$$\tag{15}$$

$$L_{ijkl} = l_{ijkl}$$

$$- \frac{nAE_0\left(\dfrac{\sigma_e}{\sigma_0} - 1\right)^{n-1}\left(\sigma_{ij} + \dfrac{v}{1-2v}\delta_{ij}\sigma_{mm}\right)\left(\sigma_{kl} + \dfrac{v}{1-2v}\delta_{kl}\sigma_{mm}\right)}{\sigma_e E^2 \sigma_0 \left(\dfrac{1+v}{E}\right)^2\left[1 + \dfrac{nAE_0\left(\dfrac{\sigma_e}{\sigma_0} - 1\right)^{n-1}}{(1+v)E\sigma_e\sigma_0}\left(\sigma_{mn}\sigma_{mn} + \dfrac{v}{1-2v}\sigma_{pp}^2\right)\right]}$$

2.3 Time dependent model

As in section 2.1 we consider an isotropic material but in the present case the mechanical properties change with time 't'. Using this formulation it is also possible to address fatigue damage by simply relating the number of fatigue cycles, N, to t through $N = f \cdot t$ where f is the frequency of load cycles during the fatigue test. In the remaining part of this paper we will state the equations in the form appropriate for fatigue problems. We assume that the time or fatigue dependent damage parameter β is related to the mechanical properties according to Equation (1) where E_0 and v_0 are the Young's modulus and Poisson's ratio before fatigue testing. The model assumes the following form for the rate of damage evolution

$$\dot{\beta} = A\left(\frac{\varepsilon_e}{\varepsilon_0}\right)^n \tag{16}$$

where A and n are constants and $(\dot{\ }) = \frac{d}{dN}( )$. Equation (16) expresses that at an effective strain level, ε_e, equal to a reference strain level, ε_0, damage increases linearly with the number of cycles at a rate of 'A', 'A' having the unit of N^{-1}. The strain measure ε_e corresponds to the peak strain level during a fatigue cycle. The exponent 'n' is included in order to accommodate a possibly very strong influence of the fatigue strain level. The parameters A and n can be determined from uniaxial fatigue tests of specimens with no stress raisers. Equation (16) implies the damage evolution function

$$\beta = \int_0^N \left(\frac{\varepsilon_e}{\varepsilon_0}\right)^n dN \tag{17}$$

An effective strain ε_e is assumed in Equation (16). We assume the following definition of ε_e

$$\varepsilon_e = \sqrt{\varepsilon_{ij}\varepsilon_{ij}} \tag{18}$$

Various other forms could be suggested for the effective strain that governs damage evolution in bone. A discussion of these other forms would follow along the same lines as the discussion in connection with the effective stress criterion suggested in Equation (3).

2.4 Time dependent numerical formulation

As in the static case we consider an isotropic elastic solid for which the stress strain relationship is described by Equation (4). Also the expression for the strain increment, Equation (5), can be used only now we differentiate with respect to N rather than with respect to stress. The form of Equation (8) is valid in the fatigue case as well. Therefore, inserting Equation (16) and Equation (8) into Equation (5), we get analogous to the derivation of Equation (9)

$$\dot{\varepsilon}_{ij} = M_{ijkl}\dot{\sigma}_{kl} + A\frac{E_0}{E^2}\left(\frac{\varepsilon_e}{\varepsilon_0}\right)^n \sigma_{ij} \qquad (19)$$

As mentioned in Section 2.2 the inverse expression of Equation (19) is often more convenient. Retracing the steps leading from Equation (10) to Equation (15) but using Equation (19) instead of Equation (9) we get for the fatigue case

$$\dot{\sigma}_{ij} = l_{ijkl}\left(\dot{\varepsilon}_{kl} - A\frac{E_0}{E^2}\left(\frac{\varepsilon_e}{\varepsilon_0}\right)^n \sigma_{kl}\right) \qquad (20)$$

2.5 Finite element analysis

The above formulations of constitutive behavior, in particular Equations 15 & 20, can be incorporated into the finite element programs ABAQUS or MARC using the user subroutines UMAT or HYPELA2, respectively. Given an increment in strain, $\dot{\varepsilon}_{kl}$, the procedure is to find the corresponding increment in stress, $\dot{\sigma}_{ij}$, using Equation (15) for static problems or Equation (20) for time dependent problems. At the beginning of each increment β and subsequently E and v are updated using Equations (1), (2), (16) and (17). The current accumulated stress, σ_{kl} and strain, ε_{kl}, are known as is everything else on the right hand side of Equations (15) & (20). Therefore, the stress increment, $\dot{\sigma}_{ij}$, can be calculated. With the stress and strain increments determined the accumulated stress and strain tensors, σ_{ij} and ε_{ij}, are updated and the solution scheme proceeds to the next increment.

2.5.1 Numerical results

The application of the model to predicting bone behavior is still in its preliminary phase. However, a few results demonstrating the capabilities of the model is presented in the following. In Figure 1 is shown the two-dimensional finite element model of a specimen with an inclined crack (or elliptical hole). Plane stress isoparametric elements were used. To simulate a bone specimen we assumed a Young's modulus of 20 GPa and a Poisson's ratio of 0.33. In Figure 2 is shown the calculated damage contours around the crack tip after 30.000 cycles at an applied farfield stress of 73 MPa. Previous models are unable to address this problem of combined non-uniform stress fields and evolving damage contours. As mentioned earlier, previous models

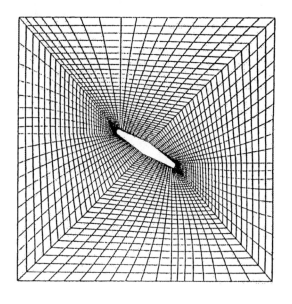

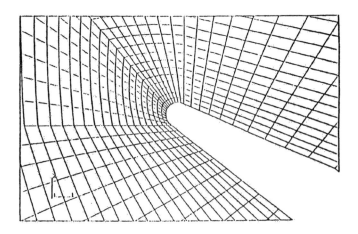

Figure 1. FEM mesh of specimen with an elliptical hole. The top figure shows the complete mesh. The bottom figure shows a close-up around the crack tip.

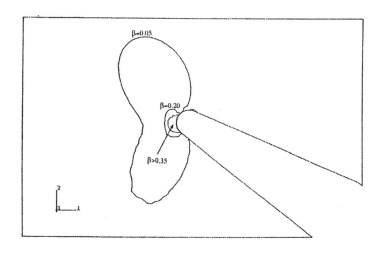

Figure 2. Damage contours around elliptical hole at 30.000 cycles,
$\sigma_{applied}$=73 Mpa, A=2*10^{-6} N^{-1}, n=2.

are unable to take into account that because of yielding, the original elastic
stress will be redistributed as damage progresses. In Figure 3 we have
presented a comparison of the predicted shape of the process zone around a
sharp horizontal crack at initial yield with the predicted shape of the process
zone at an increased load. The process zone being the area in which the
material is 'yielding'. The results shown in Figure 3 is the result from a model
of a specimen with a sharp notch and subjected to static loading. Figure 3 is
therefore a demonstration of the use of the static formulation presented in
sections 2.1 and 2.2.

3 CONCLUSION

The application of the proposed model to bone is still at an early stage,
Never-the-less, the model has obvious advantages compared to previous
models in particular its ability to model progressive damage in non-uniform
stress fields. In its current isotropic form the model is probably best suited for
bones such as the skull bones, which are approximately isotropic in the plane.
Furthermore, the adopted form of the damage parameter, β, implies that
damage should occur with no preferred directions which is unlikely for bone
in general. These shortfalls the present model shares with most previous
models. To obtain increasingly better models of bones' mechanical response
future work incorporating anisotropic material properties and a tensorial
damage parameter is contemplated.

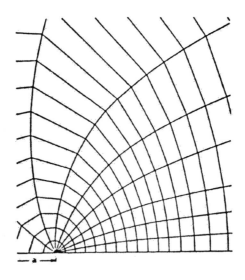

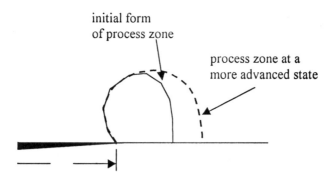

initial form
of process zone

process zone at a
more advanced state

Figure 3. Comparison of the predicted shape of the process zone around the crack-tip of a sharp horizontal crack at initial yield and at a later stage in the damage progression. The top figure shows the FEM-mesh and the bottom figure shows the predictions.

4 REFERENCES

1   CURREY, J.D. and BREAR, K. – Tensile yield in bone. Calcif.Tiss.Res., Vol. 15, pp. 173-179, 1974.

2   CARTER, D.R. and HAYES, W.C. – Compact bone fatigue damage: a microscopic examination. Clin.Orthop, Vol. 127, pp. 265-274, 1977.

3   ZIOUPOS, P., CURREY, J.D., MIRZA, M.S. AND BARTON, D.C. – Experimentally determined microcracking around a circular hole in a flat plate of bone: comparison with predicted stresses. Phil.Trans.R.Soc.London.B, Vol.347, pp.383-396, 1995.

4   KACHANOV, L.M. – On creep rupture time. Izv.AN SSR, Otd.Tekhn, Nauk, No.8, pp.26-31, 1958 (in russian).

5   BROCKENBROUGH, J.R. and SURESH, S. – Constitutive behavior of a microcracking brittle solid in cyclic compression. J.Mech.Solids, Vol. 35, No. 6, pp. 712-742, 1987.

6   HANSEN, U. – Damage development in woven fabric composites during tension-tension fatigue. J.Comp.Materials, Vol. 33, in print.

7   ZIOUPOS, P. and CURREY, J.D. –The extent of microcracking and the morphology of microcracks in damaged bone. J.Mater.Science, Vol. 23, pp. 978-986, 1994.

8   ZIOUPOS, P., XIAO, T.W. and CURREY, J.D. – Experimental and theoretical quantification of the development of damage in fatigue tests of bone and antler, J.Biomechanics, Vol. 29, No. 8, pp. 989-1002, 1996.

BIOLOGICAL MATERIALS AND STRUCTURES
Professor G. Jeronimidis, Reading University,UK

INTRODUCTION
Most biological materials and structures are composite systems where optimisation of performance has been achieved through good design rather than through the use of esoteric materials. Shape optimisation and integration of functions exploiting fibre architectures and hierarchies are the most striking aspects of efficient load-bearing solutions in Nature. As well as being relevant to biology, the study of such systems can provide examples and inspiration for multifunctional materials, optimisation strategies, smart technologies and engineering design. These aspects form the basis for the current developments of biomimetics which provides an interdisciplinary approach to these problems.

MATERIALS SCIENCE AND ENGINEERING
OF BIOLOGICAL SYSTEMS
A major difficulty in the study of the mechanical properties of biological systems is that the traditional boundaries between the "materials" and the "structures" are far more difficult to establish. To a large extent the real materials of biology are the chemical substances used in the synthesis of more complex and organised "elements". Taking wood as an example, well known and familiar to all of us, what we often consider a material is however an extremely heterogeneous solid which, at any one of the various anatomical levels can be considered a material or a structure. The representative volume elements (RVP) range from linear dimensions of the order of 10^2 m at the upper end (ring structures, grain direction, knots, etc.), to 10^{-9} m at the lower end (cellulose protofibrils). In between there is the assembly of cells structure (RVP = 10^{-3} m), the cells themselves (RVP = 10^{-5}), the composite cell walls (RVP = 10^{-6}) and the cellulose fibres themselves (RVP =10^{-8}). This situation is typical in all kinds of biological systems such as bone, tendons, insect exoskeletons, crustacean and mollusk shells, etc.

Mechanical properties depend therefore on interactions between the various sub-levels of structure. A major difficulty in predicting the response of such systems to external and internal loads is that traditional analytical techniques used in materials science, solid mechanics and engineering are not always applicable, owing to limitations imposed by the morphological complexity of the biological tissues and organs and by the fact that it not always possible to obtain measurements of physical and mechanical properties from individual sub-components or sub-elements.

The hierarchical arrangement of load-bearing fibres (cellulose, collagen, chitin), integrated with meso-structures such as cellular and laminated architectures imply highly heterogeneous and anisotropic systems designed to extract maximum structural benefit. Very often the response of is very non-linear, either because of large strain behaviour (as in soft tissues such as skin, tendons, muscles, etc.) or because large displacements are involved (leaves and petals in plants, for example). In several instances, very interesting interactions between solid sub-structures and liquid water exist as in turgid plant tissues.

FINITE ELEMENTS AT THE SERVICE OF BIOLOGICAL SYSTEMS
There is little doubt that, owing to the aforementioned complexity, Finite Elements methods offer one of the most powerful tools for an effective engineering analysis of such systems. If properly used, they can provide useful modelling techniques for the various structural sub-levels, with representative volumes appropriate to the scale of the relevant elements, and for their interaction and integration into structural components. There is an almost natural correspondence between the observed subdivisions of biological systems and the discretisation techniques used in Finite Elements. Matching the "numerical" element to the biological one for modelling purposes con provide extremely effective and "realistic" simulation techniques. Substucturing and condensation techniques used in Finite Element algorithms are conceptually similar to the interactions of sub-systems in the biological hierarchies. Moreover, the trend towards Finite Element codes which can take full advantage of parallel architecture computing obviates to some extent to problem of size of model which can be solved using a one to one correspondence between biological and numerical elements. There are several examples of this approach in the literature, covering bone, wood, skin, turgid plant tissue, insect cuticles, etc.

An interesting feature of most biological systems is the exploitation of shape to achieve minimum energy content structures. Optimisation, structural functionality (static and dynamic), integration of functions (such as the strain sensing) stems from the way in which things are put together rather than from attempting to extract high performance from the basic materials. The power of FE methods to explore the limits of response by allowing simulation of complex shapes and internal hierarchical interactions provides a unique tool for a better understanding of the biological systems with benefits in a wide range of applications such as medicine (prosthetic devices, tissue engineering, tissue mechanics), biology (biomechanics, biomaterials), smart technologies (sensors actuators, etc. based on biological ideas), composites (deformable structures, fibre architectures).

A Proposed Materials Properties Data Management System

Mr. T. T. Wong

Member of Technical Staff

1. Abstract

Material property information is fundamental to the design and analysis of any component. Inaccuracies in the material properties used for design or analysis will produce results that are also inaccurate. Even if the loads on a part are known with exact precision, its life duration cannot be determined accurately if the materials information is flawed. Therefore, meaningful results are obtained only when the material information used in both the design and analysis processes represent accurately the material in question.

Rocketdyne Propulsion & Power, a part of The Boeing Company, has recently implemented an electronic materials database system in order to maintain a library of high quality materials data. That data can now be disseminated and used by engineers very quickly and consistently. The Boeing Company, to take advantage of the knowledge collected by its many and varied divisions, is attempting to create a database system that will allow materials information to be shared between locations. This paper will examine the benefits that Rocketdyne has seen in implementing a materials properties database, the benefits that a large corporation may expect to reap in implementing a corporate wide materials database system, and also the steps needed to be taken to implement such a corporate wide system.

2. Introduction

The traditional method for increasing confidence in materials data is to run a series of tests on the material in question. This does increase confidence. Unfortunately, this may also result in unnecessary test duplication. Test duplication occurs when important information from previous tests cannot be retrieved. In these cases, previous test data, though perfectly good, are considered unreliable. Testing must, therefore, be duplicated. Though testing will never be completely eliminated, nor should it ever be, test duplication represents an unwanted burden both in cost and time. Managing materials information to prevent test duplication will benefit any company that desires to produce products with high quality at a low cost.

In addition to avoiding test duplication, managing materials information is an essential responsibility for any company. Realizing this, many companies have implemented a computerized materials database system. Various locations in The Boeing Company, for example, currently either have an electronic materials database system implemented or are in the middle of creating such a system. The Boeing Company Rocketdyne Propulsion and Power (Rocketdyne) uses a third party database software product for their materials properties database [1]. The database that Rocketdyne uses runs on a UNIX operating system and is therefore only accessible through the UNIX workstations at Rocketdyne. At the time that this paper is being written, there are at least 5 locations in The Boeing Company that are either using or in the process of developing a materials properties database system for data distribution and retention.

The exact reasons that each location felt the need to create a materials properties database system and the exact benefits that resulted vary from location to location. However, there is enough commonality between all the locations that an examination of Rocketdyne as a case study will yield useful information.

2.1 The Drive to Develop a Materials Properties Database at Rocketdyne
The dynamics of the aerospace industry are ever changing and create an ever-increasing competitive environment. The many mergers in this sector have only made competition for each contract available more intense. The emergence of foreign aerospace companies has also added to the already competitive environment. Not only is there pressure from competing aerospace companies, but customer demands require that aerospace companies be more sensitive to their needs. Aerospace companies that wish to survive and prosper can no longer conduct business today as they did 15 years ago. Companies must be able to develop and manufacture a product faster, better, and cheaper than before.

These market dynamic factors caused Rocketdyne to realize that to be the leader in its field, Rocketdyne must adopt highly efficient engineering practices. In 1993, Rocketdyne determined that it would introduce tools into the engineering process that will help its engineers towards this goal. One of the tools identified as needed to aid in the goal was a materials properties database.

2.2 The Benefits of a Materials Properties Database for Rocketdyne.

An overview of the engineering process at Rocketdyne reveals that materials properties data touch each of the three major engineering disciplines (Stress, Materials and Design). Figure 1 shows that materials engineers interact with materials properties by being the primary developers of allowables data. Designers use the properties developed by the materials engineers to determine suitable materials for the parts being designed. Stress analysts also need materials properties values to determine if the materials used can withstand the stresses that may be seen during application.

Disciplines That Interact with Materials Properties Data

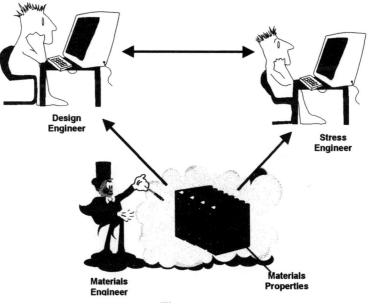

Figure 1

Rocketdyne looked at the overall engineering process and concluded that any tool that can help manage and distribute materials properties data will aid in that whole process. The tool that Rocketdyne decided to implement was a materials properties database. After several years of developing the structure of the database and populating the database, the database system was released for official use in late 1997.

1185

A more detailed examination of the handling of materials properties, Figure 2, revealed many hurdles that got in the way of having an efficient and high quality process. For example, the process of developing materials allowables data showed that the final "product" was stored in 3 places, the materials properties manual, a final report and in a file cabinet. This translated into a serious quality issue. Having three sources of materials allowables data means that these three sources must be kept up to date. This

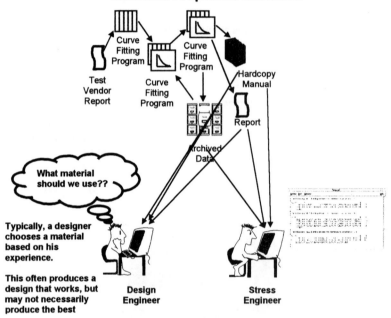

Figure 2

is a difficult maintenance issue for a company the size of Rocketdyne and one that if not performed will lead to many problems. This "before database" process also presented an area where cycle time could be improved. Under the "before database" process, the transfer of data from hardcopy format to a computerized file format that programs such as ANSYS [2] could understand was all performed by hand. This involved the stress analyst picking points from a hardcopy curve and then typing the data into a format that was suitable for ANSYS. Figure 2 highlights another opportunity for quality improvement. The designer must determine what material best fits the part that he is designing. Often, instead of poring over many volumes of hardcopy data, the designer chooses a material based on his experience. The choice of materials in this manner may not result in the "best" selection in terms of cost and properties.

The engineering process is greatly improved with the introduction of a materials properties database (Figure 3). Comparing the engineering process "before database" (Figure 2) to that with a database implemented (Figure 3) shows that the overall engineering process is greatly simplified.

Engineering Process with a Materials Properties Database

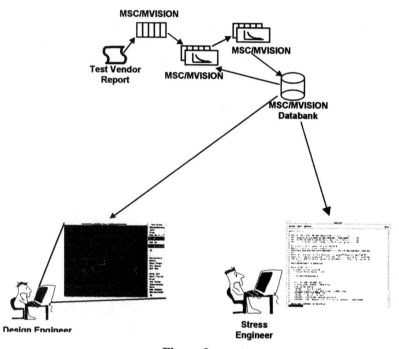

Figure 3

Notice that there are no longer three sources from which allowables data may be obtained. With a database system implemented, Rocketdyne now has a single source of materials allowables data. This greatly improves the quality of the materials properties used at Rocketdyne by ensuring that every engineer uses the same data and the most current data. Though illustrated in Figure 3, it may not be explicitly clear that stress engineers and designers now have direct access to materials properties data electronically. Designers now have access to properties data from within their CAD software such as Pro-E. Stress engineers now no longer need to

1187

tediously pick off points from a hardcopy curve and then type out a file that stress programs such as ANSYS can read. Now all they need to do is enter the database and ask it to write out a file with the properties that they need in a format that ANSYS can use. Having a materials properties database in place helps Rocketdyne towards its goal of being faster, better, and cheaper. (Reference 4 contains a more detailed description of the benefits that resulted in implementing a materials database system at Rocketdyne.)

2.3 The Benefits of a Materials Properties Database for a Large Company
  The same benefits that an individual location such as Rocketdyne sees in having a materials properties database can also be realized when such a system is implemented for a large company with many divisions. Though the "bigger is better" proverb is not always true, a large company that decides to implement a company wide materials properties database system for all of its divisions will realize other benefits on top of those already mentioned.

  "Bigger is better" for a company that uses a corporate wide materials database system because that company reaps the benefits of sharing (Figure 4). Sharing knowledge between various locations allows a company to have

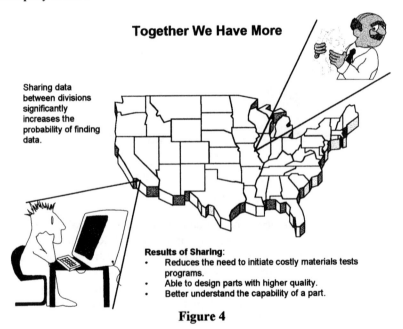

Figure 4

a wider breadth and a deeper depth of materials data than it would have if each location acted as isolated entities. The wider breadth of materials data means that there is a greater chance that data for any particular material has already been generated by one of the locations of the company. This eliminates the need to initiate a costly materials test program. There is not only a savings in terms of dollars with the elimination of a test program but there is also a savings in terms of time. Much time is consumed in obtaining the material, fabricating it into test specimens, performing the tests themselves and analyzing the data. The time saved helps keep the project on schedule and therefore keeps the customer happy. The wider breadth of data also gives designers a greater variety of materials to choose from. The designer is now given the information needed to choose a material of higher performance or of better value. The designer is also not relegated to choosing a material whose allowables properties are predicted. Instead, the wider breadth of allowables data increases his chances of finding a curve that was developed from actual testing. This eliminates the risk associated with using predicted properties.

The deeper depth of materials data allows a materials engineer to better understand the characteristics and behavior for certain materials where significant data exists. Armed with additional data, a materials engineer may be able to predict the behavior for that material (or for a similar material) outside of already existing parameters. For example, an engineer in Ohio supplied with a wealth of data on 321 Stainless Steel generated by a colleague in California may be able to use that information to predict the properties for 301 Stainless Steel.

Of course all of the above benefits of sharing may be realized without the implementation of a materials properties database system. Large companies may choose to just distribute a list of engineers who are experts on certain key materials. So when someone needs information on Alloy 718, all he/she has to do is check that list for the name of an expert. Or companies may choose to require that each location publish enough copies of all their materials data so that the other locations can have a copy. Both of these alternatives will allow corporations to realize the benefits of sharing. However, neither of these two alternatives provides a means of distributing up-to-date materials information that is cost effective and timely. The engineers who help their colleagues in other locations by answering requests for materials properties data will have to take time away from their own projects to locate the data, format it so that the requester can use it and then send it to the requester. This time consuming process must be eliminated. The best process for distributing the latest materials properties quickly and cheaply is through a company wide materials database system.

3. Requirements and Features

There are many requirements that must be met in order for the benefits of "bigger is better" to be realized. These requirements, of course, will vary from company to company. The requirements listed in this paper are not exhaustive and are only a highlight of some key features. Corporations that decide to implement such a system must involve members from its various locations in order to have a successful database system. Without such involvement, a corporation runs the risk of dividing a company rather than the intended goal of uniting for the purpose of sharing.

3.1 Database Software Requirements and Features

The first and foremost requirement for the database software is that it is accessible from any location. Since the intranet system at most companies are already established and very mature (or close to this point), a database system that connects to the intranet would be ideal. Such a system will allow every engineer to run the database either on their UNIX based workstation or the PC on their desk.

The database chosen to store the materials properties data must be able to directly interface with other programs. This interface should include two modes of operation: 1) Retrieval of information for a particular material or set of materials (programmatic access), And 2) provide a user interface to interact with the user to select a material or the type of properties required and then return this data to the "calling" program (interactive access). In other words, the database must allow other programs access to the properties in the database and the database must be able to transfer properties to other programs. So a designer who uses Pro-E must be able to access materials properties data that resides in the database from within his Pro-E environment. Similarly, a stress analyst must be able to run a program like ANSYS using a data input file generated by the materials properties database software. Though the materials data interchange may be performed by direct, custom interfaces between specific programs, a much-preferred means of transfer and access would use a standard, generic materials exchange form established by an international standards organization such as the ISO STEP standards. This eliminates the need to create a direct, or point-to-point interface for each specific program that uses materials information.

In addition to having a generic interchange of materials data between the database and other programs that require materials data, the database must be able to work with a Product Data Management (PDM) system. In such an environment, materials are electronically associated

1190

with parts and materials information is electronically version controlled. This is important since allowables may, for various reasons, change over time. It is therefore important that version control be maintained since, for example, a stress report generated 15 years ago may have used different allowables than those used today. The materials database must satisfy requests for information from both the PDM system and engineering applications (CAD, etc.). Likewise, a user interface for selecting and comparing materials data must be available in both the PDM system and engineering applications (CAD, etc.). For example, a request will be made by an application such as Pro-E. This request will be sent to a Materials Service (that resides on the network). The Materials Service (MS) exists to answer requests for materials information made on the network. The MS will draw upon a materials database for the information requested. The material selected by the Pro/E user is then "registered" in the PDM system. The PDM system can then be used to track materials use in various ongoing or released designs. If details of a material are required, the PDM system requests those details from the MS.   Figure 5 shows a schematic outline of the process just described.

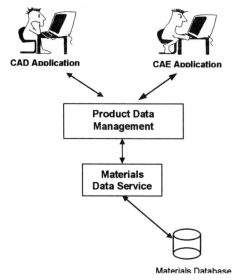

CAD Application   CAE Application

Product Data Management

Materials Data Service

Materials Database

Figure 5

It is important to note that this process is not a description of the data flow at any single location, instead it is independent of the locality of the user and the materials database.

Assumed in the above discussion of the data flow is the fact that there are multiple materials properties databases that reside on the

1191

company's network system. An alternative to having multiple databases is to have all the materials properties data in the corporation reside in one database file. At this time, this alternative seems very impractical and difficult to manage. It is more likely that successful implementation will entail multiple databases. With a multiple database system implemented, the software used must be able to search across multiple databases. This allows the engineer to make one search rather than doing multiple searches for each database that exists on the network. The database software chosen must, therefore, be able to perform searches across multiple databases that are located at various locations.

As with other software that resides on a network, the security of the materials properties data must be addressed. The data in the materials properties database will very likely represent millions of dollars in testing and analysis. The data, therefore, must not be accessible by those outside the company, unless given permission by the owners. Under this scenario, the outside company given permission will most likely be given access to only a certain portion of the data in the system. The software selected needs to be able to restrict access to certain parts of the database to certain people. In other words, access to the entire database cannot be permitted simply because someone got onto the database system. This restriction, of course, applies to those outside of the company, but also may be used to restrict access to those who are engineers within the same company. This latter feature will allow a materials engineer to access raw data that are used to develop allowables but restrict access to a designer to only allowables.

And lastly, the database must be able to store pedigree information, data points, curves, tables and pictures. The database must be able to store every piece of information that is needed to validate an allowables curve, that makes the creation of an allowables curves traceable and reproducible, and that makes any curve or set of curves searchable. Included in all this is that the database must be able to store a "range" field type. This will be needed if a curve that is stored in the database is applicable for a range of temperature. Often, for example, a high cycle fatigue curve may be good for a temperature range from 70°F to 250°F. In this case, a search for a curve at 150°F must result in finding a curve that is good from 70° to 250°F. A materials properties database must be able to store a varied amount of data.

3.2 Requirements from Each Location

For the company wide database system to be successful, each location in the corporation in question must create and maintain a database of materials properties data that they own. The benefits having such a system in place will never be fully realized until each location creates,

1192

maintains, and shares a database of their own materials properties. This is essential and this simple, obvious truth cannot be over emphasized. No matter how great the database software used and no matter how stable the network is, a company-wide database system must have materials properties data in it.

Summary of Requirements

1. Software chosen must be accessible from intranet.
2. Generic interface with CAD and CAE programs.
3. Must be able to work in a PDM environment.
4. Must be able to search across multiple databases.
5. Must have security system in place that restricts record-level access to only designated people.
6. Must be able to store tables, curves, and pictures. Related, the database must be able to store "range values" and search on them.
7. Each division/location must develop and maintain database of materials properties that they call their own.

Table 1

4. Issues to be Discussed Before Implementation

Even if all the criteria above were met, there are still issues that must be resolved and decided upon before implementation. These issues can be classified into two groups; one dealing with issues concerning input into the database and the other with the distribution of data throughout the corporation. More than likely, two groups will need to be formed to make these decisions, taking engineers from throughout the corporation. An easy way of picturing the function of these two groups is illustrated in Figure 6. The first deals with making sure that "garbage" doesn't enter into the database. And the 2nd group ensures that the data in the database is distributed to everyone who requests it.

Function of Two Resolution Groups

The first group will ensure that the data going into the data base is "clean".

The other group will ensure that the data in the database is delivered to everyone who needs it.

Figure 6

4.1 Issues concerning input to the database

The mission for the first group will be to develop standards to judge what data may be shared and what data needs to be excluded. It is not the job of this group to review every piece of data to determine if that specific piece of data should be shared or restricted. Rather, this group is responsible to set-up a "filter" to determine what may pass into the database and what must be kept out. Some of the standards may be the number of tests performed, making sure tests are performed to agreed upon standards, tests are documented sufficiently and other criteria. Once these standards are determined, this group may meet annually to determine if there needs to be any modification or deletions to the current standards.

4.2 Issues concerning data distribution

The purpose of this second group will be to develop the infrastructure needed in order to share data between the various locations in a large company. They will need to determine which database software best suits their purposes, help establish a basic schema which each division will build upon, aid in the development of each locations' own database(s) and deal with any other issues that will ensure that data may be shared efficiently. As with the previous group, this group will meet periodically to make sure that the database system is getting needed data to the requesters correctly and efficiently.

4.3 Effort Required

These two issues concerning data input and data distribution will very likely take up the bulk of the time required to implement a data management system. It's not uncommon for the various divisions in a large company to use different standards for testing, different specification for materials and different means to produce design allowables. Much time will need to be spent to come to a consensus on these issues. No doubt there will be many disagreements on the way to a consensus, but the quality of the materials data management system and the acceptance of it will depend greatly on the resolution of these two issues. It is important that adequate time and budget be allotted for these two issues.

5. Conclusions

Implementing a materials database system at any local site will help the entire engineering process to proceed faster, better, and cheaper. Rocketdyne is now experiencing many of the benefits of having such a system in place. Implementing a materials database system across the many locations of a large company will result in many other benefits. One benefit will be to provide an optimal way to take advantage of the abundance of materials properties data that is often available from the various locations of a large corporation. Data sharing is optimized with a database system implemented. For that data sharing to be efficiently shared and useful, standards must be set for the pedigree of the data itself and the manner in which it is shared. The suggested standards described in this paper are one means by which data sharing may be realized. Whatever standards are finally adopted, it is clear that materials properties data sharing must occur if a large corporation wants to take advantage of its size in order to realize the "bigger is better" proverb.

6. References

[1]   *MSC/MVISION User's Manual, Version 3*, The MacNeal-Schwendler Corporation, Los Angeles, CA, August 1996.

[2]   *ANSYS Basic Analysis Procedures Guide, Version 5.3*, Swanson Analysis Systems, Houston, PA, June 1996.

[3]   *Pro/Engineering Fundamentals Manual Version 18*, Parametric Technology Corporation, Waltham, MA, 1997.

[4]   Wong, Terry, "Optimizing the Engineering Process at Rocketdyne Using MSC/MVSION," *1997 MSC Aerospace Users' Conference Proceedings*, The MacNeal-Schwendler Corporation, 1997.

7. Acknowledgments

It goes without saying that an ambition plan like the implementation of a corporate wide materials properties database system is a task that is only undertaken with help from many people. This paper is a reflection of the help of many people. Mark Freisthler provided much insight into pitfalls that need to be avoided to have a successful system. Tony Eastland and Aryeh Meisels provided me with the managerial leadership that made this project one of the key priorities in my plateful of work. Words of wisdom from John Halchak taught me how to work within a large corporation. And Nancy Abesamis deserves special recognition for her wise counsel that can be seen in every paragraph of this paper.

Recent Developments in Materials Data Management

Edward Stanton, Thomas Kipp and Eric Lantz
The MacNeal-Schwendler Corp.

1.0 INTRODUCTION

In 1982 materials experts met at the Fairfield Glade Workshop [1] hosted by the U.S. National Bureau of Standards, now NIST, to plan the computerization of materials data for use in the then emerging CAE/CAD/CAM product design environment. A lot has happened since that Workshop and the European Materials Database Demonstrator Programme [2]. The environment now includes PDM systems, ISO/STEP standards and of course the Web. This paper is a report on recent progress made in providing unambiguous product data exchange of computerized materials data. The focus is on industry led efforts to meet life cycle material data management requirements in a "virtual engineering" environment. Participation in a multiyear NIST collaborative program, the Rapid Response Manufacturing (RRM) [3] program, in ISO/STEP PDES Inc. pilot projects and in the DARPA Agile Infrastructure for Manufactured Systems (AIMS) [4] project are the basis for the recent developments reported in the paper. They cover PDM and Bill of Materials issues, ISO/STEP material Units of Functionality and the Web enabled AIMS project that designed, analyzed and manufactured a rocket motor faster, better and a lot cheaper.

The accuracy and accessibility of materials data during the entire product life cycle is critical to manufacturers. The RRM project demonstrated PDM Web access to existing materials databases using a CORBA server and a JAVA client architecture. This architecture enables enterprise wide PDM access to Bills of Materials, test data and application specific materials data. The ISO/STEP projects are creating a collection of material Units of Functionality that capture specific material data requirements for the aerospace, automotive, electronics, and shipbuilding industries. Intranet linking of the design, analysis and manufacturing engineering data for an actual component was accomplished on the DARPA/AIMS project led by Boeing Rocketdyne. Versioning control of interactive design data exchanges at remote sites was a key to producing the small rocket motor. Design and manufacturing costs measured by a third party were reduced by more than a factor of ten and the schedule by a factor of three.

2.0 MATERIAL INFORMATION USE REQUIREMENTS

Today CAE/CAD/CAM and PDM systems are in widespread use by major manufacturers and their top tier suppliers. This has caused use scenarios and use environments for geometry and materials information to change significantly. In the case of geometry, the exchange of CAD product data using emerging ISO/STEP standards is now in production at GM and at several large aerospace companies. This change was necessary to deal with solid modeling and the variety of CAD systems used by their suppliers. ISO/STEP Application Protocol (AP) 203 is used for CAD geometry data exchange as well as Bill of Materials and PDM product data exchange.

However, materials property information, even when computerized, is rarely in a form directly useable by CAE/CAD/CAM applications and is almost never exchanged directly by these applications even when it is available. Material analysis information requirements are being addressed by the ISO/STEP New Work Item MATINF [5] that seeks to define materials information for analysis methods beyond just linear FEA. That work is described in another NAFEMS paper based in part on the earlier ESPRIT program for a Generic Engineering-analysis Model (GEM) [6]. Here we describe use scenarios for materials at different stages of the product life cycle as described in the ISO/STEP Industrial Framework Model [7].

Conceptual Design: Materials information is needed for screening candidate materials based on functional performance requirements and prior experience with candidate materials in similar products. At this stage a specific material object is **not** being designed. The use scenario is a relatively small team of advanced design engineers working in a Virtual Product Development (VPD) environment, which may include Tier one suppliers.

Design: Materials information is needed for specific stock materials, their grades and finishes, standard components and joining materials as specified in the engineering Bill of Materials. This includes material object property information for all the product's intended use environments. At this stage specific material objects are designed and analyzed for end use environments as well as other life cycle environments e.g. manufacturing, assembly, service, maintenance and disposal.

Make: Materials information is needed for specific material object fabrication processes, tooling, assembly fixtures, joining materials, dunnage, and in-process materials consumed during manufacture. These often appear in the manufacturing Bill of Materials and can include up to several hundred "touch parts" for every part in the finished product.

Operate: Materials information needed for in-service operation which includes data for health monitoring, maintenance, repair, and other material specific operating information provided by the manufacturer to the end user. The level of detail varies greatly among industries in proportion to the initial cost of the product and its service life.

Dispose: At the end of a product's service life material information about its composition, procedures for recycling and other environmental impact data are required. ISO 14000 is one source of these requirements and regulatory agencies are increasingly strict about compliance with environmental regulations.

IDEALIZED PRODUCT LIFE CYCLE

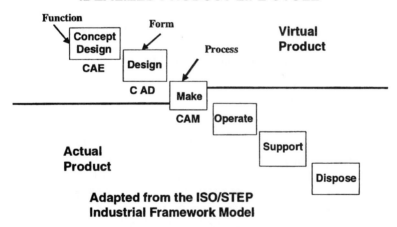

Adapted from the ISO/STEP
Industrial Framework Model

Material Designation for Web Access

Distributed Design-Build teams are beginning to use the Web in many industries to facilitate a "design anywhere, build anywhere" business model. Objects, including material objects, need to be Web accessible to these teams and that can be a challenge today. What follows is a heuristic discussion of possible designation standards that expose/publish only basic source information for a material object in a STEP product data model. The goal is a coherent link via the designation to all bills of materials data and property models needed by a design-build team at each stage in the product life cycle.

The ISO and book publishers worldwide created the ISBN numbering schema, ISO 2108, over twenty years ago with a Registry service administered by the International Standard Book Number Agency in Berlin. It's a very simple schema, the standard is three pages, with the 10 digit ISBN number parsed into four parts separated by hyphens. If you have bought a book lately, over the Web or over the counter, you know the ISBN number is likely to be the first designation requested.

The 10 Digits of an ISBN are divided by hyphens into four parts:

Part 1: Country or Group of Countries Identifier
Part 2: Publisher Identifier
Part 3: Title Identifier
Part 4: The Check Digit

It would not be difficult to transcribe this schema for manufactured materials and make changes that material manufacturers and ISO/STEP Committees require. Note that the ISO 2108 standard uses hyphenation rules to expand the number of unique ISBN designations to a very large number. Interestingly music publishers have already taken "ISMN" and other publishing groups have approached ISO TC 46/SC9 with similar requests. The suggestion made in this paper is that ISO TC184 develop a STEP designation schema for manufactured materials and perhaps other designation schemas for Part libraries to meet industry specific product designation needs in a Web integrated design-build environment.

A manufactured material product, like a book, has a form and other specific attributes that need to be linked to the designation of a manufactured material product. This is in keeping with ISO/STEP Part 45, the fundamental material resource, which defines a material as a manufactured object with associated properties in the context of its use environment. This is in contrast with the UNS materials designation, which does not link the designation to product form and other detailed attributes for a specific material product. UNS is a designation for generic material objects organized by composition that includes different material products with similar chemical composition, i.e. many similar material products have the same UNS number.

To illustrate the potential benefit of Web access to materials information we note its benefit to biochemical product design engineers. There are more than a billion DNA base pairs registered in public databases that drug designers access daily over the Web using designations based on the DNA code. In this case Web access is routinely accomplished using the BLAST DNA algorithm [8] to search a universe of Flat-files, Relational, and Object-oriented databases. Web access to highly technical data for very complex material objects is available to biochemical product designers worldwide via a common designation system.

In contrast there are hundreds of different engineering material designation schemas used by trade associations, standards agencies, government agencies, and individual companies. Traceski [9] in 1990 cataloged more than 2000 specifications and standards for plastics alone and there are probably even more for metals. Virtually all were developed in an era before Web connected computers and they implicitly require a natural language capability to search the universe of material databases. Access requires a natural language capability because no standard designation is available for engineering materials. This makes Web searches a challenge even when the same database software system built all the databases because each database schema uses different designation systems.

It is recommended that ISO TC184 Committees, material manufacturers, and other interested groups work together on an ISO/STEP designation for manufactured materials for better Web access. The benefit could be huge for users of PDM/CAE/CAD/CAM systems who are nearly all planning to use the Web to access information during product development. Existing systems for designating materials need not change if they are linked to the

ISO/STEP designation system for manufactured materials. The goal is rapid Web access to all material product information via ISO/STEP identifiers like that ISBN numbers provide to publishers and readers. The expectation is that sources of materials and materials data would index their information for Web access by design-build teams.

ISO/STEP Part 45 Material Standard

The ISO/STEP view of materials is from an industrial design perspective. The fundamental resource for materials, ISO/STEP Part 45, defines a material as a manufactured object with properties in the context of the use environment, Swindells [10]. This is in contrast to the textbook view of materials with limit point properties abstracted from microscopic representative volume elements without representing the actual property distributions in a finite volume material object as manufactured. The latter robust definition is needed for designing robust products; ones that don't run up warranty costs from local failures not predicted by oversimplified material models. The textbook definition is adequate for conceptual design (screening) and is obviously a subset of the Part 45 definition.

Unlike the ISO/STEP geometry standard (Part 42) there are no math forms in Part 45 to represent linear or nonlinear material properties. Linear property models for finite element mechanical and thermal analyses are in the finite element standard (Part 104) and these are used in AP 209 for modeling metallic and composite structures. The comprehensive AP (MATINF) for materials information in work (Leal, et. al. [4]) will require several enhancements to existing STEP standards to support the nonlinear math forms commonly used by materials engineers. Also, efforts need to be made to harmonize the information models proposed for MATINF with the AP 209 STEP model, Hunten [11]. A schematic shown here illustrates the STEP "product backbone" in AP 209 that connects materials models and finite element product models.

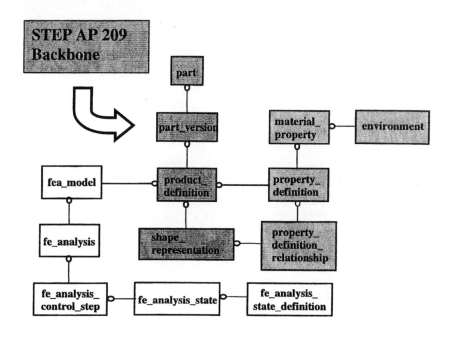

ISO/STEP Material Units of Functionality

Several STEP Application Protocols have Units of Functionality for product information related to the Bills of Materials typical of an industry. For example the automobile industries AP 214 and other industry AP's have such Units of Functionality. In several cases the AP's which are Part 200 series STEP standards have not yet reached International Standards (IS) status. The information described here may change as the result of the ballot process required to advance from Committee Draft (CD) to Draft International Standard (DIS) to IS. In the following description of UoF Application Elements text in UPPERCASE indicates the object is specific to the AP.

Part 203 (IS): Configuration controlled design
AP 203 Bill of Materials UoF – Table 3
  Application Element
      ALTERNATE_PART
      COMPONENT_ASSEMBLY_POSITION
      Transformation
      ENGINEERING_ASSEMBLY

Security_code
Engineering_assembly to planned_effectivity
ENGINEERING_MAKE_FROM
ENGINEERING_NEXT_HIGHER_ASSEMBLY
As_required
Component_quantity
Reference_designator
Unit_of_measure
Engineering_next_higher_assembly to component_assembly_position
ENGINEERING_PROMISSORY_USAGE
SUBSTITUTE_PART

Part 209 (CD): Composite and metallic structural analysis and related design
AP 209 Materials UoF – Table 15
  Application Element
    ANISOTROPIC_MATERIAL
    DISCONTINUOUS_FIBER_ASSEMBLY
    FILAMENT_ASSEMBLY
    HOMOGENEOUS_MATERIAL
    MATERIAL_DIRECTION
    Material_orientation
    MATERIAL_PROPERTY
    Property_name
    Property_value
    MATERIAL_SPECIFICATION
    Material_designation
    Document
    STOCK_CORE
    STOCK_MATERIAL
    property
    Reference_direction
    Relate_to
    Specified_material

Part 210 (CD): Electronic assembly, interconnect and packaging design
AP 210 Bill_of_Material UoF
  Application Element
    ALTERNATE_PRODUCT
    ALTERNATE_SELECT_PRODUCT

ASSEMBLY_ALTERNATE_PRODUCT
ASSEMBLY_COMPOSITION_RELATIONSHIP
ASSEMBLY_MAKE_FROM
INTERCONNECT_MODULE_STRATUM_ASSEMBLY_
RELATIONSHIP
ENGINEERING_MAKE_FROM
MATERIAL_ASSEMBLY_RELATIONSHIP
MATERIAL_COMPOSITION_RELATIONSHIP
NEXT_HIGHER_ASSEMBLY_RELATIONSHIP
PRODUCT_ASSOCIATION
PROMISSORY_USAGE_RELATIONSHIP
SUBSTITUTE_PRODUCT
TEST_SELECT_PRODUCT

Part 214 (CD): Core Data for Automotive Mechanical Design Processes
AP 214 Material_Property UoF – Table 27
Application Element
ITEM_PROPERTY
Item_property to design_discipline_item_definition
Item_property to dimension
Item_property to fea_model
Item_property to mated_item_association
Item_property to shape_item
Item_property to material
Item_property to material_property
MATERIAL
Description
Name
Material_property to property_value
PROPERTY_VALUE
Lower_limit
Specified_value
Upper_limit

1.0 CAE/CAD/CAM Access to Materials Information

As a value proposition, access to materials information requires more than computerized data. It requires referential integrity among the CAE/CAD/CAM systems used to produce a product and coherent representations of the materials data. Attention to this level of detail can be

1205

daunting, witness the materials communities' difficulty with something as basic as designation, but it is within the capabilities of current information technology. A framework for CAE/CAD accesses to STEP material object data under PDM version control is illustrated here to make the connection between design and analysis functional requirements for a typical industrial use scenario.

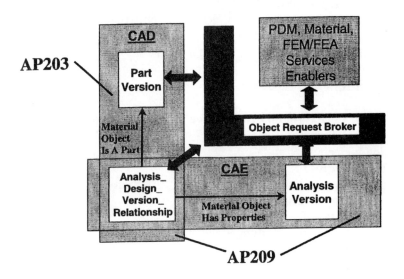

RRM Material Property Client Server Project

The NIST ATP Rapid Response Manufacturing program had numerous pilot projects. One was the Interoperable Materials Property Data Services Provider project managed by Lantz [3] with requirements defined by Ford, Boeing Rocketdyne and Texas Instruments. Initial requirements centered on CAE data management in a use environment that included PDM services. This led to a portable materials client requirement that,

- Could be invoked from different applications
- Provided for material selection and property display
- Could re-link to a material record via a materials object
- Could return a material object & properties to an application

As is often the case in collaborative programs, the initial pilot project identified new industry requirements needed to advance an R&D solution to a production solution. A client-server architecture similar to that shown here was developed using a Java client and Corba server and tested successfully at Boeing Rocketdyne. This demonstrated Web access to material databases via a portable Java client in a pilot project. It remains to scale up this result to production level projects.

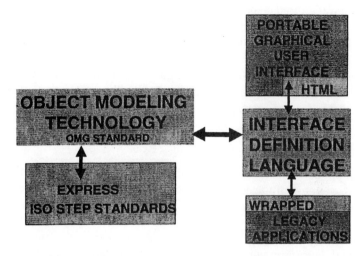

AIMS Collaborative Design/Manufacture Pilot Project

The DARPA Agile Infrastructure for Manufactured Systems (AIMS) Simple Low Cost Innovative Engine (SLICE) 2 project had a 10x development speedup and 100x cost reduction goal for the design and manufacture of a prototype rocket engine. Although materials data management was not the driver on this project the team leader, Boeing Rocketdyne has developed material databanks to support its engineering process, Wong [12]. The Web was a very important key to linking team members around the country to form a virtual product and process development environment. Boeing Rocketdyne under the leadership of Bob Carmen managed the project.

The ability to conduct CAE over the Web was demonstrated on the project with the aid of the Multimedia Environment for Collaborative Engineering (MECE) system. It allowed any team member to access the project electronic notebook over an Intranet created for the project via a Web browser. They were able to markup screen images of various CAD and CAE models to highlight areas of concern on applications at any site on the

Intranet (AIMSNet). It wasn't necessary to have every application at every team members' site. Rocketdyne maintained the vault for project data and provided version control services.

Desktop components like EXCEL and the electronic Intranet Notebook were also important to effective and open communications on the project, which designed and manufactured a polyurethane prototype rocket engine in ten (10) months. The development cost was reduced from $4.5M to $47,000. The USC Marshall School of Business audited project financial performance [4], and Boeing Rocketdyne produced a videotape documenting the program for DARPA. Interestingly the metrics kept on the program were used to track not control events. They were used to help make intelligent changes when unexpected events were encountered on this R&D project.

5.0 SUMMARY

Yes, a lot has happened to enhance material data management since Fairfield Glade in 1982 but even more needs to happen to keep pace with advances in product data management. The test community now has materials database standards for computerization used by several large companies and commercial material database publishers. Awareness of metadata requirements for determining what data are valid for a particular intended use has increased but there are few design teams that directly connect material properties for analysis to metadata or to the Bills of Material for a product. Verification and validation of materials data still occurs late, often at design verification long after critical design decisions have been taken. Handbook and other paper sources are beginning to give way to material databases but we remain close to the calculator stage in materials data for design use.

Pilot projects from the STEP, RRM, and AIM programs mentioned in this paper are beginning to integrate materials services into virtual product design processes. Java clients and CORBA servers for material information have been demonstrated at Ford and Boeing that provide materials data access to CAE/CAD/CAM applications. However, enabling software tools are only one factor in the implementation of IPPD design methodologies that include materials as active design variables. Changes in organizational culture, changes in legacy design to manufacturing processes and reductions in the complexity of materials data access are necessary as well. Materials experts need to become part of the virtual product development process.

The modularization of ISO/STEP standards and the implementation of the STEP PDM schema into CAD and CAE standards are important milestones in meeting referential integrity requirements among the several Bills of Materials found in product development. The accuracy and accessibility of materials data during the entire product life cycle is a dynamic requirement and standards are critical for products with very long life cycles. More work is required to enable Web access to coherent materials data across all the disciplines required by manufactured products. A good first step would be a common designation system for manufactured material objects.

REFERENCES

1. Computerized Materials Data Systems, J. H. Westbrook and J. R. Rumble, Eds., Proceedings, Workshop held at Fairfield Glade, Tennessee, 7-11 Nov. 1982.
2. Materials Information for the European Communities, N. Swindells, N Waterman and H. Krockel Eds. Proceedings, Concluding Workshop of the Materials Demonstrator Programme held at Petten, The Netherlands, 6-8 December 1989.
3. Rapid Response Manufacturing Final Technical Teams Report, RRM Document 500D035.4, National Center for Manufacturing Sciences. 1998.
4. Allen, G. A. and Jarman, R., Collaborative Research & Development: Manufacturing's New Tool, John Wiley & Sons, Appendix D, 1999.
5. ISO/NWI 10303-N737, Materials information for design and verification of products (MATINF), ISO TC184/SC4 Nominator, David Leal, 1998.
6. ESPRIT CIME 8894, Generic Engineering-analysis Model, Core Model, David Leal, et.al., 1996.
7. ISO 10303-N219, Industrial Framework Model, ISO TC184/SC4 Editor, Christopher Vaughan, 1999.
8. Studt, T., Open, Object-Oriented Software Deals with Drug Data, R&D Magazine, August 1998, pg 67-68.
9. Traceski, F. T., Specifications & Standards for Plastics & Composites, ASM International, 1990.
10. ISO 10303 Part 45, Materials, ISO TC184/SC4 Editor, Norman Swindells.
11. ISO 10303 Part 209, Composite and metallic structural analysis and related design, ISO TC184/SC4 Editor, Keith Hunten.
12. Wong, T., Optimizing the Engineering Process at Rocketdyne Using MSC/MVISION, 1997 MSC Aerospace Users' Conference Proceedings.

LOW-COST FIBERGLASS COMPOSITE BRIDGE DECK AND DISASTER-RELIEF HOUSING

Ohanehi[1], D. C., Hayes[2], M. D., Lesko[3], J. J., and Cousins[4], T. E.
[1] Res. Scientist., Engr. Sci. & Mech. Dept., Virginia Tech, Blacksburg, VA 24061
[2] Research Engineer, Michelin Company, 515 Michelin Rd, Greenville, SC 29605
[3] Asst. Prof., Engr. Sci. & Mech. Dept., Virginia Tech, Blacksburg, VA 24061
[4] Assoc. Prof., Charles E. Via, Jr. Dept. of Civ. & Env. Engr., Virginia Tech, Blacksburg, VA 24061, USA

ABSTRACT

Cost-effective analysis methodologies are critical for the development of low-cost infrastructure systems. Two industrial applications were used to demonstrate cost-effective finite element analysis (FEA) for infrastructure development. The applications were a low-cost pultruded fiberglass bridge deck and recycled-plastic reinforced composite housing structures.

ANSYS 3-D FEA models were generated for a prototype pultruded fiberglass composite bridge deck. The developmental bridge deck was a phase in an experimental effort to install and monitor test sections of composite decking in the Schuyler Heim lift span bridge in Long Beach, CA, USA. The deck was fabricated using low-cost, off-the-shelf, glass/polyester square tubes and flat cover plates bonded together using an epoxy adhesive. The deck was fatigue tested to 3 million cycles at 3 Hz and then statically loaded to failure. The initial FEA was done using Cosmos/M 1.75A software before moving on to ANSYS 5.3 to meet additional modeling requirements. This paper highlights modeling lessons that should be of interest to FEA software users and developers. The lessons include helpful but less-publicized software features like the extrusion method and associated usage tips.

The same modeling method was applied to a basic structural housing unit. A structural unit concept based on recycled plastics was developed for alleviating problems of homelessness, and providing housing for disaster relief, refugee shelters, as well as low-cost housing primarily in Third-World countries. The structural material was recycled high-density polyethylene (HDPE) reinforced with glass fibers or locally available fillers such as sand and talc. An ANSYS 3-D FEA was employed for evaluations of deflections and stresses under self-weight, snow, and wind loading conditions. The FEA procedure was validated using a scaled plexiglas physical model. Design decisions were made using the model.

Because the applications were deflection controlled and were large composite structures, they lent themselves easily to accurate analysis using fairly coarse meshes. More accurate stress and strain analyses constitute a next step and may be implemented using global-local or sub-modeling analysis tools.

1. INTRODUCTION

Plastic reinforced plastics are finding increasing applications in a variety of civil engineering infrastructure. The applications generally are large composite structures with special requirements for FEA modeling. Bridge decks, plastic composite homes, and tractor trailer rails are recent or current applications at Virginia Tech. A unique feature of the deck and home applications is cost-minimization strategies.

Bridge decks appear to be one of the most promising applications of reinforced composites in infrastructure applications because of the weight and durability advantages of composites over concrete. Several composite bridge deck developments are underway [1-3]. The primary subject of this paper is FEA work in support of Strongwell's development of a low-cost composite deck using off-the-shelf pultruded products [4].

The same FEA methodology was applied to a second low-cost structure [5]. The Daedalus World Shelter concept is a contribution to a simultaneous solution of two world problems, namely waste plastics recycling, and shelters for disaster relief, the homeless, and those without adequate shelter as well as low-cost housing. The material was recycled HDPE reinforced with glass fibers or locally available filers like sand or talc. The use of recycled plastics and the resultant low cost advantage made the Daedalus structural unit concept unique among widely publicized plastic homes. No FEA work has been used in related development and FEA work in support of this program constitutes the second half of this paper.

2. BRIDGE DECK APPLICATION

Strongwell's composite deck employed pultruded sections from their standard, off-the-shelf, EXTREN product line. Exclusive use of standard sections decreased the cost of the deck, and made it readily available and adaptable to many deck depths. The deck concept was part of an effort to install and monitor test sections of composite decking in the Schulyer Heim lift span bridge in Long Beach, California, USA [6]. The prototype deck was 1.22 m by 4.27 m by 121 mm (4 ft by 14 ft by 4.75 in.), was tested statically to shearing "punch-through" failure after 3 million cycles of fatigue testing at 3 Hz, and was modeled using FEA.

2A. Composite Bridge FEA Literature
Most composite bridge deck development included FEA modeling. While Zureick [3] gave only sketchy details on composite deck FEA, he covered most pre-1995 development work. He ended with a debatable conclusion that "approximations to the load distribution characteristics and boundary conditions over the bridge girders as done in earlier works are not warranted if the reluctance to use this technology by bridge designers is to be overcome." This study was based on the assumption that approximations were required and that the best approximations should be sought. Burnside and Barbero's approximations were required by severe wavefront limitations of the dated ANSYS 4.4 package but yielded valuable insights on the buckling response of some geometries and fiber orientations [7]. The ANSYS optimization option was invoked for minimizing the bridge deck volume (and mass) and there was a discussion on ANSYS Poisson's ratio conventions, a potentially confusing issue that is still of current interest for ANSYS composite materials applications. Sotiropoulos's work included ANSYS composite deck modeling and popular themes in composite deck FEA literature namely, composite action in deck cross sections and the effectiveness of deck adhesive joints [2]. Sirisak's FEA on steel decks [8] was very relevant to composite decks. Using simple trusses, bars and beams in ANSYS 5.0, a shakedown analysis defined the bridge's reserve capacity by a rating factor.

It was generally agreed that composite deck design was deflection controlled. While detailed stress analysis was required for joint designs and predictions on initiation of edge delamination, the bulk of the modeling centered on FEA to generate deflections and serviceability limit states. This paper describes FEA work for assessing the structural deflections and local strains.

2B. Bridge Deck Description
The prototype deck was formed from twelve 102 by 102 by 6.35 mm (4 x 4 x 0.25 in.) square tubes sandwiched between two 9.53 mm (0.375 in.) thick plates, all made from polyester/glass (Fig. 1). The plates and tubes were Strongwell's standard EXTREN 525 pultruded shapes. The tubes were connected using both an epoxy adhesive and Fibrebolt® studs which ran transversely through the webs of

Fig. 1– Side View of Composite Deck in Test Configuration

each tube. The top plate was bonded to the square tubes using the epoxy adhesive. The deck was intended for a design load of 92.5 kN (20.8 kips), an AASHTO HS20-44 wheel load with a 300% Impact Factor included [9]. In the test configuration modeled, the deck was supported by W16x40 steel beams spaced 1.22 m (4 ft) apart and was connected to each steel stringer using three steel bolts spaced 4 or 5 tubes apart. Holes were drilled through the deck and steel beam flanges with standard bolts and washers bearing directly on the top composite plate. Wood inserts inside the fiberglass tubes at the hole locations were used to provide additional transverse integrity under bearing load. The deck was oriented so that square tubes ran transverse to the steel girders (Fig. 1).

2C. FEA Model For Deck

All models in this study were generated on a PC with the Windows NT operating system. Cosmos/M 1.75A [10] was used in the initial models. Later models used ANSYS 5.3 and 5.4 [11] on a Pentium 300 MHz PC.

The models were created to study the response of the deck in a connected deck-stringer configuration and allow for future parametric studies of the deck at other stringer spacings and deck thicknesses.

Shown in Figures 2a and b are deck panels, used in the testing program and modeled by FEA. The models in Figures 2a and b are respectively center- and end-patch loaded. A 222.4 kN (50,000 lb) loading was distributed over a 508 by 305 mm (20 x 12 in.) loading patch area and deflections at the design loads were calculated by scaling.

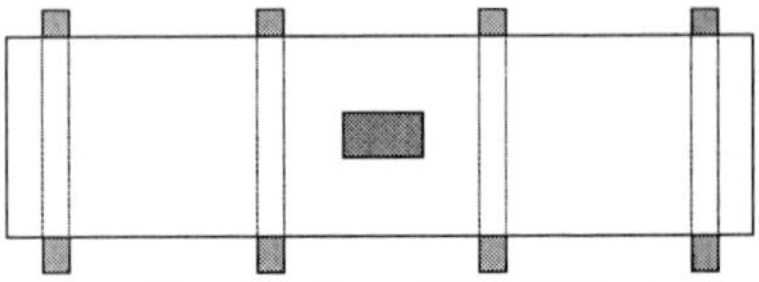

2a - Service Load Test (Center Span)

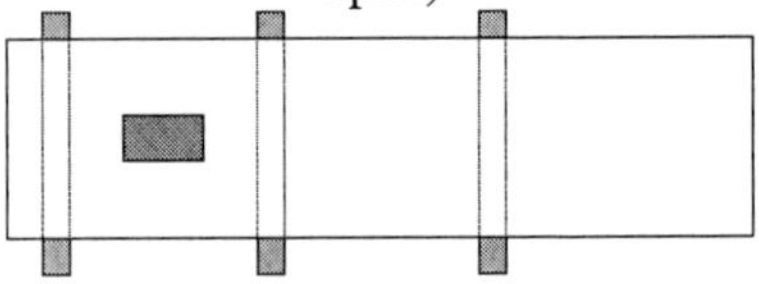

2b - As Received Strength (Left Span)

Fig. 2 – Deck Schematic Showing Test Load Locations

Deck tubes and plates had both uni-directional and continuous strand mat (CSM) glass fibers in an isophthalic polyester resin. Deck material properties used in the models are listed in Table 1, and were minimum "homogenized" material properties for polyester/glass structural shapes from Strongwell's design manual [12].

To facilitate the meshing of the large structure, the "extrusion" method was applied extensively using a meshed 2-D cross section as a template for generating brick elements. The models generated had approximately 8700 3-D elements with 28,700 nodes. SOLID45 brick elements were used to model the deck (tubes and plates) and steel support I-beams, and were 8-node elements with an orthotropic material property option. Each node had three translational degrees of freedom. The element had creep, stress stiffening, large deflection and large strain capabilities that were not employed here.

Initial models used pin supports for the I-beam support. Effects of a variety of boundary conditions were studied. The final deck model included a 3-D model of the 3 support I-beams and facilitated the use of more realistic boundary conditions. The model connected the deck to the support wide -flange beams with shared nodes at the bolt locations.

Tube-tube and tube-plate adhesive bonds were represented as rigid connections. Mottram estimated a 10% reduction in the flexural rigidity of pultruded beam assemblies due to adhesive bonding flexibility [13]. The rigid connection assumption gave accurate deflection results in the current study potentially because of bond quality. For detailed stress analysis, the adhesive bonds may require a plate element on elastic foundations.

Mechanical Property	Structural Tube 102 mm x 102 mm (4 in. x 4 in)	Plate 9.53 mm (0.375 in.)
Ultimate Flexural Stress Length Wise, MPa (ksi)	207 (30)	207 (30)
Ultimate Flexural Stress Cross Wise, MPa (ksi)	69.0 (10)	124 (18)
Flexural Modulus of Elasticity Length Wise, MPa (ksi)	11034 (1600)	13793 (2000)
Flexural Modulus of Elasticity Cross Wise, MPa (ksi)	5517 (800)	9655 (1400)
Ultimate Shear Stress, MPa (ksi)	31.0 (4.5)	41.4 (6)
Shear Modulus Length Wise, MPa (ksi)	2931 (425)	NA

Table 1 – Material Properties of Tube and Plate

The Fibrebolts® and the wooden inserts at bolt locations were not included in the model for modeling convenience and gave good deck deflection results. The effect of bolt holes will best be introduced through a sub-model of a small section of the deck including the bolts.

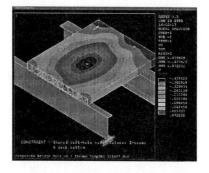

Fig. 3 – Vertical Deck Deflections

2D. Comparison Of Deck Model Results With Test Data
Table 2 summarizes the results from the finite element models of the deck and compares these results with those from the testing program. Deck deflections and strains were compared at the maximum design load, 92.5 kN (20.8 kips). The model provided a good prediction for vertical deflections (Fig. 3) with the deflections predicted by the

1215

model at 92.5 kN (20.8 kips) of load approximately 10% higher than those measured during the tests. Slight over-prediction of the deck deflection was expected since minimum design material properties were used in lieu of actual material properties, and over-compensated for the stiffening effect of the coarse mesh.

The general stress and strain contour patterns were reasonable. In general, the longitudinal strains on the bottom surface of the deck under the load patch were within 7% of those measured before the fatigue tests. Not shown in Table 2 were strains from some high strain areas of the deck panel with poor correlation to the test results. The mesh used was fairly coarse and was selected primarily because of ease of modeling, and hardware and limited-version software capabilities. Given the non-uniformity in the deck cross-section and material properties, fairly complex strain distributions were reasonable. Both test and FEA model values of transverse strains showed similar bands of alternating compressive and tensile patches (Fig. 4). It was demonstrated, through variations in boundary conditions, that the bands were controlled by the boundary conditions. For better strain predictions, refined sub-models are needed in areas of high strain (Sec. 4). Improved strain data may also be generated from a full-scale model with detailed (and time consuming) non-uniform meshing.

The results were qualitatively consistent with the damage data. There was a lot of stress "activity" around the loading patch area and in particular, around tubes 5 and 8 (load/no-load transition area). Vertical deck members (flanges) did not show large stresses and support the absence of fracture or delamination in the vertical members during the test-to-failure. Better correlations require buckling analysis.

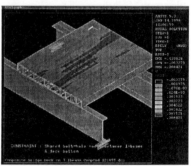

Fig. 4 – Longitudinal Strains
(Across sq. tube direction)

2E. FEA Lessons
The bridge deck model highlighted the use of the "extrusion" method for generating higher-order elements from lower-order geometries. In addition to the deck model, one of the authors had used extrusion for modeling active magnetic bearing hardware, truck rails, and low-cost housing structures, the subject of the next section. Extrusion is clearly a powerful technique that has not been publicized until recently. Texas Instruments' applications of extrusion in modeling semiconductor packages were described in [14,15] where ANSYS macros were used to automate the process. Understandably, extrusion may not be used conveniently in many complex structures. The deck model demonstrated

moderately complex structures that were good candidates for extrusion. ANSYS 5.5 has added enhancements to the extrusion options including VSWEEP for both extruding hex's and hex-meshing existing volumes. However, ANSYS had very limited documentation on extrusion but had much more than Cosmos/M. Cosmos extrusion time was very long and extrusion was apparently not actively promoted.

A few pitfalls were apparent in applying ANSYS extrusion. Numerous redundant nodes and keypoints were generated during extrusion. In applications where redundant nodes and keypoints were not advantageous, redundant entities may be merged before the solve step. However, caution should be exercised in merging the entities. In one ANSYS 5.3 run, the merging step resulted in a

Load Configuration	Load kN (kips)	Vertical Deflection under load, mm (in.)		Longitudinal Strain under load, microstrain	
		Model	Tests	Model	Tests
Middle Span loaded (see Figure 2 (a))	92.5 (20.8)	4.32 (0.170)	3.81 (0.150)	1130	1210
End Span Loaded (see Figure 2 (b))	92.5 (20.8)	4.57 (0.180)	3.81* (0.150)*	1298	1270*
			4.32** (0.170)**		1060**

* as received condition   ** post fatigue

Table 2 – Comparison of FEA and Test Results

corrupted database. There was also confusing documentation on acceptable 2-D elements for use as templates in extruding 3-D entities. One may argue about the benefits of expending the effort to generate uniform elements by extrusion as opposed to using automatic meshing to generate mostly tet elements. The current study did not explore theoretical issues related to hex's versus tet's. However, practical experience did demonstrate models with smaller numbers of elements that ran fast (10-minute run time), gave good results for deflections and were visually appealing. In areas of high strains, sub-modeling may be used for detailed meshing to improve accuracy.

3. EMERGENCY HOUSING APPLICATION

The second example of FEA extrusion was the low-cost housing structure.

3A. Basic Concept   Fig. 5 – Unit Panel FEM

The Daedalus structural housing unit employed a basic building block, a ribbed plate panel (Fig. 5), for all parts of the plastic composite home namely, walls, roofing, flooring, doors, and windows. The structural analysis consisted of beam-theory-based parametric studies and FEA of the unit panel, in addition to a complete structure FEA. In all analyses, brick elements (SOLID45) were used in place of shell or 2-D element. Bricks offered all the advantages listed for the bridge example while maintaining physically meaningful models for the primary users of the results namely, architects, building contractors, plastics extruders and civil engineers who may not be familiar with FEA.

3B. Unit Panel Optimization
A simple beam-based parametric study was used to calculate initial "optimal" parameters for a 1.25 x 1.25 m (49.2 x 49.2 in.) panel. The initial parameters were fed into a panel FEA model and used to generate an "optimal" combination of panel parameters. The selected "optimal" parameters would be evaluated by mold analysis and were used for the FEA evaluation of a complete structural unit.

The analysis methodology used here was a simplification of published procedures. By using equivalent unribbed flat thickness for ribbed sections, Lifshey and Campo provided simplified design equations for ribbed plates [16,17]. Throne and Progelhof developed an iterative method, based on the ribbed-plate moment of inertia equations, to generate plate optimum rib designs [18]. Trantina et al. discussed a software package, Ribstiff, for "quick, approximate solutions for the stiffness of laterally-loaded rib-stiffened plates" and included geometric nonlinearities [19]. Sherman et al. included detailed geometrical definitions and employed the ADINA nonlinear FEA package [20]. Cases examined using the non-automated optimization method included diagonal rib structures for an automotive hood. Kwak et al. combined the ABAQUS FEA for a composite hood with diagonal ribbing with automated optimization using IDESIGN based on recursive quadratic programming [21]. Chung and Lee's method was an automated optimization of a ribbed plate structure using ANSYS and a topology optimization technique [22]. In seeking to exclude thickness-varying plate results that were difficult to manufacture, Chung and Lee's method became complex.

Simplicity was emphasized in the methodology employed in this study. An initial dimensional guess was generated from the beam-based stiffness optimization and multiple ANSYS runs were used to generate an acceptable optimum. ANSYS automated optimization option was not invoked.

The panel FEA used bricks with a coarse but uniform mesh (Fig 5). The panel mesh was generated by extruding a unit cell and then replicating the cell to form a panel. The constraints on the panel were vertical "pinned" support around the perimeter of the panel. Self weight was the first loading. A standard ASCE 7 snow load [23] was next applied and calculated deflections were marginally

1218

acceptable (0. 3 in. or 0.76 cm or L/ 160, where L is the span length). Significant roof deflections were not acceptable for flat roofs with snow or rain loading because roof deflection and the resulting ponding would grow out-of-control. Therefore, a pitched roof was recommended and flat roofs excluded from further consideration.

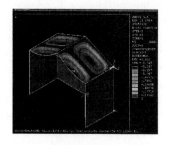

Fig. 6 - Deflections Under Snow Loading

3C. Whole-Structure FEA
To keep the FEA size manageable, the roof and walls were modeled using equivalent flat plates for the ribbed-plate panels. A unit panel may have been represented with a super-element but the equivalent flat plate method was chosen for simplicity. Flat-plate thicknesses were selected to achieve equivalent ribbed-plate deflections at the connections. The loading consisted of the roof dead weight and snow loads. The structure's base was constrained in 3 translational directions. Deflections under snow loads were shown in Fig. 6 for a half-symmetry model. The maximum roof deflection of L/134 was acceptable for a pitched roof, where L is the roof span.

A one-seventh scale model of the structure was constructed from 2.5 mm (0.1 in.) thick plexiglas sheets. The plexiglas sheets did not include stiffening ribs as in the structure's panel but were used strictly to validate the FEA procedure. The sheets were assembled using an adhesive, attempting to most accurately represent FEA boundary conditions. Deflections were monitored while the model was loaded with sandbags to simulate roof snow loading. Maximum measured roof deflections were 0.4 mm (0.017 in.) for 0.2 kPa (5.1 lbs/sq. ft) loading and the plexiglas FEA calculated a maximum deflection of 0.3 mm (0.010 in.). The 0.2 mm (0.007 in.) discrepancy between the tests and FEM results was acceptable.

For the wind-load model, standard ASCE 7 loads were used [23]. Use of 90 or 135 m/s (200 or 300 mph) wind speeds was unrealistic because at those speeds a host of issues not addressed in this analysis became dominant. Such issues included damage caused by projectiles transported by the wind and window-joint strengths. For an assessment of wall and roof deflections, and wall-wall and wall-roof connections, a typical wind speed of 45 m/s (100 mph) was selected for the continental U. S.A. and other countries of interest for a mean recurrence interval (MRI) of 50 years. For Exposure C (flat open terrain), and Category I building (certain temporary facilities), and applying a knock-down factor for a 10-year MRI, a preliminary worst-case wind load of 0.5 kPa (10 lbs/sq. ft) was assumed on the roof and all walls. All pressures but the windward pressure were suction (vacuum) pressures. FEA deflections are shown in Fig. 7. A more realistic but less severe load case will be run using lower suction pressure values on the roof, leeward and side walls. However, the ASCE 7 code was aimed at excluding excessive wall deflections that may damage the wall plaster and other layers that were not relevant to the structure. Therefore, larger structure's deflections for ASCE 7 loads may be acceptable. The model was run for a panel with plate and rib thicknesses of 3.5 mm (0.14 in.). Panels with 5.0 mm (0.2 in.) thicknesses were planned and FEM runs with these dimensions were expected to give low deflections easily meeting wind load requirements.

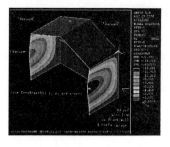

Fig. 7 - Deflection Under Wind Loading

3D. Analysis Summary

The structure's design was clearly deflection controlled. Standard snow loads gave acceptable deflections for the 3.5 mm (0.14 in.) thick panel design whose performance will be enhanced through the usage of 5 mm (0.2 in.) thick panels as planned. Standard wind loads were expected to give acceptable deflections for the 5 mm (0.2 in.) thick panel design. It was assumed that wind loads represented the controlling design constraint and will be used to address earthquake loads. Other key conclusions derived from the analysis included the need for a pitched roof and for fiber reinforcement. Design loads for connections and foundations were expected to emerge from more detailed analyses using sub-models. The unit panel was optimized for stiffness and cost using an FEA model. Thus, several key design decisions were made on the basis of the FEA models.

4. STRATEGIES FOR LIMITED-VERSION FEA SOFTWARE

It has become clear that there is a significant application pool for FEA where only limited hardware and limited-version FEA software are available. The need has been experienced in active magnetic bearing and specialist electric motor industries and in university research applications including composite bridges, truck rails, and low-cost housing structures. Burnside acknowledged such a need in connection with composite bridge analysis [7].

Thus, the FEA procedure demonstrated for the deck and home applications may be a strategy for small companies and university environments. Where resources were not as limited, the procedure may be used in modeling very large structures. Extrusion would be used to generate a fairly uniform mesh with minimal element count and sub-modeling used to get detailed stress-strain maps. A companion method, sub-structuring or super-elements, has been used for large structures like a composite body of a bus [24] and composite bridge decks [3]. These strategies are global-local methods of analysis [24,25] that need to be considered by FEA software and third-party developers for further development for easy implementation. There are tradeoffs in applying any of these methods or in using a very large model with a fine mesh but judicious application leads to useful results.

5. CONCLUSIONS

It was shown that FEA extrusion modeling of structures with deflection-controlled design was effective. The coarse mesh was adequate for a major design phase to be followed with sub-modeling for improved meshing and stress results. The procedure was demonstrated for a low-cost composite deck and recycled-plastic reinforced composite homes where applications of limited software and hardware resources contributed to improved product design and cost minimization.

6. RECOMMENDATIONS

The composite deck model may be extended to predict modes of failure and associated failure loads. Improved composite-materials finite elements and estimates of interlaminar stresses in ANSYS and Cosmos/M will assist in improved modeling of the deck and structural housing unit.

7. REFERENCES

1. YIN, T Y - An Overview of the Design of Composite Decks for Steel Highway Bridges, Fiber Composites in Infrastructure, ICCI'98, Eds., Saadatmanesh, H. and Ehsani, M. R., Tucson, pp. 593 – 602, Jan 1998.
2. SOTIROPOULOS, S N, ET AL. - Theoretical and Experimental Evaluation of FRP Components and Systems, J. of Structural Engineering, Vol. 120, No. 2, pp. 464 – 485, Feb. 1994.
3. ZUREICK, A-H ET AL. - Fibre-Reinforced Polymeric Bridge Decks, Structural Engrg. Review, Vol. 7, No. 3, pp. 257 – 266, 1995.
4. HAYES, M D, OHANEHI, D, LESKO, J J, COUSINS, T E, AND WITCHER, D – Static and Fatigue Performance of a Square Tube and Plate Type Fiberglass Composite Bridge Deck System, J. of Composites for Construction, ASCE, 1999 (submitted).
5. OHANEHI, D C, LESKO, J J., and EASTERLING, W S. - Recycled Composite Structures for Third-World Housing Applications, Materials and Construction - Exploring the Connection, Procs. 5th Construction Matls Congress, ASCE, Matls Engrg Div, Cincinnati, OHIO, May, 1999.
6. CALTRANS - Experimental Fiber-Reinforced Plastic Deck Panels for the Schuyler Heim Bridge, CALTRANS RFQ/RFP #59A0028, 1997.
7. BURNSIDE, P H AND BARBERO, P J - Design Optimization of a Composite Short Span Bridge, Procs, ANSYS Conf. and Exhib., pp. 8.19 – 8.29, May 1992.

8. SIRISAK, S ET AL. - Evaluation of Bridge Structures Using ANSYS, <u>Procs, ANSYS Conf and Exhib</u>, pp. 1.139 – 1.147, May 20 – 22, 1996.

9. AASHTO - <u>Standard Specifications for Highway Bridges</u>, American Assoc of State Highway and Transportation Officials, Wash, DC, 1992.

10. STRUCTURAL RESEARCH & ANALYSIS CORP. - <u>Cosmos/M Users Guide 1.75</u>, Santa Monica, CA, 1996.

11. ANSYS INC. - <u>Structural Analysis Guide</u>, Rev5.4, Houston, PA, 1998.

12. STRONGWELL, INC. - <u>Strongwell Extren® Design Manual</u>, Bristol, VA, USA, 1998.

13. MOTTRAM, J T - Short- and Long-Term Structural Properties of Pultruded Beam Assemblies Fabricated Using Adhesive Bonding, <u>Composite Structures</u>, Vol. 25, pp. 387 – 395, 1993.

14. BHUVARAGHAN, B, ET AL. - Specialized Macros for Model Construction and Loading of Semiconductor Packages, <u>Procs., ANSYS Conf. and Exhib.</u>, May 1998.

15. BHUVARAGHAN, B, ET AL. - Geometry Import Techniques of Semiconductor Packages from Pro/Engineer and AutoCAD, <u>Procs., ANSYS Conf. and Exhib.</u>, May 1998.

16. LIFSHEY, A L - Designing Ribbed Parts, <u>Plastics Des. Forum</u>, Mar/Apr, pp. 59–64, 1980.

17. CAMPO, E A - Calculating Deflection and Stress of a Ribbed Structure, <u>Plastics Design Forum</u>, Nov./Dec., pp. 55–57, 1982.

18. THRONE, J L AND PROGELHOF, R C - Optimized Ribbed Plate Design, <u>ANTEC'89</u>, pp. 1637–41, 1989.

19. TRANTINA, G G, ET AL. - Selecting Materials for Optimum Performance, <u>Plastics Engrg</u>, pp. 23-26, Aug. 1993.

20. SHERMAN, K C, ET AL. - Engineering Performance Parameter Studies for Thermoplastic Structural Panels, <u>ANTEC '89</u>, pp. 640-44, 1989.

21. KWAK, D-Y, ET AL. - Optimal Design of Composite Hood with Reinforcing Ribs through Stiffness Analysis, <u>Composite Structures</u>, Vol. 38, No. 1-4, pp. 351-59, 1997.

22. CHUNG, J AND LEE, K - Optimal Design of Rib Structures Using Topology Optimization Technique, <u>Proc. Instn. Mech. Engrs.</u>, Vol. 211, Part C, pp. 425–37, 1997.

23. ASCE 7-95 - <u>Minimum Design loads for Buildings and Other Structures</u>, ASCE, 1995.

24. REDDY, J N AND MIRAVETE, V - <u>Practical Analysis of Composite Laminates</u>, pp. 257 – 281, CRC Press, 1995.

25. HARYADI, S - <u>Global/Local Analysis of Laminated Panels With Cutouts and Cracks</u>, Ph. D. Dissert, VPI&SU, Blacksburg, VA, Oct1996.

8. ACKNOWLEDGEMENTS

Strongwell Corp., Bristol, TN, Daedalus Project, Inc., Alexandria, VA, and Center for Innovative Technology, VA, USA.

Design of Composite Structures Under Variable Loading

Dr. José Sancho Rodriguez

MTorres Diseños Industriales, S.A. Ctra. Huesca Km.9 31119 Torres de Elorz (Navarra). Spain

SUMMARY

Finite Element Method has demonstrated to be a very useful tool for performing calculations on non isotropic laminated materials. Nevertheless even the simplest linear static calculations presents more difficulties than with their counterpart non laminated isotropic materials. These calculations need to consider the orthotropic material properties as well as the orthotropic strength (nine values for each one). Additional problems arise when free edges, open holes, or discontinuities are needed because of the interlaminar shear stresses that can cause the total failure of the component.

Besides the previous problems, for accurately designing any component or structure, both dynamic effects as fatigue life must be taken into account to ensure the right behaviour of the structure under real life loading.

The present paper presents the experience of designing a large composite structure (windmill blade) taking into account all the previous problems (orthotropic laminated material), performing also dynamic and fatigue analysis.

Several additional computer programs to evaluate fatigue data and cumulative damage following Palmgreeen-Miner and Manson rules are developed. Some other ones are also used and introduced as subroutines to make the fluid flow calculations needed to obtain accurate pressure distribution around the structure for different shapes and angles of attack.

Besides these separate codes, user subroutines to introduce loading on the structure, calculate user failure criterion, capture additional data on selected elements and plot additional data, are developed and linked with the main code of the finite element program used for the design, MARC.

Basic testing for obtaining fatigue life data and damping ratios required for the analysis are performed on coupons for the different materials employed in the design and for the joining between different parts.

Along the whole work, results of the calculations have been compared with real simplified testing (simple loading to check deflections and stresses in the critical parts of the structure) to check the accuracy of the finite element analysis. Simplifying hypothesis have also been checked with different options trying to ensure their accuracy.

For performing design of structures and parts made out with composite materials, several conventional theories have been used in the past. The most traditional and well known method is the Classical Laminated Theory (CLT), based on the Kirchoff [1-3] hypothesis for shear behaviour that offers a good approach to the design of simple slender parts because some of the stresses and deformations are lost (ε_{zz}, ε_{xz}, y ε_{yz}). The equations of this method are the following ones:

$$u(x,y,z) = u_0(x,y,z) - z\frac{\partial w_0}{\partial x}(x,y)$$

$$v(x,y,z) = v_0(x,y,z) - z\frac{\partial w_0}{\partial y}(x,y)$$

$$w(x,y,z) = w_0(x,y)$$

If the part being analysed is not so slender, or the structure is loaded in such a way that stresses out of the plane of the laminae are important, some other assumption must be included in the CLT to correct it for this out of plane behaviour. A common way to take this fact into account is just by including non linear terms in the equations of CLT that introduce a non linear variation of the deflections with the thickness of the laminate. Next equation shows basically the way of introducing these non linear terms.

$$u(x,y,z) = u_0(x,y,z) - z\frac{\partial w_0}{\partial x}(x,y) + u_3 z^3$$

$$v(x,y,z) = v_0(x,y,z) - z\frac{\partial w_0}{\partial y}(x,y) + v_3 z^3$$

$$w(x,y,z) = w_0(x,y)$$

In any case, even with the inclusion of the higher order contribution to the equations, there are some discrepancies between theory and experimental results. These discrepancies can be seen in the next figures [4] for different higher order theories, being the 3D the full three dimensional approach or exact solution of the problem.

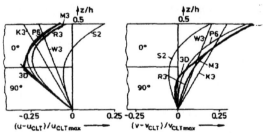

Fig. 1. Comparison between in plane displacements in a 0/90° laminate with an slenderness ratio (length/thickness) of 5 [4]

To avoid these problems, a full three dimensional calculation of the composite structure should be performed, the problem is that this kind of calculation is almost impossible to be carried out in most of the practical applications[5].

In the case of a blade of a wind energy converter (windmill), the blade root is specially critical because of the adhesive joint and the flanges or similar devices required to translate all the loading from the blade to the hub. Three dimensional stresses appear in this area and the calculation must take that into account, being simultaneously time effective as non linearities involved are very heavy. Besides that fatigue and dynamic loading are driving issues for the total design of the machine. Material used is AS4/8552

WORK PERFORMED FOR SIMPLIFICATION

As pointed out before, interlaminar stresses σ_z, τ_{xz}, and τ_{yz} are likely to be responsible of the failure of the root (the most critical part). This means that the finite element model to be performed must be capable of calculating these stresses.

Unfortunately, classical simplifications of the problem (plain strain or axisymmetrical calculations) may not be used in this case, as interlaminar stresses are lost due to the nature of the simplifying hypothesis assumed in both cases.

To overcome this lack of information, the only possibility is performing a full three dimensional analysis using eight noded brick elements and using, initially at least one element in thickness per layer of the laminate. This approach implies the use of a large amount of computer resources as the number of degrees of freedom of the problem is very large.

Simplification is required as the analysis to be run is iterative due to the nonlinear nature of the contacts arising between the flanges of the composite plates and between the composite plates and the steel inserts themselves. Besides that, the effect of the bolts should be introduced in the same model, not using external links that can make some energy to be released out of the joint to the clamping of the link. All these reasons make the problem non lineal.

To investigate possible simplifications of the finite element mesh, a part of the model is calculated under lineal considerations. So, the model is just a composite plate with flange and a steel insert glued to it subjected to a tensile load and restricting displacements in two points of the outer face of the steel insert. The sketch of the calculation is in the following figure (fig. 2).

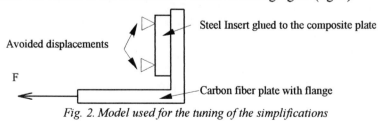

Fig. 2. Model used for the tuning of the simplifications

In all the calculations loads have been introduced in the model as displacements to ensure that the behaviour is the same that the blade root will have during both real life and testing stage.

Full model

This is the first calculation performed on the simplified lineal model. On it, each layer of the laminate is simulated with one eight noded brick element in thickness direction. Each layer is given the appropriate orientation and the material properties of each layer with orthotropic material properties.

With this model, and using a displacement of .1 mm. in the left border of the flanged plate, the stresses obtained in the elements of the fillet (the worst area for interlaminar stresses), are in the following ranges

	σ_x	σ_y	σ_z	τ_{xy}	τ_{yz}	τ_{yz}
Minimum	-185	-29.47	-39.7	-33.49	-9.895	-17.66
Maximum	201.4	40.07	25.55	38.51	16.72	45.94

First Simplification

The first simplification of the previous model is made diminishing the number of elements in the thickness direction, by simulating two real layers with a single element in thickness direction. These theories are called equivalent single layer theories[6], and, in this case, they are direct approximations as $0°$ plies are included in each single layer.

With this model, and using a displacement of .1 mm. in the left border of the flanged plate, the stresses obtained in the elements of the fillet (the worst area for interlaminar stresses), are:

	σ_x	σ_y	σ_z	τ_{xy}	τ_{yz}	τ_{yz}
Minimum	-187	-25	-36.61	-31.56	-12.02	-15.22
Maximum	180.7	32.16	27.06	32.22	16.40	39.72

Final Simplification

If results of the full model and the first simplification are checked, small differences can be seen between them while the number of degrees of freedom of the problem is greatly reduced. With this aim, a larger simplification is performed, assuming that each element in thickness includes the behaviour of 4 layers with $0°$ orientation. The layers with $\pm45°$ are kept in the model as delamination failure is foreseen to start in the interface between these layers and the first $0°$ one. No additional calculation is required to obtain equivalent single layer properties as these properties are the same for a group of unidirectional plies.

1226

With this model, and using a displacement of .1 mm. in the left border of the flanged plate, the stresses obtained in the elements of the fillet (the worst area for interlaminar stresses), are:

	σ_x	σ_y	σ_z	τ_{xy}	τ_{yz}	τ_{yz}
Minimum	-200.4	-27.5	-33.24	-32.43	-12.01	-17.78
Maximum	178	33.17	25.56	39.7	16.86	43.63

By comparison of the three calculations showed in this point, it is clear that simplification can be performed without loosing accuracy in the results, including interlaminar stresses.

Next table shows results comparison for the three models. Maximum and minimum stresses for the three cases reported may be checked. The last column also includes a percentage comparison between the calculation with the full model (the most accurate one) and the final simplification. This provides a measure of the accuracy and validity of the simplification performed in the last and definite model.

Stress	Max Full model	Min. full model	Max 1st simplif.	Min 1st simplif.	Max final model	Min. final model	Δ_{max}1-3	Δ_{min} 1-3
σ_x	201.4	-185	180.7	-187	178	-100.4	11.6%	1%
σ_y	40.07	-29.47	32.16	-25	33.17	-27.5	17.2%	6.7%
σ_z	25.55	-39.7	27.06	-36.61	25.56	-33.24	0.4%	16.3%
τ_{xy}	38.51	-33.49	32.22	-31.56	39.7	-32.43	3.1%	3.1%
τ_{yz}	16.72	-9.895	16.4	-12.02	16.86	-12.01	0.8%	21%
τ_{zx}	45.94	-17.66	39.72	-15.22	43.63	-17.78	5%	0.7%

Maximum differences between models are not important, and, if checked the annex, patterns of the stresses seen in contour plots are very similar. For these reasons, and taking into account that next calculations are non lineal, the model with the final simplification is used.

ABOUT MIXED FINITE ELEMENT MODELS

Another way to perform detailed three dimensional analysis of areas of interest in composite structures, is the mixture between shell analysis with a laminated theory implemented on their formulation, and eight noded brick elements to solve for the real 3D stress field. The approach is very interesting although there are still some items left to investigate before tuning the finite element model. The first aspect is the number of elements. Again, to simulate the real 3D behaviour of a structure with brick elements, at least one row of bricks are required per each layer of material. This means that if the part being analysed is moderately thick, the number of elements can be increased very fastly up to an unacceptable level. To avoid this problem, equivalent singer layer theories are used, being capable of decreasing the number of elements in thickness direction of the laminate by simply grouping several layers of the material into a single row of finite elements with equivalent orthotropic properties as pointed in the previous points of this article. Of course, interlaminar response inside this group of layers is lost, so that grouping should be made with caution to avoid loosing important stress components of the response of the structure.

In MARC, this is an straightforward procedure, as there are automatic generation of coupling equations between three dimensional shell elements and brick elements[7]. This means that shell can be used in areas far away from the critical zones while 3D brick elements with equivalent single layer theories can be used for modelling details.

Next figure (fig. 3) shows an sketch with results of one of the calculations performed with the simplifications included in this article. Procedure used for the calculations is explained later.

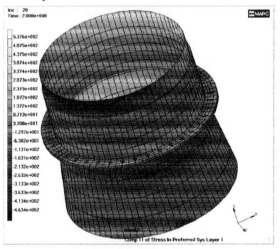

Fig. 3. Model used for the blade root

CALCULATIONS PERFORMED ON THE BLADE

Finally, the procedure used for the calculations of the blade is shown. This way of performing all the analysis is quite embarrassing but is one of the only possibilities to be capable of dealing with the real problem in which aerolasticity, dynamics and fatigue behaviour are of maximum importance and influence on the accuracy of the design.

The first problem is always how to calculate the loads appearing due to the operation of the wind energy converter (WEC). Here there are three key aspects to deal with. The first one is the random nature of the wind. Variability appears both in time as in space, so that, there is a necessity of obtaining time series of wind for different points of the rotor[8]. The wind is generated using controlled random numbers to keep the acceleration of it inside certain margins known by the experience. Next figure (fig. 4) shows the variation of the wind at hub height for one of the cases analyzed.

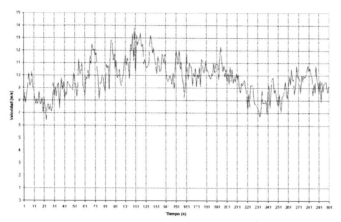

Fig. 4. Example of wind variation at hub height (speed vs. time)

This variation is introduced into a large user subroutine implemented in MARC to load the blades[9]. This user subroutine what makes is just to calculate the loads along different stations of the blade taking into account the wind at each moment, the rotating speed of the blade and the pitch (angle of attack) of the blade at every moment. This aerodynamic calculation is made based on the well known blade element theory that can be corrected to take into account blade losses either with the Prandtl or with the Goldstein method. The development of this aerodynamic code named EOLO[10] is outside the scope of this article.

An important aspect is the control algorithm that must be included in the user subroutine as the pitch variation of the blade depends on the wind, the actual power being produced and the rotating speed of the rotor. Besides that, control should not be either too fast or too slow. The first one can introduce very high fatigue loading on the blades while the second one can be dangerous because of the problems if overspeed in the rotor. Overspeed can cause higher loads due to centrifugal forces and, what is more important, can induce vibration in the tower of the WEC as it is getting closer to its first natural frequency.

All these aspects are introduced in a simplified model made out with MARC using a large deflection algorithm with follow for option in the loads (always perpendicular to the blades) and using an updated Lagrange procedure for the formulation of the equations. Following this procedure the advantage is that the model is analysed taking into account all the dynamic phenomena appearing due to the rotation of the blades and to the wind variation[11].

Next figure (fig. 5) shows the simplified model used in MARC.

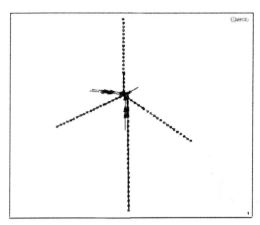

Fig. 5 Model used for loading calculation

From these calculations, a whole set of loadings can be obtained for each blade station and each mean wind speed with following aspect.

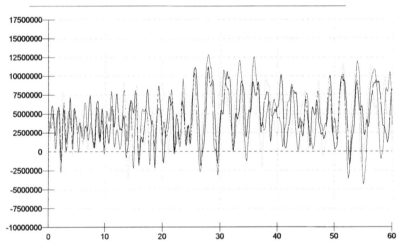

Fig. 6. Bending moments in one station of the blades

Once the loads are available for different wind speeds and for all the cases specified by the international regulations, a rainflow cycle counting is performed. This counting technique is based on the hypothesis that the wind distribution along a year fits a two parameter (C and α) probability (P) Weibull distribution based on mean wind speeds (V_m) as stated in the next formula.

$$P = e^{-\left[\frac{V}{V_m}\bigg/C\right]^{\alpha}}$$

The application of this formula to an average wind speed gives the following figure (Fig. 7).

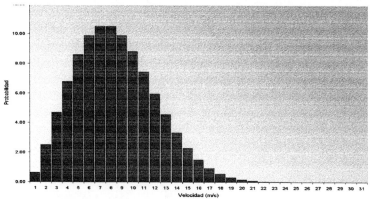

Fig. 7. Weibull distribution of a windfarm location

With the loads, the Weibull distribution and the rotating frequency (speed) of the WEC, the number of repetitions of each set of load calculation is stated and a rainflow cycle counting can be performed. Results of this rainflow cycle have the following aspect [12] (fig. 8).

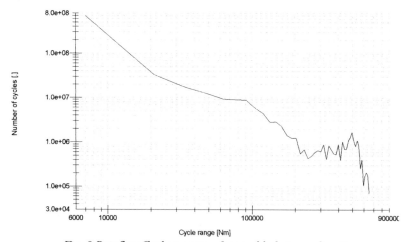

Fig. 8.Rainflow Cycle counting for one blade in one location

The calculation procedure presented up to this point is required to obtain the most accurate loading values possible without using large supercomputers. Loads are finally presented as the result of a rainflow counting. With these values, some further modification is required because they must be imposed to different stations of the blade. Besides that, the rainflow procedure used by BLADED[12], the additional software used for performing aerolastic and fatigue calculations, looses average values. To take into account not only ranges but also average values, an additional code was developed in

1231

house[13]. This code obtains an average loading for each range based on a cumulative damage approach based on Miner's rule. By using this code, the following kind figures are obtained (fig. 9) and their values are applied to each station of the blades.

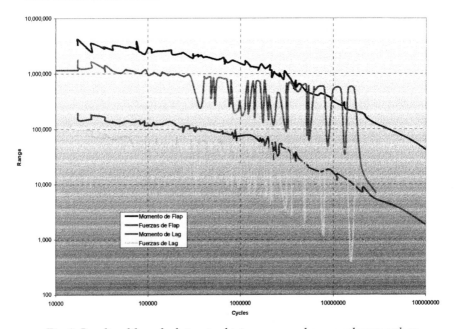

Fig. 9. Results of the calculation to obtain ranges and averaged mean values

One of the most important advantages of this kind of calculations is that the detailed FEM calculations of the complicated parts of the WEC (blade root, hub, main axle, tower, etc.) can be solved with static algorithms, saving time in the development of the machine. Nevertheless, it should be noted that of course some information is lost and some errors are inherent to this methodology. So, the effect of nonlinearities in the adhesive joint of the blade root, effect of the bolting of the blade, possible plastic joints in some parts are not taken into account. There are some bibliography [14] where the effect of simplifying hypothesis as using time series or quasi static calculations for fatigue analysis on WEC parts are analysed.

For calculating each part, a FEM is made for that part and some further extension to introduce all the loading in the extension avoiding dummy stresses introduced by local loading of the numerical model. Figure 3 shows how to make a model of the blade root, some part of the blade is included in the FEM model. To see how the simplifying theories with which this article has started, next figure (Fig. 10) shows a detail of the FEM model of the blade root. Unfortunately, it is not possible to see in that model the adhesive layer but it is included as a layer of brick elements.

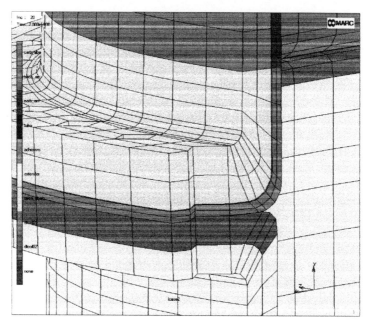

Fig. 10. Detail of the blade root

CONCLUSIONS

Along this article, a way for performing detailed dynamic and fatigue analysis of large structures has been shown. Simplifying techniques to decrease the time required for solving the problem, without loosing too much accuracy are presented for a real industrial application.

Procedures shown here can also be applied to many other industrial fields taking the benefits of a cut down in the time to market period as well as decreasing the cost of new developments. All the procedures must be certified on a prototype to ensure the total foreseen life of the machine is achieved with a reasonable safety margin.

BIBLIOGRAPHY

1 TSAI, S. W. AND HAHN, H. T. - Introduction to Composite Materials. Ed. Technomic Westport CT. 1980.

2. JONES, R. M. - Mechanics of Composite Materials. McGraw-Hill Book Company (Scripta Book Company). 1975

3. REDDY, J. N. AND MIRAVETE, A. - Practical Analysis of Composite Laminates. Ed. CRC Press 1995

4 ROHWER, K. - Computational models for laminated composites. Zeitschrift für Flugwissenschaften und Weltraumforschung 17 (1993) pp 323-330

5 SANCHO, J. - A Procedure for Studying Delamination in complex structures. Proceedings of the 11th International Conference on Composite Materials. Ed Murray L. Scot. Woodhead Publishing Limited. 1997

6 SANCHO, J. - PhD. Dissertation. University of Zaragoza. 1996.

7 MARC. - User Information. Volume A. 1994 MARC Analysis Research Corporation

8 SPERA, DAVID A. - Wind Turbine Technology, Fundamental concepts of wind turbine engineering. ASME Press, New York 1994.

9 MARC. - User Subroutines. Volume D. 1994 MARC Analysis Research Corporation

10 GÜEMES, A. - EOLO, Manual del Usuario y Teoría. PhD Work. Universidad Politécnica de Madrid 1987.

11 EGGLESTON, D. M. STODDARD, F.S. - Wind Turbine Engineering Design. Van Nostrand Reinhold, New York. 1987

12 Bladed for Windows. V3.1 Garrad Hassan & Partners Ltd.

13 DIGITAL. - Digital Fortran, Language Reference Manual. Digital Equipment Corporation. Massachusetts 1997.

14 RAHLF, U. OSTHORST, R. Numerical Fatigue life prediction for improved reliability and cost effectiveness of wind turbines components. Proceedings of the European Wind Energy Conference. Ed Rick Watson. Irish Wind Energy Association. 1997.

FINITE ELEMENT MODELLING OF DELAMINATION INDUCED FAILURE OF FLAT AND CURVED COMPOSITE PANELS

G J Short, F J Guild and M J Pavier

Department of Mechanical Engineering, University of Bristol, Queens Building, University Walk, Bristol, UK, BS8 1TR

SUMMARY

A delamination is an area of de-bonding between adjacent plies in a layered composite panel. Since delaminations reduce the compressive strength of the panel, it is useful to be able to understand the mechanism of the strength reduction. The mechanism consists principally of buckling of the delaminated plies resulting in bending of the remaining panel.

A series of post-buckling finite element analyses have been carried out to predict the stress distribution in delaminated flat composite panels under uniaxial compressive load. The effect of delamination size and through thickness position has been studied. It is found that the buckling of the delamination causes a stress concentration, which may be used to predict failure using a maximum stress criterion. Failure stresses have been compared with experimental measurements with reasonable agreement.

A further series of finite element analyses consider delaminated panels with an initial curvature. The initial curvature effects the buckling characteristics of the delaminated plies and therefore the resulting strength reduction.

1. INTRODUCTION

An important type of damage in a layered fibre reinforced plastic composite panel is an area of de-bonding between adjacent plies. This type of damage is known as a delamination and may be caused by an accidental impact or a manufacturing defect. The presence of a delamination in a panel supporting compressive load may reduce considerably its strength.

Previous work [1] has shown the mechanism of strength reduction results from the plies above the delamination buckling out of plane. Once this buckling occurs, the remaining plies become subject to bending in addition to the in-plane compressive load. The stresses in the remaining plies are therefore higher than would exist in an undelaminated panel leading to a reduced failure load.

In this paper a series of finite element and experimental studies are described of the compressive strength reduction for flat composite panels containing delaminations. The geometry of the delamination, its size and position through the thickness of the panel, is varied to investigate the strength reduction.

Following the studies on flat panels, a series of finite element analyses are described of the strength reduction for panels with initial curvature. Although the curvature is quite small, it is sufficient to effect the buckling behaviour of the delaminated plies and hence the compressive strength reduction.

2.   MODELLING OF DELAMINATIONS IN FLAT PANELS

The geometry of the delamination considered in this paper is shown in Figure 1. The square delamination of dimension a by a is placed centrally in a square panel of dimension b by b and thickness T. The delamination is located a distance t from the top surface of the panel.

The panel is taken to be laid up using unidirectional plies of E-glass fibre reinforced epoxy 7714 with a lay-up of $[0, +45, -45, 0]_s$. and properties $E_{11} = 46.0$ GPa, $E_{22} = E_{33} = 13.0$ GPa, $G_{23} = 4.6$ GPa, $G_{13} = G_{12} = 5.0$ GPa, $v_{23} = 0.42$ and $v_{13} = v_{12} = 0.3$. The through-thickness position of the delamination must be in integer values of the ply thickness of 0.3mm. In this work, three positions were used: position A between the first and second, 0 and +45 plies, position C between the third and fourth, -45 and 0 plies and position D centrally between two 0 plies.

The panel size was taken to be 50mm by 50mm, constrained so that the loaded edges are simply supported and the unloaded edges are built in. Three in-plane sizes of delamination were used: 10mm by 10mm, 15mm by 15mm and 25mm by 25mm.

The finite element mesh is shown in Figure 2. A layer of 20 node orthotropic brick elements were used for the plies above the delamination while another layer was used for the plies below the delamination. The finite element system used for the analyses was ABAQUS version 5.7

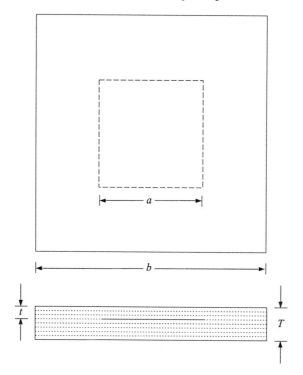

Figure 1   Geometry of delamination in flat panel

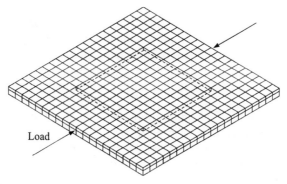

Figure 2   Finite element model of delaminated panel

The elements representing the delaminated plies were given a small initial out if plane displacement so as to initiate buckling. Compressive load was then applied to the model in increments, continuing beyond the load at which the delaminated plies buckle. Figure 3 shows the out of plane displacement of the centre of the delaminated plies and the centre of the remaining plies of the panel for a 25mm by 25mm delamination in position

C. As can be appreciated, the delaminated plies buckle at a load corresponding to an average compressive stress of about 27 kN. Once the delaminated plies buckle, the remaining panel also begins to deform out of plane.

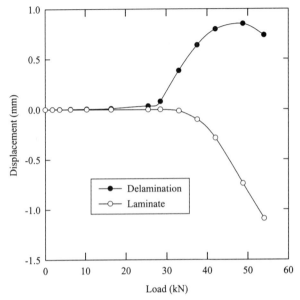

Figure 3   Out of plane displacement versus load for 25 C panel

Failure of the delaminated panel is predicted by the maximum fibre direction stress reaching a limiting value. The stress distribution was first examined to locate the position of maximum value, determined to be along the centre of the loading direction and either in the centre of the delaminated area or near the edge of the delamination.

Figure 4 shows the maximum fibre direction stress versus load for the 25mm by 25mm delamination in position C. A value of 550 MPa is taken from experiments on undelaminated panels for the fibre direction failure stress in compression. Therefore failure of the panel is predicted at a load of about 35 kN.

3.   COMPARISON WITH EXPERIMENT

A programme of experimental work was carried out to measure the failure loads of composite panels containing delaminations. The panels were the same 50mm width as the geometry of the finite element models but were 200mm long to include space for the fitting of end tabs. The panels were held in an anti-buckling guide preventing gross buckling but allowing out of plane displacements to occur in the central part of the panel so as to give

1238

edge conditions comparable to the finite element model. Details of the anti-buckling guide may be found elsewhere [2].

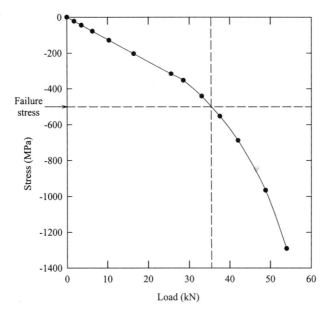

Figure 4   Fibre direction stress versus load for 25 C panel

The failure loads of panels with a range of delamination sizes and through thickness positions were measured. In all cases failure was sudden and was preceded by buckling of the delaminated plies and progressively increasing out of plane displacement.

Figure 5 shows the failure loads of a series of panels with a varying through thickness position of delamination. All panels had the same delamination size of 25mm by 25mm. The through thickness position is normalised with respect to the thickness of the panel, so that a delamination with a normalised position of 0.5 is in the centre of the panel, that is in the D position. The failure loads are also normalised with respect to the failure load for an undamaged panel of 45 kN. The finite element predictions are also shown on Figure 5 and show reasonable agreement with the experimental trends.

The failure loads for panels with a varying size of delamination are shown in Figure 6. All delaminations were in the through thickness position A. The delamination size is normalised with respect to the width of the panel. Again, finite element predictions show reasonable agreement with the experimental trends.

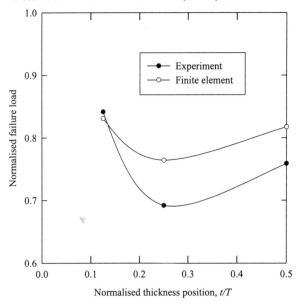

Figure 5   Failure loads for delaminations of varying thickness position

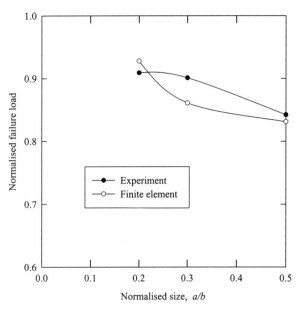

Figure 6   Failure loads for delaminations of varying size

4. MODELLING OF DELAMINATIONS IN CURVED PANELS

The strength of a delaminated panel is controlled partially by the buckling load of the delaminated plies. It is therefore expected that the strength of a panel will depend on its initial curvature, since initially curved panels will have a higher buckling load.

Finite element models have been developed for panels with an initial curvature, where the plane of curvature is normal to the loading direction. The geometry of the curved panel is shown in Figure 7. Two different curvatures were considered of radius 75mm and 125mm. For all cases the delamination size was 25mm by 25mm in through thickness position C. Note that in the through thickness position of the delamination is measured from the outer surface of the panel.

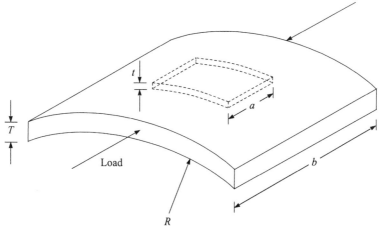

Figure 7   Geometry of delamination in curved panel

Again, compressive load was applied to the model in increments, continuing beyond the load at which the delaminated plies buckle. Figure 8 shows the out of plane displacement of the centre of the delaminated plies and the centre of the remaining plies for the two cases of curvature, compared to the flat panel results. It can be seen that the buckling load for the delaminated plies increases with curvature.

Figure 9 shows the maximum fibre direction stress versus load for the curved panels compared to the flat panel. It can be appreciated that the predicted effect of curvature on the failure load will depend on the value chosen for the failure stress. For example a value of failure stress of about 700 MPa will coincide with a cross-over point, therefore little effect of curvature will be predicted. For a value of 550 MPa used previously, a small effect is predicted.

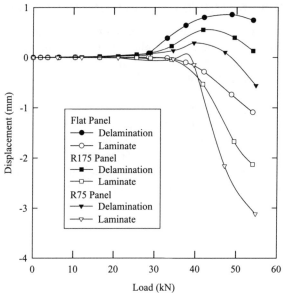

Figure 8   Out of plane displacement versus load for curved panels

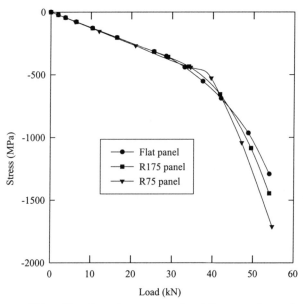

Figure 9   Fibre direction stress versus load for curved panels

5.   CONCLUSIONS

Finite element models have been developed to analyse the behaviour of flat composite panels containing delaminations. The models predict buckling of the delaminated plies followed by the post-buckling behaviour. Predictions of failure load have been made based on the maximum fibre direction stress and using a value derived from experiments on undelaminated panels.

Experiments have been carried out on panels with varying delamination size and through thickness position. The experimentally measured failure loads show reasonable agreement with the finite element predictions.

Finite element analysis has been carried out of curved composite panels containing delaminations. Out of plane displacements and maximum fibre direction stresses have been predicted. The results obtained seem to suggest that the effect of curvature on failure loads will be small. Work is continuing to provide experimental measurements with which to compare the finite element predictions.

ACKNOWLEDGEMENTS

Mr G J Short is supported by an EPSRC Studentship Award No. 97700381.

REFERENCES

1. PAVIER, M J AND CLARKE, M P, Finite element prediction of post impact compressive strength in carbon fibre composites, <u>Composite Structures</u>, Vol. 36, 141-153, 1996

2. PAVIER, M J AND CLARKE, M P, Experimental techniques for the investigation of the effects of impact damage on carbon-fibre composites, <u>Composites Science and Technology</u>, Vol. 55, 157-169, 1995

AN OVERVIEW OF NON-LINEAR BOUNDARY ELEMENT FORMULATIONS IN ENGINEERING PROBLEMS

A.A. Becker[1], H. Gun[1] and C. Chandenduang[1]

ABSTRACT

A brief review of non-linear Boundary Element formulations is presented, dealing with material non-linearity involving plasticity and creep, geometric non-linearity and boundary non-linearity. A number of examples are shown in order to demonstrate the accuracy of the BE formulations in non-linear analysis.

1.   INTRODUCTION

The Boundary Element (BE) method is well established as an accurate numerical tool particularly well suited for linear elastic problems. Due to its high resolution of stresses on the surface, the BE approach has been shown to be accurate in problems involving stress concentration, fracture mechanics and contact analysis. However, its extension to non-linear problems involving material and geometric non-linearities is not widespread and is under-developed when compared to the Finite Element (FE) method. In many non-linear BE formulations the interior of the solution domain has to be discretised, thus losing the main BE advantage of surface-only modelling. Another difficulty encountered in the non-linear BE formulations is the accurate evaluation of strongly-singular integral functions.

This paper presents an overview of different non-linear BE formulations with particular emphasis on material, geometric and boundary non-linearities. Brief details of how such non-linearities are tackled analytically and numerically by the BE method are presented. A number of non-linear BE problems are presented including elasto-plasticity in a pressurised thick cylinder, creep of a plate with a circular hole and frictional contact of a pin and connecting rod. The BE solutions are shown to be accurate and in good agreement with the FE solutions.

---

1   Department of Mechanical Engineering, University of Nottingham, Nottingham, UK

2. BE ANALYTICAL NON-LINEAR FORMULATIONS

Non-linearities in solid mechanics have been traditionally classified into four categories, as follows:
(i)   Time-independent material non-linearity such as elastoplasticity,
(ii)   Time-dependent material non-linearity such as creep and viscoelasticity,
(iii)   Geometric non-linearity in which the changes in the geometry of a structure due to its displacement under load are taken into account, and
(iv)   Boundary non-linearity which occurs in most contact problems.

In this section, a brief review of the analytical non-linear BE formulations is presented for each type of non-linearity.

2.1   Plasticity BE Formulation

In linear problems, the BE formulation reduces the dimensionality by one by transforming the variables from volume to surface values. Therefore, unlike the FE method which requires whole body discretisation, only the surface (boundary) of the domain requires discretisation in linear BE problems. The basis of the BE formulation is a boundary integral identity for displacements, relating the displacement at an interior point p to the displacements and tractions at a boundary point Q, over the surface S, as follows (see e.g. the textbooks by Brebbia et al [1], Becker [2] and Banerjee [3]):

$$u_i(p) + \int_S T_{ij}(p,Q) \, u_j(Q) \, dS(Q) = \int_S U_{ij}(p,Q) \, t_j(Q) \, dS(Q) \qquad (1)$$

where u_i and t_i are the displacement and traction vectors respectively, and U_{ij} and T_{ij} are the displacement and traction kernels, respectively, which are functions of the positions of points p and Q and the material properties. The above equation ignores the effects of body forces and non-linear material non-linearity.

To include the effect of the non-linear material behaviour, it becomes necessary to model either the whole of the interior solution domain or only the part which exhibits a non-linearity. Elastoplastic behaviour can be incorporated in an additional volume integral term based on the work done by the strain rate multiplied by the stress at the load point as follows (see e.g. Lee and Fenner [4]):

$$\int_V W_{kij}(p,q) \, \dot{\varepsilon}_{ij}^p(q) \, dV \qquad (2)$$

where V is the volume of the solution domain and q is an interior point. The dot above the plastic strain ε_{ij}^p indicates the rate of change of strain with respect to time. The kernel W_{kij} can be interpreted as the stress at point q due to a unit

1246

load at the load point p in the kth direction. This approach is referred to as the initial strain method. Alternatively, the volume integral can be written in terms of the initial stress which can be used as the primary unknown [4].

To calculate the plastic strain rates, the von Mises flow rule can be used. The following expression can be used for the plastic strain increments in terms of the total strain increments:

$$\dot{\varepsilon}_{ij}^{p} = \frac{3}{2}\left(\frac{\dot{S}_{kl}\,\dot{\varepsilon}_{kl}}{1+H/3\mu}\right)\frac{\dot{S}_{ij}}{(\sigma_{eq})^{2}} \tag{3}$$

where S_{kl} and σ_{eq} are current deviatoric and equivalent stresses respectively, μ is the shear modulus and H is the slope of the uniaxial plastic stress–strain curve. Alternatively, the plastic strain increments can be expressed in terms of the stress increments.

2.2   Creep BE Formulation

The BE formulation for creep problems can be based on the initial strain approach which has the same form as that of elastoplasticity by replacing the plastic strain rates in equation (2) with the creep strain rates and expressing the displacements and tractions in rate form (see, for example Ref [5]).

The constitutive material law can be expressed as a power law based on the Norton-Bailey uniaxial creep equation [6]. The multi-axial creep strain rate based on the von Mises flow rule can be written as follows:

$$\dot{\varepsilon}_{ij}^{c} = \frac{3}{2}\,mA\,(\sigma_{eq})^{(n-1)}\,S_{ij}\,t^{(m-1)} \tag{4}$$

where A, m and n are material constants dependent on the temperature. m=1 is used for the secondary creep stage and m<1 is used for the primary creep stage.

2.3   Geometric Non-Linearity BE Formulation

An integral equation based on the reciprocal work theorem can be derived to account for geometric non-linearity by including an additional integral over the volume (see, e.g. [7, 8]). BE formulations for geometric non-linearity problems have used the updated Lagrange approach in which the solution is tracked from time t to t + Δt, i.e. the updated state is used as the reference state [9, 10]. By differentiating the reciprocal energy expression with respect to time, an integral equation for the displacement rates (velocities) and the traction rates can be obtained, similar to the elastoplastic formulation. To account for geometric non-linearity and rigid boy rotations, an extra volume integral, usually called the 'geometric correction' integral, can be derived to allow for finite deformations and rotations within the body. In cases where the deformation is compressible, e.g. in soil problems, additional volume integrals are necessary to account for the compressibility of the material (for details, see Ref [11]).

2.4   Contact BE Formulation

In multi-domain or contact problems, the elements on the contact interface do not have prescribed displacements or tractions, but must satisfy three conditions:
(i)   Equilibrium conditions (equal and opposite tractions)
(ii)   Continuity of displacements (no overlapping of elements)
(iii)   Friction slip conditions (shear stress must not exceed the friction limit).
These conditions can be expressed as algebraic relationships between the displacements and tractions of the bodies in contact and directly incorporated in the equation solver. This direct coupling of the contact variables means that it is not necessary to use any form of gap or interface elements between the contacting surfaces.

3.   BE NUMERICAL NON-LINEAR IMPLEMENTATION

3.1   Discretisation of the Surface and the Volume

Since partial or full interior modelling is required in non-linear problems, both surface (boundary) elements and volume (internal domain) cells are necessary in order to perform the integrals arising in the BE formulation. Figure 1 shows a typical three-node boundary element and a typical eight-node quadrilateral domain cell for two-dimensional problems. For the boundary and interior cell element, the geometry can be described in terms of linear, quadratic or cubic shape functions in a local coordinate axes system (see e.g. Becker [2]). For isoparametric elements, the displacement and traction vectors can be expressed in terms of the appropriate shape functions.

It should be noted that in non-linear applications, where full domain discretisation may become necessary, the BE meshes will be similar to FE meshes. However, the interior BE cells do not have to be inter-connected to the boundary elements, and the size of the BE solution matrix is proportional to the number of the boundary nodes.

3.2   Numerical Integration Schemes

The integrals appearing in the boundary integral equations have to be numerically calculated in order to obtain the coefficients needed to form the set of algebraic equations. The kernel functions contain singularities of the order of $1/r$ or $1/r^2$ where r is the distance between the load point P and the field point Q or the interior point q. Therefore, the integrals become singular when P coincides with either Q or q. It is important to devise accurate numerical integration schemes to evaluate the integrals in such cases, as it has a direct influence on the accuracy of the solutions. When P and Q do not coincide, the standard Gaussian quadrature formulae can be used, even if p and Q are in the same element. However, when P coincides with either Q or q, special integration schemes have to be devised. A number of approaches have been devised for dealing with this singularity (see for example Refs [2, 3, 12]).

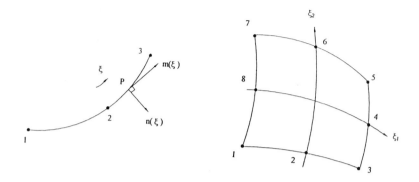

Figure 1 : 2D quadratic boundary elements and domain cells

3.3 Assembly of the System Equations

After performing the integrations in equations (1) and (2) by taking each node in turn as the load point, the coefficients of the displacements and tractions can be assembled in matrices. The linear algebraic equations obtained from the discretised integral equations can be formed as follows:

$$[A] [\dot{u}] = [B] [\dot{t}] + [W] [\dot{\varepsilon}^{NL}] \tag{5}$$

where the matrices [A], [B] and [W] contain the integrals of the displacement, traction and domain kernel functions, respectively. The matrix $[\varepsilon^{NL}]$ represents the non-linear strain rate vector. For two-dimensional problems, if the total number of boundary nodes is N and the total number of the domain cell points is H, then the solution matrices [A] and [B] will be square matrices of size 2N x 2N, whereas the matrix [W] will be a rectangular matrix of size 2N x 3H. Unlike the FE method, all BE matrices are fully populated.

Boundary conditions are specified as either displacement rates or traction rates on the boundary of the domain, or a linear relationship between them. The boundary conditions are considered to be in incremental form.

To be able to implement a standard equation solver, the matrices [A] and [B] must be rearranged by exchanging the relevant columns such that the unknowns are placed in one vector [x] on the left hand side, and the prescribed known values are placed in another vector [y] on the right hand side, as follows:

$$[A^*] [\dot{x}] = [B^*] [\dot{y}] + [W] [\dot{\varepsilon}^{NL}] \tag{6}$$

The non-linearity of the solution procedure arises because the non-linear strain increments on the right hand side of equation (6) are unknown and are functions of current stress state or time according to the material constitutive equations. Therefore, iterations have to be performed to ensure that the solution is acceptable, i.e. the error in the approximations is reasonably small. In practice, this means that the loads have to be applied in small increments such that the associated non-linear strain increments are also small. By using approximate values of the non-linear strains, the right hand side of equation (6) can be combined as follows:

$$[A^*][\dot{x}] = [b] \qquad (7)$$

where [b] now contains the contribution of the plastic strain increments as well as the prescribed displacement and traction values. Since the solution matrix [A*] is not symmetric and fully populated, the Gaussian elimination technique is usually used to solve the equations.

It is worth emphasising that the size of the solution matrix [A*] is either 2N x 2N (2D problems) or 3N x 3N (3D problems) where N is the total number of surface (boundary) nodes, rather than the total number of nodes on the boundary and in the domain.

3.4   Incremental/ iterative solution procedures

In most BE or FE numerical treatments of non-linear problems, the total load is applied in small increments and iterations are performed within each load increment to ensure that the approximations are acceptable within a specified tolerance. In the initial strain elastoplastic BE approach, it is possible to obtain the plastic strain increments by using the flow rule in equation (3), which can handle perfectly plastic material behaviour.

In BE formulations, there is no significant difference between the initial stress approach and initial strain approach, because the integral equations in both approaches include the effect of plasticity. The initial strain formulation is considered more suitable for traction-control problems, because the first approximations for the stress increments are usually reasonably accurate.

A robust convergence criterion must be used in order to ensure that the error in the approximations is acceptably small. Convergence criteria used in BE formulations are usually based on either the norm of changes in the primary unknown vector [x], or the change in non-linear strain increments.

3.5   Time Stepping Scheme

In elastoplastic problems, the material behaviour is assumed to be independent of time, and the variables are expressed as rates simply to indicate the order of events. At the end of the iteration process, a suitable 'pseudo-time' step is chosen and multiplied by the variables (rates) to obtain the actual increments of the variables.

For geometric non-linearity problems, small time steps are used over which iterations are performed to ensure convergence of the solution. Iterations are necessary because the unknown velocity gradients occur both on the surface as well as the volume. In the iterative process, a first approximation for the velocity gradients is obtained at time t=0, and the equations solved to obtain a second approximation for the velocity gradients. The process is repeated as required until convergence within an acceptable tolerance is achieved. The time history of the variables can be obtained by updating the geometry and marching forward with time. The updating of the geometry can be either done at each time step, or every few time steps to save execution time. Details of the BE numerical implementation for geometric non-linearity, including modelling conservative and non-conservative (follower) loads can be found in Ref [11].

In time-dependent material non-linearity, the total creep time is divided into small time steps where the creep strain increments are assumed to be small. A typical creep time stepping algorithm can summarised as follows:

(i) At time t=0, solve the equations for the displacements and tractions.

(ii) Calculate the creep strain rates over a small time step using the creep law and include the creep term in the vector [b] in equation (7).

(iii) Solve the system equations for the unknown displacement and traction rates at the boundary. From the boundary values, calculate the displacement rates at the internal nodes.

(iv) Calculate the stress and strain rates at the boundary using Hooke's law and calculate the strain rates at the internal nodes by differentiating the interior displacements over each cell via the shape functions.

(v) Check convergence, i.e. the change in creep strain or stress should be reasonably small using a suitable convergence criterion. If no convergence is obtained, reduce the time step and repeat the analysis until convergence occurs.

(vi) Update the variables and repeat the procedure for the next time increment until the final creep time is reached.

Note that since the matrices [A*] and [W] in equation (6) are not changed during each time step, they need to be calculated only once and stored. Refinements to the above procedure are possible by using an automatic time stepping scheme in which the time step is adjusted during the analysis.

3.6   Contact Algorithm

In contact problems, each domain is treated separately to form equation (5), and the resulting matrices [A] and [B] are coupled together according to the relevant contact conditions, with the number of unknowns remaining equal to the number of equations. The compatibility and equilibrium equations are exactly modelled by coupling the relevant variables directly. Independent mesh discretisation of the contacting surfaces can be used to eliminate the need for node-on-node contact within the interface. The independent mesh discretisation is implemented either by a length mapping procedure or by creating fictitious nodes in the contact area [13].

The BE implementation of the contact constraints in the global system of equations is relatively more efficient and accurate than the FE method because both the tractions and displacements exist as nodal variables whereas the FE equations involve only the displacements as the nodal degrees of freedom.

In most practical contact problems, the extent of the contact area and the stick-slip conditions are initially unknown and must be determined by an iterative procedure and, in history dependent frictional problems, load incrementation. An efficient automatic iterative scheme can be employed in which interpenetration of the contacting elements or overlap just outside the contact area is prevented, and elements with tensile stress are released from the contact area. The iteration procedure also allows the nodes with the shear stress exceeding the Coulomb friction limit to slip in the next iteration with a limit imposed on the value of the tangential traction.

4.   EXAMPLES OF N0N-LINEAR BE APPLICATIONS

To demonstrate the applicability and accuracy of the BE method in non-linear applications, a number of examples are presented in this section. The examples include elasto-plasticity, creep and contact applications using the BEACON software package [2]. The BE solutions are compared to the corresponding analytical and FE solutions obtained using ABAQUS [14].

4.1   Plasticity in a Pressurised Thick Cylinder

This problem concerns a thick cylinder under an internal pressure in which the diameter ratio is taken to be 2. The analytical solution of this problem was presented by Hodge and White [15]. The material is assumed to be elastic-perfectly plastic with the following material properties: $\sigma_y = 200$ N/mm^2, E = 200×10^3 N/mm^2 and $v = 0.33$.

The boundary element discretisation is shown in Figure 2 where a 15° sector is used to represent the cylinder with 4 cells. The BE solutions for the hoop stress distribution across the wall thickness are shown in Figure 3. The BE solutions show very good agreement with the analytical and FE solutions.

4.2   Creep of a Plate with a Circular Hole

This 2D plane stress problem concerns a plate with a circular hole subjected to a tensile stress of 50 N/mm^2 in the x-direction. The dimensions of the plate are 20 mm x 36 mm with a hole radius of 5 mm. The creep material properties are

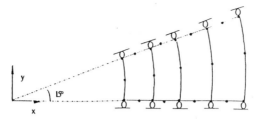

Figure 2 : BE mesh for the thick cylinder problem

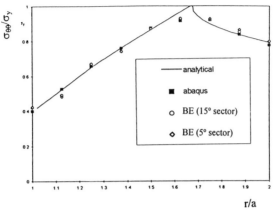

Figure 3 : BE solutions for plasticity in a thick cylinder problem

as follows: A = 3.125×10^{-14} per hour, m = 1, n = 5, E = 200×10^3 N/mm^2, v = 0.3. The test is conducted for 1 hour. The BE mesh of a symmetric quarter of the plate is shown in Figure 4 where 24 boundary elements and 32 cells are used. An automatic time stepping scheme is used with an initial time of 10^{-3}. Figure 5 shows a comparison of the BE and the corresponding FE solutions of ABAQUS, where it can be seen that a very good agreement is obtained.

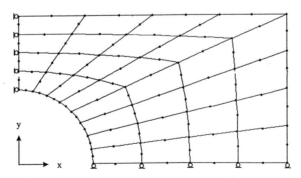

Figure 4 : BE mesh for the creep of plate with a hole problem

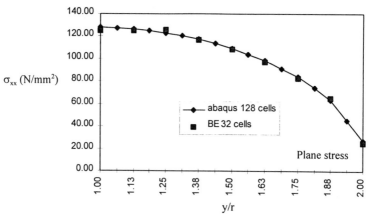

Figure 5 : BE solutions for creep of plate with a hole problem

4.3   Frictional Contact of a Pin and a Connecting Rod

This example concerns the contact of a pin inserted into a connecting rod with a small initial clearance. An external load is applied on the inner circumference of the pin, and the effect of friction between the contacting surfaces is examined. The BE mesh of 84 elements used to model a symmetrical half of the problem is shown in Figure 6 where independent mesh discretisation is used for the contact surfaces [16]. The BE solutions and the corresponding FE solutions using ABAQUS for the normal contact pressure (Po=2.98 x 10^6) for μ =0 and 0.1 are shown in Figure 7. Both solutions are in close agreement.

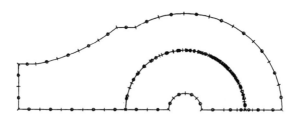

Figure 6 : BE mesh for the pin and connecting rod problem

1254

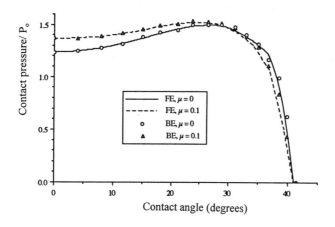

Figure 7 : BE solutions for the pin and connecting rod problem

5. CONCLUSIONS

A review of the BE approach in non-linear solid mechanics problems, including plasticity, creep, geometric non-linearity and contact, is presented. Full or partial modelling of the interior solution domain becomes compulsory in most non-linear BE applications. The BE method is shown to be an accurate alternative technique to the FE method in non-linear problems.

REFERENCES

1. BREBBIA, C.A., TELLES, J.C.F. and WROBEL, L.C., Boundary element techniques, Springer, Berlin, 1984.

2. BECKER, A.A., The boundary element method in engineering, McGraw-Hill, London, 1992.

3. BANERJEE, P.K., The boundary element methods in engineering, McGraw-Hill, London, 1994.

4. LEE, K.H. and FENNER, R.T., A quadratic formulation for two dimensional elastoplastic analysis using the boundary integral equation method, J. Strain Analysis, 21, 159-175, 1986.

5.   MUKHERJEE, S., Boundary element methods in creep and fracture, Applied Science Publishers, London, 1982.

6.   KRAUS, H., Creep analysis, John Wiley, New York, 1980.

7.   NOVATI, G. and BREBBIA, C.A., Boundary element formulation for geometrically nonlinear elastostatics, Applied Mathematical Modelling, 6, 136-138, 1982.

8.   CHANDRA, A. and MUKHERJEE, S., Application of the boundary element method to large strain large deformation problems of viscoplasticity, J. Strain Analysis, 18, 261-270, 1983.

9.   MUKHERJEE, S. and CHANDRA, A., Nonlinear solid mechanics, in Boundary Element Methods in Mechanics, edited by D.E. Beskos, Elsevier Science, London, 1987.

10. ZHANG, Q., MUKHERJEE, S. and CHANDRA, A. Shape design sensitivity analysis for geometrically nonlinear problems by the boundary element method, Int. J. Solids & Structures, 29, 2503-2525, 1992.

11. CHANDRA, A. and MUKHERJEE, S., Boundary element methods in manufacturing, Oxford University Press, 1997.

12. HUANG, Q. and CRUSE, T.A. Some notes on singular integral techniques in boundary element analysis, Int. J. for Numerical Methods in Engineering, 36, 2643-2659, 1993.

13. OLUKOKO, O.A., BECKER, A.A. and FENNER, R.T. A review of three alternative approaches to modelling frictional contact problems using the boundary element method, Proc. Royal Society of London, Series A, 444, 37-51, 1994.

14. ABAQUS Version 5.5, HKS Inc., Rhode Island, 1995

15. HODGE, P.G. and WHITE, G.N. A quantitative comparison of flow and deformation theories of plasticity, J. Applied Mechanics, Trans. ASME, 17, 180-184, 1950.

16. OLUKOKO, O.A., BECKER, A.A. and FENNER, R.T., A new boundary element approach for contact problems with friction, Int. J. for Numerical Methods in Engineering, 36, 2625-2642, 1993.

A Contact Model for Rough Surfaces

Ulf Sellgren, Stefan Björklund and Sören Andersson

Dep. of Machine Design, Royal Institute of Technology, Stockholm, Sweden

ABSTRACT

Engineering surfaces can be characterized as more or less randomly rough. Contact between engineering surfaces is thus discontinuous and the real area of contact is a small fraction of the nominal contact area. The stiffness of a rough surface layer will thus have an influence on the contact state as well as on the behavior of the surrounding system. A contact model that takes the properties of engineering surfaces into account has been developed and implemented in a commercial FE software. Results obtained with the model have been compared and verified with results from an independent numerical method. Results have shown that the height distribution of the topography has a significant influence on the contact stiffness but that the curvature of the roughness is of minor importance. The contact model that was developed for determining the apparent contact area and the distribution of the mean contact pressure could thus be based on a limited set of height parameters that describe the surface topography. By operating on the calculated apparent pressure distribution with a transformation function that is based on both height and curvature parameters, the real contact area can be estimated in a postprocessing step.

1. INTRODUCTION

The local state between interacting bodies is determined by the behavior of the system and the characteristics of the engineering surfaces. In some cases, the actual topographies of interacting surfaces have a significant influence on the behavior of an entire system. There is, as an example, a direct relation between the thermal and electromagnetic resistivity between interacting bodies and the real area of contact between the mating surfaces. The level of shielding from electro-magnetic fields and the actual thermal heat flux in a design thus depend strongly on the contact condition. The reliability of high performance mechanical components can be severely reduced by fatigue failures. Subsurface crack initiation and propagation in such machine elements depend on the load sequences as well as on the apparent contact area and pressure distribution between the interacting surfaces.

A proper numerical treatment of all important physical phenomena that occur in a contact zone requires a detailed discretization of the actual topography and a thorough material model. To obtain a good estimate of the forces that are transferred through the contact zone, a reasonable portion of the actual assembly must also be represented in the numerical model. To keep the problem at a manageable size, it is thus necessary to keep the model as simple as possible.

Elastic contacts between rough surfaces have been treated numerically by Björklund & Andersson [1]. Their numerical method is useful in many studies of contact mechanics and wear. It is though limited to contact situations where a half space approximation is valid. Since the FE method can describe the state of an object with arbitrary shape it is interesting to develop a surface model that consider the properties of engineering surfaces and that is suitable for implementation in a commercial FE software. Such a model is presented in this work. The relationships between different surface parameters have been studied with the Björklund & Andersson numerical method. It has also been used to compare and verify results obtained with the FE based surface model.

2. CONTACT MECHANICS AND TOPOGRAPHY

Despite the relatively simple mechanics theory used for modeling contact problems, the practical simulation of this class of problems is computationally difficult, mainly because the boundary conditions of the bodies under consideration depend on the solution variables [2]. A general contact algorithm consists of two distinctive parts. The first part is a contact search algorithm that identifies penetration, i.e. $\lambda > 0$, between the different sub-domains Ω_1 and Ω_2 (see figure 1). Examples of different types of contact search algorithms that have attracted a broad attention are the master-slave type [3], the single surface type [4], the "pinball" type [5], and the hierarchy type [6][7]. The second part of a general contact algorithm satisfies the kinematic contact condition on Γ_c :

$$\lambda \le 0 \tag{1}$$

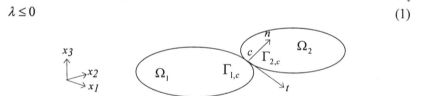

Figure 1. Two bodies in contact.

The impenetrability constraints on the displacements are usually enforced either approximately with the penalty method [8] or more exactly with the Lagrange multiplier method [9]. The two methods are basically complementary [10], which is one of the reasons for the growing popularity of augmented Lagrangian methods.

The topography of engineering surfaces can be characterized by a random height distribution. Contact between such surfaces is, in general, discontinuous and the real area of contact A_r is a small fraction of the nominal contact area A_n. An idealized kinematic contact condition for real engineering surfaces that takes the surface topography into account is :

$$\lambda \le \xi \tag{2}$$

where ξ is the distance between the mean planes of the two contacting surfaces at initial contact (see figure 2). The second part of a general contact algorithm also calculates the pressure p and the tangential traction σ_t on the interacting surfaces:

$$p \geq 0 \qquad (3)$$

$$\sigma_t = f(p) \qquad (4)$$

Figure 2. Rough surfaces, and the concept of mean surfaces.

The topography of a surface can easily be measured with a profilometer. It is usually described with statistical height parameters. The average roughness R_a and the standard deviation R_q of the height of the surface from the center-line, are parameters that are frequently used. A statistical roughness value such as R_a or R_q gives however no information about the height distribution. The first attempt to do this was the bearing area curve [11], sometimes referred to as the Abbott-curve, which expresses the fraction of the nominal area as a function of the height parameter z. A topography model that captures the essential characteristics of the bearing area curve, with a very limited set of parameters, is the linearized five parameter model [12] that is shown in figure 3.

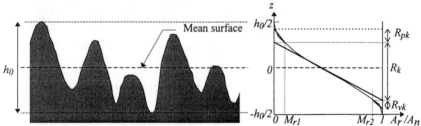

Figure 3. The bearing area curve.

3. A TOPOGRAPHY BASED SURFACE MODEL

3.1 The Contact Stiffness and the Apparent Contact Area

Although engineering surfaces are rough, they are rather flat. The contact between such surfaces is thus discontinuous and distributed. As the applied load is increased the contact spots will grow. The developed model is based on the assumption that the deformation of any point in the roughness layer is independent of its neighboring points. This assumption is increasingly in error for higher contact loads.

It has been shown, e.g. [13] and [14], that contact situation between an elastic sphere with radius R and a slab, when the friction is neglected, can be described with two non-dimensional parameters α and μ.

$$\alpha = R_q/\delta_0 = R_q\left(4RE^{*2}/9F_n^2\right)^{1/3} \tag{5}$$

$$\mu = \eta R_q\left(2R/\kappa\right)^{1/2}8/3 \tag{6}$$

where δ_0 is the compression of the bulk material, η is asperity density, κ is the mean summit curvature of the asperities, which is of the same order as the root-mean-square (r.m.s.) curvature of the surface. For two bodies with rough surfaces pressed against each other, composite values can be used, e.g.:

$$R = \left(1/R_1 + 1/R_2\right)^{-1} \text{ and } R_q = \left(R_{q,1}^2 + R_{q,2}^2\right)^{1/2}$$

The relative size of the apparent contact area A_a, compared with the Hertz solution, depends primarily on α, i.e. the load and the roughness. The second non-dimensional parameter μ has only a secondary influence on A_a. It is though crucial for a proper evaluation of the real contact area A_r.

The approximation that any point in the surface layer is independent of its neighboring points is a justification for utilizing an elastic foundation model (see figure 4), with one-dimensional springs acting in the normal direction z to the mean surface with properties as:

$$F_{z,i} = k_{z,i}u_{z,i} \tag{7}$$

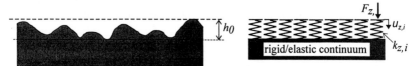

Figure 4. The non-linear contact stiffness as an elastic foundation model.

A common approach is to include the stiffness of the bulk material in the elastic foundation model. The chosen approach here is to combine an elastic foundation model that describes the stiffness of the surface topography with an ordinary FE model of the bulk material. The "topography stiffness" $k_{z,j}$ of a surface area segment A_i is a non-linear property that is determined by integrating the contribution from the distribution of the material according to the Abbott-curve for the topography layer.

$$k_{z,i}(z) = \frac{1}{h_0}\int_{-h_0/2}^{h_0/2}EA_{r,i}(z)dz \tag{8}$$

where $A_{r,i}(z)$ is given by the bearing area curve.

3.2 Finite Element Implementation

In a FE model, the influence on the physical behavior from the surface topography, can be treated at three levels of abstraction - a detailed model with the actual asperities included, a semi-detailed topography layer with a nonlinear constitutive model, or as an abstracted model with an explicit nonlinear contact stiffness. All three approaches have their specific strengths and weaknesses. A detailed FE model will be very large in the general 3D case, and it is thus highly impractical when the behavior on a systems level is required from a simulation model. A detailed model of a single asperity may though give valuable insight for simpler, i.e. more abstracted, models. In an abstracted model, the micro-mechanical effects can be modeled as a non-linear contact stiffness, and a state-dependent non-linear coefficient of friction. Such a model can be implemented by adapting a standard penalty-based master-slave type of general contact element [15]. In a semi-detailed model, which is the chosen approach here, the surface topography can be modeled as a thin non-linear elastic layer.

A first implementation was to connect nonlinear elastic springs with a stiffness according to equation 8 to each surface node on the target surface (see figure 5). The nominal area for each spring was calculated by adding the surface contribution from the closest integration points on the faces of the neighboring elements. The nodal contact area and the nodal pressure distribution that are output from the FE analysis are treated as apparent values. The real contact area is interpreted in a postprocessing step.

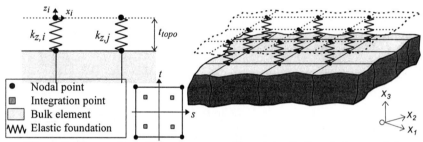

Figure 5. The contact stiffness as a non-linear elastic foundation model.

In a second implementation, the topography was modeled as a nonlinear elastic material model for a continuum layer with the thickness $t > t_{topo}$ (see figure 6). In the topography layer, the relation between the stress σ_z and the engineering strain ε_z in the normal direction z to the contact surface is $\sigma_z = E_{topo}\varepsilon_z$, where $\varepsilon_z/\varepsilon_c = \left(-z + h_0/2\right)/h_0$ and $\varepsilon_c = h_0/t_{topo}$.

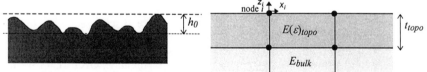

Figure 6. Surface roughness as a discretized nonlinear elastic layer.

E_{topo} is determined from the bearing area curve for the surface according to:

$$E_{topo}\left(\varepsilon_z\right)/E_{bulk} = A_r(z)/A_n \qquad (9)$$

For $\varepsilon_z \geq \varepsilon_c$ we have $E_{topo} = E_{bulk}$. Figure 7 shows such a transformation from a multi-linear bearing area curve to a multi-linear elastic model.

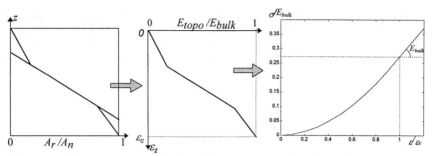

Figure 7. The non-linear contact stiffness as a transformed bearing area curve.

3.3. FE Analyses of the Contact Stiffness and the Apparent State

For comparison with analytical Hertzian solutions of contact between bodies with idealized geometries and smooth surfaces, the contact state between a steel sphere with a radius R of 25 mm pressed against a slab was studied with the presented FE method. The sphere was smooth but the surface of the slab was modeled with a roughness that was defined with a procedure developed by Patir and Cheng [16]. Some examples of rough surfaces that were generated are given in figure 8. Scaling of the topography of a generated surface with a specific wavelength produced surface variants with different roughness. The fastest decay autocorrelation length S_{al} was used for describing the wavelength. Surface 7 in figure 8 is a synthetic bearing area curve, which was used to study the stiffness effect from of a very rough surface.

Surf. #	S_{al} (μm)	R_q (μm)	γ_q (1)	κ_q (1/μm)	R_k (μm)	R_{pk} (μm)	R_{vk} (μm)	M_{r1} (%)	M_{r2} (%)
1	16	1.0	0.14	0.020	2.54	0.9	0.9	9.80	90.37
2	16	0.5	0.07	0.010	1.27	.45	.45	9.80	90.37
3	16	0.1	.014	.0020	0.25	0.09	0.09	9.80	90.37
4	158	1.0	0.04	.0048	2.54	0.7	0.9	6.97	88.31
5	158	0.5	0.02	.0024	1.27	0.35	0.45	6.97	88.31
6	158	0.1	.004	.00048	0.25	0.07	0.09	6.97	88.31
7	-	2.5	-	-	6.35	-	-	-	-

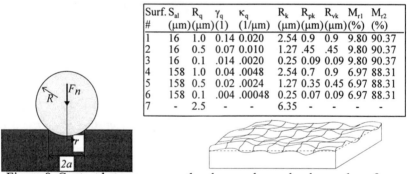

Figure 8. Contact between a smooth sphere and a randomly rough surface.

A large portion of the slab and the sphere were discretized with linear isoparametric elements with the commercial code ANSYS. The bulk material in both geometric domains were given linear elastic generic steel properties, and the general master-slave type of contact elements CONTAC48 were created between the two mating surfaces. The coefficient of friction was zero. Numerical tests with the two implemented models of the topography layer showed a negligible difference in the results. All results that are presented in the following sections are from simulations that were performed with the nonlinear continuum model.

The contact pressure distribution between smooth surfaces calculated by ANSYS is plotted as a function of the applied normal load F_n in figure 9A.

The apparent contact pressure distribution for the same case but with a rough slab surface R_k=6.35μm is shown in figure 9B. The roughness was modeled as a multi-linear continuum layer with a thickness t_{topo} equal to 2xR_k. The absolute difference between the pressures for the smooth case and the rough case is shown in figure 9C. We can clearly observe that roughness increases the apparent contact area and decreases the maximum contact pressure with a value that is almost independent of the load level. For comparison, the roughness was also modeled as a multi-linear linear elastic foundation model. The results for the pressure and the contact area were almost identical for the two models.

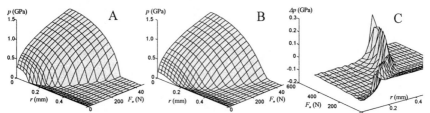

Figure 9. Sphere against a plane slab test case.

The contact state between smooth surfaces of linear elastic bodies with idealized shape and no friction forces present, was solved by Hertz in 1882. The Hertz solutions for the contact radius and the maximum contact pressure for a sphere pressed against a plane slab are:

$$a_{Hertz} = \left(1.5\,FR/E^*\right)^{1/3} \text{ and } \hat{p}_{Hertz} = 1.5\,F_n/\pi \cdot a_{Hertz}^2$$

where $E^* = \left(0.5\left(\left(1 - \upsilon_1^2\right)E_1^{-1} + \left(1 - \upsilon_2^2\right)E_2^{-1}\right)\right)^{-1}$ is the composite modulus.

In figure 10, the apparent contact pressure has been normalized with $\hat{p}_{Hertz}$ and the radial coordinate r has been normalized with a_{Hertz}.

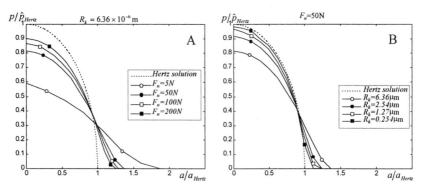

Figure 10 Normalized contact pressure for different loads (A) and roughness (B).

The FE calculated contact area between rough surfaces is larger than for a smooth surface, i.e. it is the apparent contact area that is calculated. In figures 10 and 11, we can clearly see that an increased roughness/load ratio, i.e. an increased α, will increase the calculated area and consequently reduce the apparent contact pressure .

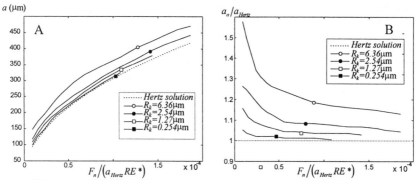

Figure 11 Apparent contact area (A) and in normalized form (B) as a function of the normalized normal load for different roughness values.

4. THE REAL AREA OF CONTACT

4.1 The Relation Between Real Area and Apparent Area

The real contact area approaches the apparent contact area when the load is increased. For a single sine shaped corrugation with roughness amplitude B and wavelength L, i.e. the height $z(x)$ of the surface above the mean height is:

$$z(x) = B\sin(2\pi x / L) \tag{10}$$

The slope γ and the r.m.s. slope γ_q are thus:

$$\gamma = \partial z/\partial x = (2\pi B/L)\cos(2\pi x / L) \tag{11}$$

$$\gamma_q = \sqrt{2}\pi B/L \tag{12}$$

The general contact capabilities in ANSYS 5.3 and I-DEAS Master Series 5[1] was used to find the smallest apparent mean contact pressure $\bar{p} = \bar{p}_{a,\mathrm{limit}}$ for which $A_r = A_a$, for an idealized 2D case (see figure 12).The FE analyses showed that full contact was obtained when the mean apparent pressure reached:

$$\bar{p}_{a,\mathrm{limit}}/G^* \approx 4B/L \tag{13}$$

$$p(x) = \bar{p}_{a,\mathrm{limit}}\left(1 + \sin(\pi x/L)\right) \tag{14}$$

Since the shear stress is related to shear deformation as $\tau/G=\gamma$, equation 13 clearly indicates the limit contact state is mainly due to local shear deformation that can be represented by the r.m.s. slope γ_q. The compressive contact stresses $p(x)$ decrease the initial corrugation amplitude B. A modified corrugation amplitude $B' = f(BpE^{-1})$ can thus be used. If we relate the pressure to the composite Young's modulus for the actual material combination we get:

$$\bar{p}_{a,\mathrm{limit}}/E^* \approx \hat{p}_{\mathrm{limit}}/(2E^*) \approx \bar{\tau}_{a,\mathrm{limit}}/G^* = \sqrt{2}\pi B'/L \approx \gamma_q \tag{15}$$

---

[1] ANSYS and I-DEAS Master Series are trademarks of ANSYS INC. and SDRC respectively.

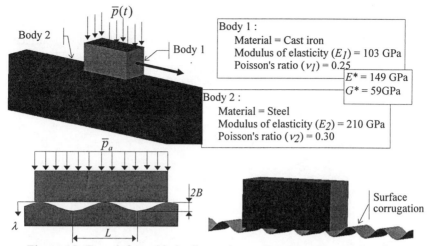

Body 2

$\overline{p}(t)$

Body 1

Body 1 :
  Material = Cast iron
  Modulus of elasticity (E_1) = 103 GPa
  Poisson's ratio (v_1) = 0.25

$E^* = 149$ GPa
$G^* = 59$ GPa

Body 2 :
  Material = Steel
  Modulus of elasticity (E_2) = 210 GPa
  Poisson's ratio (v_2) = 0.30

$\overline{p}_a$

$2B$

λ

L

Surface
corrugation

Figure 12. Two deformable bodies and one sine-shaped contact surface.

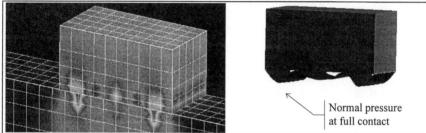

Normal pressure
at full contact

Figure 13. Contact between ideal block and corrugated slab.

It has been shown [13] that the ratio of the real to the apparent contact area for a regular wavy surface can be determined from the mean apparent pressure:

$$A_r / A_a = (2/\pi)\sin^{-1}\left(\overline{p}_a / \overline{p}_{a,\lim it}\right)^{1/2} \tag{16}$$

For a random rough surface with an exponential probability function and elastically deforming asperities, the ratio of the real contact area to the apparent contact area is directly proportional to the mean apparent pressure [14]:

$$A_r / A_a = 2\pi^{1/2}\left(R_q\kappa\right)^{-1/2}\overline{p}_a / E^* \tag{17}$$

For a Gaussian distribution this relationship would be slightly different. Replacing κ with γ_q in eq. 16 and setting the area ratio to unity gives:

$$\overline{p}_{a,\lim it} / E^* = \left(1/2\pi^{1/2}\right)\gamma_q \tag{18}$$

For each node in contact, a value for the real area of contact can thus be calculated according to equation 16, and summed to a total value for the real area of contact:

$$A_r = \sum A_r^n = \sum A_a^n \cdot g\left(R_q, \kappa, p_a^n, E^*\right) \tag{19}$$

1265

4.2 Comparison of FE Results with an Independent Numerical Method

For rough surfaces, the real contact area is smaller than for smooth surfaces. The ratio A_r/A_a approaches unity at increased loading. By combining the stress-strain state in the topography layer with the non-linear elastic material model for the layer, A_r can be calculated for each contact node. The real to Hertzian contact area ratio as a function of applied load and roughness wavelength is plotted for two roughness values in figure 14.

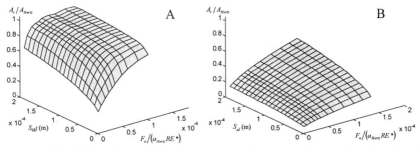

Figure 14 Real/Hertzian contact area ratio as a function of normalized load and roughness wavelength S_{al} for R_k=0.254μm (A), and R_k=1.27μm (B).

The contact state at increased loading between the set of rough isotropic surfaces and the sphere were calculated with a previously developed and verified numerical method for three-dimensional analysis of elastic contact between rough surfaces [1]. For each generated surface, the contact state was calculated at 31 different positions on the surface (see figure 15). Some calculated contact spots are plotted, at a few discrete load levels, with the FE based interpretation of the real contact area, for roughness with a relatively short wavelength in figure 16, and for a relatively long wavelength in figure 17. Results from the independent numerical method are labeled Micro-mech in figure 16.

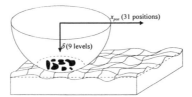

Figure 15. Contact between a smooth sphere and a randomly rough surface.

With a low load to roughness ratio, e.g. figure 17D, the wavelength is no longer small compared to the size of the apparent contact zone.

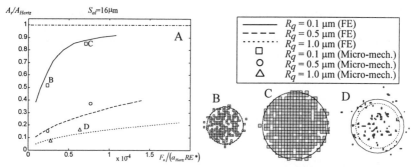

Figure 16. Real/Hertzian contact area ratio as a function of load and roughness, and three contact spots (B-D) for S_{al}=16μm.

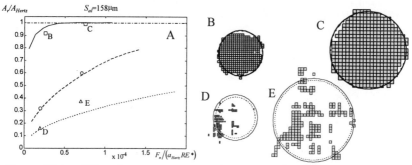

Figure 17. Real/Hertzian contact area ratio as a function of load and roughness and four contact spots (B-E) for S_{al}=158μm.

5. CONCLUSIONS

For contact between rough surfaces, the contact stiffness and thus the accuracy of the calculated apparent contact area and the resulting stress state in the interacting bodies can be significantly improved if the surface topography is taken account for. A model of the non-linear elastic stiffness of a randomly rough isotropic surface layer has been developed and implemented in a commercial FE code. A linearized bearing area curve can be used to get an improved contact stiffness and thus a better estimate of the apparent contact area and the stress state. For random isotropic surfaces the ratio between real contact area, which for rough surfaces is a fraction of the apparent contact area, can, with reasonable accuracy, be interpreted in a post-processing step from the actual roughness and curvature surface parameters when and computed stress-strain state.

6. DISCUSSION

Surface roughness increases the apparent contact area. It thus reduces the maximum apparent contact pressure and the maximum shear stress in the bulk material. This effect is inversely related the normal load and directly related to the roughness. The real contact area though depend primarily on the amplitude and wavelength properties of the interacting asperities.

1267

The surface topography that in the present model is treated as a surface layer with non-linear elastic properties can easily be implemented as a more abstracted or idealized model, where this effect is treated as a non-linear contact stiffness in the penalty or augmented Lagrangian based contact elements.

7. ACKNOWLEDGEMENT

The presented work was financially supported by grant from Volvo Research Foundation and Scania CV AB, which is gratefully ackowledged..

8. REFERENCES

1 Björklund, S. and Andersson, S. - A numerical method for real elastic contacts subjected to normal and tangential loading, Wear, Vol. 179, pp.117-122, 1994.

2 TAYLOR, RL and SACKMAN, JL - Contact-Impact Problems Volume 1. Engineering Report and User's Manual, Report No. DOT-HS-805-629, Department of Civil Engineering, University of California, Berkeley, California, 1980.

3 HALLQUIST, JO - NIKE2D - a vectorized, implicit, finite deformation, finite element code for analyzing the static and dynamic response of 2-d solids, RPT. UCRL-52678, LLNL, Livermore, 1979.

4 BENSON, DJ and HALLQUIST, JO - A single surface contact algorithm for post-buckling analysis of shell structures, Comp. Methods Appl. Mech. Eng., Vol. 78, pp. 141-163, 1990.

5 BELYTSCHKO, T and NEAL, MO - Contact-impact by the pinball algorithm with penalty and Lagrangian methods, Int. Journal Num. Methods. Eng., Vol. 31, pp. 547-572, 1991.

6 NILSSON, L, ZHONG, Z-H and OLDENBURG, M - Analysis of shell structures subjected to contact-impacts, Proc. Symposium on Analytical and Computational Models of Shells, ASME CED-Vol. 3, ASME, New York, pp. 457-482, 1989.

7 ZHONG, Z-H and NILSSON, L - A unified contact algorithm based on the territory concept, Comput. Methods Appl. Mech Eng., Vol. 130, pp. 1-16, 1996.

8 ODEN, JT and PIRES, EB - Nonlocal and nonlinear friction laws and variational principles for contact problems in elasticity, Journal of Applied Mechanics, Vol. 50, pp. 67-76, 1983.

9 BATHE, K-J and CHAUDHARY, A - A Solution Method for Planar and Axisymmetric Contact Problems, Int. J. Numer. Methods Eng, Vol. 21, pp. 65-88, 1985.

10 GUERRA, FM and BROWNING, RV - Comparison of Two Slideline Methods Using Adina, Computers & Structures, Vol 17, pp. 819-834, 1983.

11 ABBOTT, EJ and FIRESTONE, FA - Specifying surface quality", Mechanical Engineering (ASME), Vol. 55, pp. 569, 1933.

12 DIN, Measurement of Surface Roughness; Parameters Rk, Rph, Rvk, Mr1, Mr2 for the description of the material position in the roughness, German Standard, DIN 4776, 1990.

13 GREENWOOD, JA and TRIPP, JH - The elastic contact of rough spheres, Trans. ASME, Series E, Journal of Applied Mechanics, Vol. 34, pp. 153-159, 1967.

14 JOHNSON, KL - Contact mechanics, Cambridge University Press, Melbourne; Australia, 1987.

15 Sellgren, U. and Olofsson, U. - A frictional model for the micro-slip range, Proc. NAFEMS WORLD CONGRESS '97, pp. 534-545, 1997.

16 Patir, N. And Cheng, H.S. - An Average Flow Model for Determining Effects of Three-Dimensional Roughness on Partial Hydrodynamic Lubrication, ASME Journal of Lubrication Technology, Vol. 100, pp. 12-17, 1978.

F.E. Predictions of Residual Stresses in Fusion Welded Steel Pressure Vessels

G.P. Campsie, M.A. Rahim, A. Ramsay and J.D. M^{C}Vee

Self-equilibrating locked-in or residual stresses are an important consideration in the fatigue and fracture assessment of fusion welded steel pressure vessels. However, the experimental determination of residual stresses is time consuming and often requires the use of destructive techniques on full-scale structural arrangements. Moreover, for certain structural arrangements, the use of experimental techniques may not be possible.

Coupled thermo-mechanical and mechanical finite element analyses have been carried out, using both MARC and ABAQUS, to evaluate the magnitude and distribution of residual stresses induced at an internal ring stiffener to plating T-butt fillet weld. Multiple pass welding has been simulated, with thermal loading applied to the models on a pass by pass basis. Rotational symmetry has been assumed and metallurgical phase transformations are ignored. Further, no attempt has been made to follow the complex temperature history experienced by the structure during welding.

Initial comparisons with experimental results for longitudinal residual stresses acting through the plating thickness have shown reasonable correlation. This suggests that continued development and application of validated numerical modelling techniques for prediction of welding residual stresses may allow the quality of advice offered in relation to fracture and fatigue crack growth to significantly improve.

BEHAVIOR OF HELICAL PIERS IN FROZEN GROUND

He Liu[1], Ph.D., P.E.
Hannele Zubeck[2], Ph.D., P.E.
Daniel H. Schubert[3] , P.E.
Sean J Baginski[4]
Yifei Shi Hsieh[5] , Ph.D.

SUMMARY

The need for improving our understanding of the behavior of helical piers in frozen ground has prompted this analysis. Funded by a grant from the Alaska Science and Technology Foundation, this study seeks to examine the performance and behavior of helical piers including finite element methods to determine the distributions of stresses within the soil and the steel.

Four objectives were established to complete this study; 1) Analyze the immediate displacement and stress distribution in soil due to a design axial load, 2) Analyze the stresses within the pier structure when it is subjected to this load using a submodel that is created from the larger scale model. 3) Provide a detailed analysis of the stresses within the pier itself during installation using a small scale, detailed model of the spiral structure. 4) Analyze longer-term creep settlement behavior in frozen ground is investigated using a model similar to (1). ANSYS was chosen to model all four of these situations for its ability to perform complex, nonlinear, three-dimensional analyses.

Little data exists for the displacement of helical piers and the surrounding

---

[1] Assistant Professor, Civil Engineering Department, University of Alaska Anchorage, Alaska, USA
[2] Assistant Professor, Civil Engineering Department, University of Alaska Anchorage, Alaska, USA
[3] Director, Technical Services Branch, Office of Environmental Health & Engineering, US Public Health Service, Anchorage, Alaska, USA
[4] Graduate Student, Civil Engineering Department, University of Alaska Anchorage, Alaska, USA
[5] Design Analyst, WGC HSC, Inc., Buffalo, New York, USA

soil when subjected to axial, vertical compressive loads, therefore these results will be compared with physical tests to be conducted at the US Army Corps of Engineers CRREL Laboratory in the near future. Stresses within the pier indicated in the installation model compare favorably with known failure modes for these pier systems. The model results will increase understanding of these pier systems in various soil conditions as well as provide insight into design considerations to be utilized in future development of these piers.

1. INTRODUCTION

Helical piers or anchors come in many shapes and sizes. Typically they consist of a central shaft that is made from square or round sections that can be either solid or hollow. To this shaft is connected from 1 to 4 spiral plates which are designed to mobilize more soil under normal loading conditions. The advantage of helical piers is their light weight and ease of installation. Helical piers have long been used to resist tension forces generated by uplift and overturning of various structures. Current design methods for determining the capacity of helical piers are based upon very simple formulas and may or may not be conservative. More research must be performed in order to establish a more rigorous and reliable design approach. Some research has been conducted to describe the overall capacity and displacement of helical piers subjected to vertical and lateral loading Narasimha Rao et. al. [1], Narasimha Rao and Prasad [2] Ghaly et. al. [3]. Much less has been done to examine the distribution of stresses in the soil surrounding these piers, their behavior under prolonged loading in frozen ground or the stresses within a typical helical pier under normal working loads or during installation. Helical piers may present a real economical solution for arctic foundations that currently rely on drilled and slurried piers or spread footings. A more complete understanding of the stresses developed within the pier structure and the soil around it will provide valuable information for design of these foundations, as well as further the efforts of research in this area.

The purpose of this analysis is to develop a useful model for analyzing helical piers using finite element methods. The following are the main objectives of this analysis. 1) Large Model: This model will analyze the soil stresses and displacements immediately after the pier is subjected to its design load. This data is also critical for the development of more detailed analysis using submodeling techniques. 2) Small Model: The small model is a submodel of the large model. It will analyze the stresses developed within the spiral structure by using results from the large model analysis. 3) Installation Failure Model: A detailed model of the spiral structure subjected to a torsional load during installation will provide insight into the failure mechanism of helical piers during construction. 4) Creep Model: Creep analysis will be conducted to determine the long-term displacement and soil stress in frozen ground.

The data gathered from these models will provide a basis for future experiments to be conducted using helical piers. These experiments in turn will provide a check for the methods and assumptions used in this analysis. Following this check, the modeling technique will be refined in order to maximize the accuracy of the modeling results.

2. ANALYSIS

1. Large Model: The large-scale model was a cylindrical volume of soil with a helical pier at the center. To represent the half space of soil, sufficient dimensions were necessary to minimize the effects of boundary conditions on the model results. An overall diameter of 50 in. and an overall depth of 180 in. were decided on. The pier's embedment depth was 60 in. with 4 helices spaced 10 in. apart. Boundary conditions for this model include translational restraint in the horizontal directions for the sides of the cylinder and full restraint of the bottom of the cylinder. The loading condition includes a 35 kip axial load applied at the top of the pier. This was based on typical design loads for piers of similar size and shape.

The geometry chosen for the pier includes a 3-½ in. outer diameter pipe section with ½ in. wall thickness for the shaft; and a 10 in. diameter, ½ in. thick plate for the spiral. Steel for the pier system was assumed to be A572 Grade 50. For this large-scale model, the spirals were modeled using flat shell elements instead of the actual spiral structure. This was done in order to limit the overall number of elements in the model. The element and material properties used in this model appear in Table 1.

Material	E (ksi)	ν	ρ (lb·s^2/in^4)	Element Type	Nodal DOF
Soil	5	0.35	1.94 e-4	Solid	3
Steel shaft	29000	0.3	7.34 e-4	Beam	6
Steel plate	29000	0.3	7.34 e-4	Shell	6

Table 1. Element and Material Properties

The Drucker-Prager yield criterion was used in this and all other models utilizing soil elements. The Drucker-Prager yield criterion is described in Figure 1 and equations 1 and 2.

1273

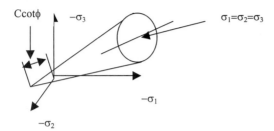

Figure 1. Drucker-Prager Circular Cone Yield Surface

The material constant is:

$$\beta = \frac{2\sin\phi}{\sqrt{3}(3-\sin\phi)}$$ Eq. (1)

ϕ= the angle of internal friction = 31°
The material yield parameter is:

$$\sigma_y = \frac{6c\cos\phi}{\sqrt{3}(3-\sin\phi)}$$ Eq. (2)

c= the cohesion value for the element = .005 ksi

The results of this analysis indicated that the reaction stresses developed within the soil immediately after placement of a 35 kip load were not distributed evenly among the plates. The reaction stresses directly below the bottom plate were much higher than in other regions. The stress in this region was 37.50 psi, which is six times higher than similar reaction regions above it. The fact that there was little significant contribution from the upper three spirals is, likely to be caused by the

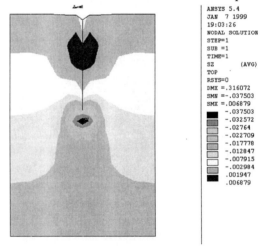

```
ANSYS 5.4
JAN  7 1999
19:03:26
NODAL SOLUTION
STEP=1
SUB =1
TIME=1
SZ        (AVG)
TOP
RSYS=0
DMX =.316072
SMN =-.037503
SMX =.006879
    -.037503
    -.032572
    -.02764
    -.022709
    -.017778
    -.012847
    -.007915
    -.002984
     .001947
     .006879
```

Figure 2. Vertical in the Soil Volume Stress Distribution

confinement of the soil between the plates, which allows the soil in that volume to move for the most part, as a unit. Vertical compressive stresses were concentrated within an area of about 3 time the radius of the spiral section and expanded outward and downward from the bottom of the pier.

1274

The vertical stress distribution can be seen in Fig. 2. Some tension stress was developed above the top plate with magnitudes up to 6.88 psi. These stresses are caused by adhesion between the soil and the pipe or plate elements in this region and were limited by the Drucker-Prager yield criteria.

The soil settlement due to the applied load, shown in Figure 3, was greatest at points near the pier. This is a result of the small cohesion value used in this model, which reduced the ability of the soil elements to transfer shear outward from the pier. The maximum settlement of 0.316 in., occurred at the top of the pier and the pier was displaced relatively rigidly. The plate deformation was not of any significant interest in this model, however it was analyzed much more closely in the detailed models.

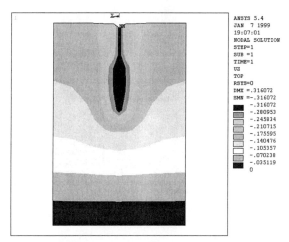

Figure 3. Vertical Displacement Distribution in the Soil Volume

2. Small Model: The small model was a submodel that used the results from the large model to generate the loads on the spiral structure. The pipe and spiral were modeled with the actual geometry and dimensions using shell elements. The submodel was dimensioned in order to avoid errors due to the simplification of the pier structure that was used in the large-scale model. The resulting model was a cylinder of soil with a 15-in. diameter and a 10-in. depth. Using the submodel routine in ANSYS, the program interpolated displacements for coordinates corresponding to the boundary nodes of the submodel. These displacements were then applied to the submodel to calculate the internal displacements and stresses within the boundaries of the volume.

The results from this analysis indicate that the largest vertical compressive stresses in the surrounding soil occur at the outer edges of the spiral. The magnitude of these stresses within the soil ranged up to 19.52 psi along a line directly below the outer edge of the spiral. The vertical stress distribution within the soil surrounding the bottom spiral is shown in Figure 4.

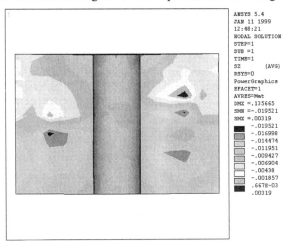

Figure 4. Vertical Stress Distribution in Soil at Bottom Spiral (spiral removed for clarity)

The spiral structure is stressed at its highest level where the plate joins the pipe section and most specifically at the inside point along the cutting edge of the spiral itself. Figure 5 shows the stresses within the spiral resulting from the stresses within the soil. Under the 35 kip maximum load, Von Mises stresses varied along the radial direction of the spiral up to a maximum of 64 ksi at the bottom most point of the weld. This is at or slightly above the yield point for most steel used for helical piers.

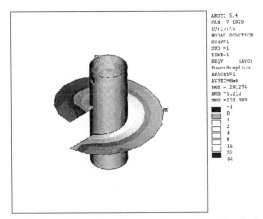

Figure 5. Von Mises Equivalent Stress on the Spiral Structure

3.   Installation Failure Model: The failure model simulates a critical failure during installation when the spiral encounters a hard rock or ice formation. In this analysis it is assumed that there are no reactions due to the soil and the entire torque is resisted at the cutting edge of the spiral. This model used the same detail of the spiral structure as the small model however it contained no soil elements. The spiral structure was restrained at the outside end of the bottom of the spiral while a torque was applied at the top of the pipe section. The bottom of the pipe was restrained against translation in the vertical direction. The magnitude of the torque was 90 kip-in, which is in the range used during installation of these piers. The spiral was modeled using the same material properties as the small model with the addition of Bilinear Isotropic yield criteria including $\sigma_y = 46$ ksi and $E_T = 1450$ ksi. The shell elements for this model allowed for large deformation and had 6 degrees of freedom at each node.

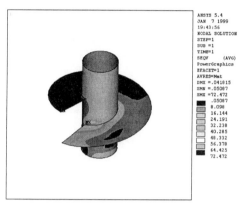

Figure 6. Von Mises Stress During Installation Failure

The results from this model show clearly that large stresses will develop at the weld between the spiral and the shaft (see Fig. 6). These stresses are sufficient to yield the steel and cause failure of the pier. This stress is highly concentrated at the bottom end of the connection and rapidly dissipates within an inch of this point.

4. Creep Model: This model used the same dimensions and material properties as the large model with the addition of creep parameters. It was assumed that dead load would be the only applied load for long term application, and therefore 20%, or 7 kips was applied axially at the top of the pier. The creep equations were determined based on previous research Ladanyi [4] and were of the form:

$$\dot{\varepsilon}_e = \dot{\varepsilon}_c \left(\frac{\sigma_e}{\sigma_{cu\theta}} \right)^n \qquad \text{Eq. (3)}$$

where: $\dot{\varepsilon}_e = .01 yr^{-1}$
$n = 3$
$\sigma_{cu\theta} = .00551$ ksi at 1°C.

The creep model was performed for a time period of 1 year at a temperature of 274 K. The displacement results provided apply only to primary settlement conditions and do not account for secondary or tertiary settlement. The initial settlement under the 7 kip load was –0.031 in. and increased nearly linearly to –0.062 in., about 200% of the initial settlement over the course of one year (see Figure 7). The stresses in the soil were shown to relax significantly at all points outside the immediate viscinity of the pier itself.

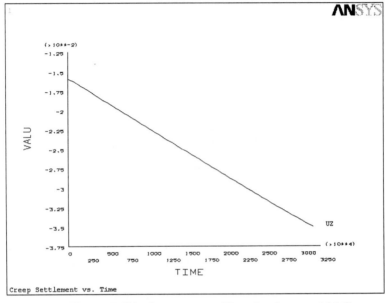

Figure 7. Displacement vs. Time for 1 year at 1° C

3. CONCLUSIONS

The large model provides valuable information regarding the distribution of stress and displacement within the soil. This stress, may or may not depend on the geometry of the pier system in question. However, it does indicate that closely spaced spiral structures will not provide significant increases in the overall capacity of the pier. As can be seen from the analysis, of the four spirals in this pier system, only the bottom one contributes to any large extent to the pier's capacity.

The small model provided a detailed examination of the stresses within the spiral during the application of design loads. The highest stresses were encountered at the top and bottom ends of the spiral welds. This information can be used to make changes to the design of these structures if necessary as well as provide a more detailed picture of the physical situation encountered by helical structures embedded in soil.

The installation failure model displayed the stresses encountered during installation of the pier system under worst case conditions. The stress concentration that occurred at the bottom of the weld was very similar to known, but rare, failure mechanisms for helical piers. This information can be used to design better connection geometry at this critical location.

Finally, the creep analysis indicated the creep behavior of helical piers in frozen ground and provides valuable insight into the magnitude of settlement that can be expected over extended periods of time.

Further research will be conducted to examine the effects of increasing or decreasing the number, spacing, diameter and thickness of the spiral plates as well as changing the overall depth of the pier and varying the soil conditions. This study has provided the groundwork for future analyses and will be useful in determining testing and design procedures for helical piers.

4. REFERENCES

1. Prasad, Yenumula V.S.N., Rao, Narasimha, "Lateral Capacity of Helical Piles in Clays", *Journal of Geotechnical Engineering*, Vol. 122, No. 11, November 1996, pp. 938-941.

2. Rao, Narasimha, Prasad, Yenumula V.S.N., "Estimation of Uplift Capacity of Helical Anchors in Clays", *Journal of Geotechnical Engineering*, Vol. 119, No. 2, February 1993, pp. 352-357.

3. Ghaly, Ashraf; Hanna, Adel; Ranjan, Gopal; Hanna, Mikhail; "Helical Anchors in Dry and Submerged Sand Subjected to Surcharge", *Journal of Geotechnical Engineering*, Vol. 117, No. 10, October 1991, pp. 1463-1470.

4. Ladanyi, Branko, "Shallow Foundations on Frozen Soil: Creep Settlement", *Journal of Geotechnical Engineering*, Vol. 9, No. 11, 1983, pp.1434-1448.

5. Ruipiper, Stan, Edwards, William G., "Helical Bearing Plate Foundations for Underpining", *Foundation Engineering Procedings Congress/SCE/CO Div*, June 25-29, 1989, pp. 1-10.

NONLINEAR FINITE ELEMENT ANALYSIS OF A PRETENSIONED CONCRETE BEAM STIFFENED BY A DOUBLE-CABLE SYSTEM.

Monique Bakker[1], Jan Kerstens[1] and Cees Kleinman[1]

[1]Faculty of Architecture, Building and Planning, Eindhoven University of Technology, the Netherlands.

SUMMARY

This paper presents a nonlinear finite element model for a girder formed by a pretensioned concrete beam stiffened with a double-cable system, and the comparison of the model results with test results. The paper starts with a description of the prototype of the girder and its posttensioning procedure, as tested in the laboratory. Then the finite element model for this girder is described. The finite element model is made in ANSYS 5.3 using beam and truss elements only. In the model elastic material properties have been assumed. The model is nonlinear due to large deflection effects (large rotations but small strains), and due to a contact problem caused by a cable running frictionless through the struts. This cable has been modelled by follower forces. The magnitude of the applied follower forces is determined from the difference in cable length in the deformed and undeformed girder geometry. This requires an iterative procedure, which was accomplished by using the ANSYS parametric design language. Finally the results of the finite element model are compared with the results of laboratory tests, performed on a prototype of the girder. The paper concludes with recommendations for improvements to the posttensioning procedure, following from the comparison of model and test results.

1 INTRODUCTION

The girder, which will be described in this paper, is a component of an innovative type of light-weight load-carrying structure, which is under development at the Faculty of Architecture, Building and Planning of the Eindhoven University of Technology in a series of projects carried out by students to obtain their master's degree. The final aim is to integrate the girder into a portal frame system also stiffened with a double-cable system. To assist the determination of the structural properties of this girder, a finite element model has been built. To verify this model, a full scale model of the girder has been built and tested in the Pieter van Musschenbroek laboratory. In this paper the girder as tested in the laboratory will be described. It should be realized that this girder is a prototype and not the final design.

2 EXPERIMENTAL RESEARCH

2.1 Description of girder

The girder consists of a pretensioned, precast, high-strength concrete beam supported by concrete struts of a double-cable system (see Figure 1). Posttensioning of the first cable causes a hogging deformation of the girder. The tensioning forces of this cable are balanced by compression forces in the concrete beam. Posttensioning of the second cable causes a sagging deformation. The tensioning forces of this cable must be balanced by the reactions at the supports. In the tested prototype both cable 1 and cable 2 consist of two cables. The struts are connected to the beam by means of pinned joints. Ideally, during posttensioning both cables should run frictionless through the struts. In that case the prestress would be constant over the length of the cables. However, to ensure the stability of the cable system during tensioning, cable 1 is fixed to the struts. Posttensioning cable 1 from one side thus results in an asymmetrical deformation mode. If both cables were fixed to the struts, it would be impossible to attain the required prestress over the entire length of the cables. A large part of the prestress would be lost due to bending deformations in the struts. After completing the tensioning procedure both cables are fixed to the struts.

The pretensioning of the concrete beam (by internal, bonded tendons) has to prevent tensile stresses in the beam (during transport, installation and posttensioning of the cables). This prestress does not influence the rigidity of the beam but increases the cracking moment , so that the beam longer retains the rigidity of the uncracked cross-section.

2.2 Posttensioning procedure

To prevent the development of tensile stresses in the beam the cables are posttensioned alternately. Tensioning a cable in one step could result in collapse of the girder. The cables are posttensioned alternately from one side, starting with cable 1. When one cable is tensioned (active tensioning) the oil circuit of the hydraulic jack of the passive cable is closed, resulting in a fixed initial (unstrained) length of the cable. Due to the deformations of the girder during active tensioning the forces in this fixed cable will also change (passive tensioning). Initially, when both cables are still without tension, the girder is supported by temporary props at mid span. In the first step cable 1 is

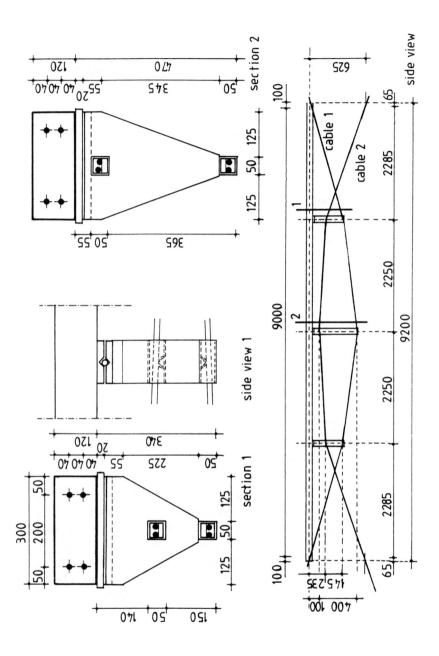

Figure 1: Dimensions of girder, as tested in the laboratory

1283

tensioned a little so that the girder experiences a small hogging deformation and the temporary props can be removed. Cable 2 is also tensioned a little, so that both cables contact the struts.

2.3 Measurements

During the posttensioning procedure the following measurements have been carried out:
- strains in the steel cables by means of strain gages,
- hogging deformations of the beam, by means of LVDTs,
- jacking forces by means of pressure registration.

step	jacking force of cable 1 [kN]	jacking force of cable 2 [kN]	hogging deformation [mm]
1	3.05	0.59	13.0
2	28.14	(7.67)	27.4
3	(37.58)	25.14	15.4
4	55.53	(31.10)	27.9
5	(58.37)	46.50	22.0
6	84.01	(53.88)	37.9
7	(83.36)	68.16	33.9
8	111.30	(75.04)	50.3
9	(110.20)	89.29	47.2
10	137.99	(88.76)	53.2
11	(137.99)	111.19	48.6

Table 1: Overview of measurements

The values between parentheses denote the cable forces in the passively tensioned cable. When the initial (unstrained) length of this cable does not change in a loading step, the jacking force of the passively tensioned cable will have to increase. However, in the tensioning steps 7, 9, 10 and 11 the jacking forces of the passively tensioned cable do not increase. This may be caused by an increase in the initial length resulting from slip between the cable and the hydraulic jack. It may also be caused by inaccurate measurements of the jacking forces, as the measurements of the cable forces indicate that these forces do increase.

3 FINITE ELEMENT MODEL

3.1 Aim and choice of elements

The purpose of the finite element model is to determine the displacements and global force distribution resulting from the tensioning procedure, that is:
- deflections of the concrete beam and rotations of the struts,
- internal forces in the cables,
- internal forces and moments in the concrete beam,
- compression forces in the connections between beam and struts.

These forces and displacements can be determined with a model consisting of beam and truss elements. To account for the influence of stress-stiffening and geometry changes in the concrete beam and steel cables a large deflection/large rotation/small strain analysis has been carried out.

3.2 Geometry

The behaviour of the 3-dimensional girder can be described by a 2-dimensional model (see Figure 2), because the girder experiences no out-of-plane deformations. Cable 1 and cable 2, which in the tested prototype consist of two cables each, are modelled by one cable. Figure 2 presents the geometry of the finite element model. The concrete beam and struts are represented by their center lines. Offsets are modelled by very stiff elements, taking care that their stiffness is still sufficiently small to avoid ill-conditioning of the stiffness matrix.

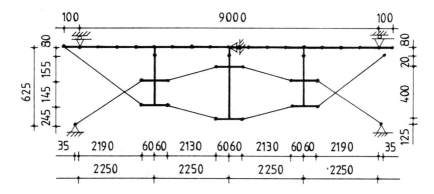

Figure 2: Geometry of finite element model

The first loading step cannot be described with the model, because the temporary props have not been included. The hogging deformation of the girder, measured in the test after the first step, is applied to the model as an initial imperfection of the concrete beam. It is assumed that this initial imperfection has a sinus shape over the length of the concrete beam.

3.3 Material properties

In the model linear elastic material properties for both concrete and steel cables are assumed. For the concrete this is justified because (during post-tensioning of the cables) the applied prestress prevents cracking of the concrete and the concrete compression stresses remain below the proportional limit.

Beam and struts: concrete class B105

measured modulus of elasticity $E_{concrete} = 40087 \ N/mm^2$

(Pretensioning of concrete beam by internal bonded tendons:
        FeP 1860, 4 strands Ø 12.9, prestress = 10.7 N/mm²
No reinforcement except for some hair-pins and stirrups near the supports)

Steel cables: FeP 1860 , 2 strands Ø 15.2 (A = 2·137.62 mm²)

measured modulus of elasticity $E_{steel} = 197155 \ N/mm^2$

3.4 Modelling of cables

The cable which is fixed to the struts can simply be modelled with truss elements. The cable running frictionless through the struts cannot easily be modelled using truss elements and contact elements, since the direction of the cable changes when running through the struts. Therefore it was decided to model the cable by forces, instead of by elements. Since the direction of the forces in the cable changes when the geometry of the girder changes, the cables are modelled by follower forces. These forces are applied as surface loads on the faces of beam elements, so that the orientation of the beam element determines the direction of the follower force. The extensional rigidity of these 'dummy' beam elements should be chosen sufficiently small to have negligible influence on the internal forces in the girder, but also sufficiently large to avoid ill-conditioning of the stiffness matrix. The flexural rigidity has no influence since the 'dummy' beam elements are connected to the struts by hinges. It would be more convenient to use dummy truss elements. However, in ANSYS no surface loads can be applied to these elements.

When a cable is actively tensioned, the magnitude of the follower forces is equal to the applied jacking force. When the cable is passively tensioned, the magnitude of the follower forces has to be determined from the extension of the cable, resulting in an iterative procedure as will be discussed in the tensioning procedure for cable 1.

3.5 Active tensioning of cable 1

Figure 3 shows the model used for active tensioning of cable 1. Since cable 1 is fixed to the struts, the cable can be modelled with a truss element, except for the part of the cable where the jacking force $F_{1;jacking}$ is applied. Cable 2, which runs frictionless through the struts is modelled by follower forces. First the initial (unstrained) cable length ℓ_0 of cable 2 is calculated from the undeformed geometry, and an estimate $F_{2,estimated}$ is made of the forces in cable 2. Then the jacking force $F_{1;jacking}$ and the estimated forces in cable 2 are applied to the model, and a geometrically nonlinear calculation is carried out to determine the deflections of the girder. From the thus obtained deformed geometry the length ℓ of cable 2 is determined. From this length the forces $F_{2,lenght}$ in cable 2 corresponding to the elongation of cable 2 can be calculated as:

$$F_{2;length} = EA\frac{\ell - \ell_0}{\ell_0} \qquad (1)$$

When it is found that $F_{2,length} > F_{2,estimated}$ then the estimated force in cable 2 was too small, when $F_{2;length} < F_{2;estimated}$ the applied force was too large. From the difference between $F_{2;length}$ and $F_{2;estimated}$ then a better estimate of the forces in cable 2 is made. This procedure is repeated until $F_{2;estimated}$ is sufficiently close to $F_{2;length}$.

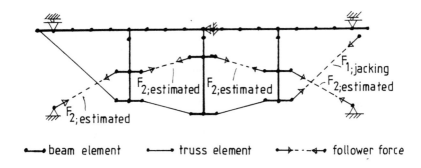

Figure 3: Model for active tensioning of cable 1

3.6 Active tensioning of cable 2

The modelling of active tensioning of cable 2 is much simpler than that of cable 1. The follower forces describing the cable are simply taken equal to the jacking force (see Figure 4). Note that in the model for active tensioning of cable 2, the part of cable 1 attached to the hydraulic jack is modelled by a truss element instead of by follower forces. To replace the follower force by a truss element a new finite element model is defined whose initial geometry corresponds to the deformed geometry of the previous model. The stress distribution is transferred to the new model by means of temperature and initial strains.

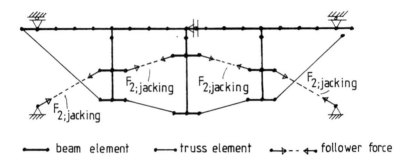

Figure 4: Model for active tensioning of cable 2

4 RESULTS

4.1 Overview of results

Figure 5 shows the hogging deformations of the girder during the various tensioning steps as calculated with three different models and measured in the test. Figure 6 shows the jacking forces during these tensioning steps. In model 1 the jacking forces in the passively tensioned cable are determined based on the assumption that the end of the cable (near the hydraulic jack) is completely fixed. In model 2 the jacking forces in the passively tensioned cable are kept constant. In model 3 it is assumed that cable 2 is not completely fixed during passive tensioning, so that the initial (unstrained) length increases. By trial and error it was found that by increasing the initial length with 1 mm in every even tensioning step, the jacking forces of cable 2 correspond very well to the jacking forces measured in the test.

After tensioning step 9 the models are no longer valid, because then the model results in a tensile force in the connection between the struts and the

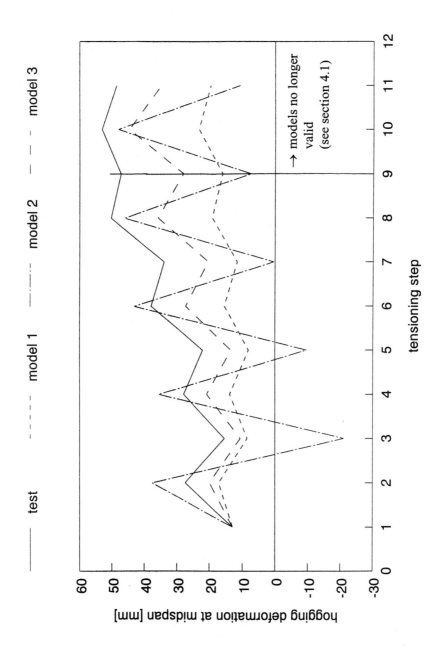

Figure 5: Hogging deformations of the girder during posttensioning

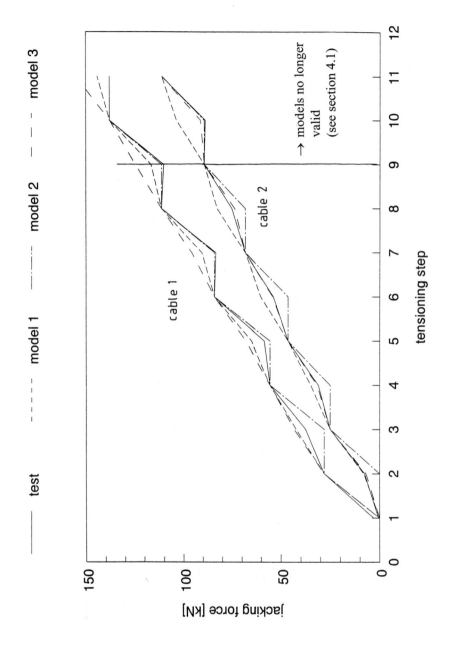

Figure 6: Jacking forces during posttensioning

beam. In the girder as tested in the laboratory this connection can carry only compression forces.

4.2 Discussion of results

When comparing model with tests results two questions need to be answered:
1 Does the model predict the cable forces correctly?
2 Applying cable forces as measured in the test, does the model predict correct deformations?

In the tensioning steps 2, 4, 6 and 8 the jacking forces in cable 2 calculated with model 3 correspond very well to those measured in the test. However, it can be seen that the hogging deformations predicted by this model are consistently smaller than those measured in the test. This is partly caused by the fact that it is assumed that in tensioning step 1 the model has the same hogging deformation as the test, while the jacking forces are not exactly identical. If we would assume a smaller initial imperfection in the model (which is included in the hogging deformation) the hogging deformations predicted by model 3 would correspond reasonably well with those measured in the test.

From the results of models 1 to 3 it can be seen that the deformations of the girder are quite sensitive to the magnitude of the jacking forces of the passively tensioned cable. An overestimation of these forces results in smaller variations of the hogging deformations in the various tensioning steps, because the forces in the passively tensioned cable resist the deformation resulting from active tensioning. Also, when the force in the passively tensioned cable is larger, it takes a smaller load increment (and thus smaller deformations) in the next tensioning step, to increase the tensioning force in this cable to the prescribed value.

It can be seen that model 1 overestimates the forces in the passively tensioned cable, while model 3 underestimates them. Model 2 shows that slip of cable 2 during passive tensioning reduces the forces in the passively tensioned cable. It seems likely that slip may have occurred during the test. It should be noted that it is difficult to predict the amount of slip which will occur during passive tensioning.

5 CONCLUSIONS

In principle it is possible to determine the forces arising during passive tensioning in the cable running without friction through the struts of the double-cable system by means of a finite element model. However, the thus determined forces are very sensitive to changes in the unstrained cable length, resulting from slip in the connection between cable and hydraulic jack. Since this slip is difficult to predict and hard to prevent totally, it is recommended that both cables of the girder are post-tensioned simultaneously so that passive tensioning need not be modelled. During posttensioning the chosen tensioning procedure has to be followed carefully, since small deviations in cable forces result in large differences in deformations.

6 ACKNOWLEDGEMENT

The research described in this paper has been carried out by R.G.W. Hofstee in a research project to obtain his master's degree.

7 REFERENCES

HOFSTEE, R.G.W - Ontwikkeling van een onderspannen beton constructie, BKO report 96.14., Eindhoven University of Technology, 1996.

Prof. P Croce
University of Pisa
ITALY

An Innovative Procedure for Non-Linear Analysis of Frames

Finite element analysis of non-linear frames depends on the mesh refinement, so that when large structures have to be studied, to obtain satisfactory results it is necessary to use a great number of elements. Consequences of the mesh refinement are an important increase of the number of d.o.f's of the structural model and of the computational time, also considering that the solution is equally obtained by step-by-step procedure.
If the paper an innovative procedure for non-linear elasto-plastic analysis of frames is presented.

The procedure is based on an analytical-numerical local solution of the non-linear elasto-plastic problem of each beam element, satisfying step-by-step the global constraints at the ends and evaluating the out of balance forces to restore the nodal equilibrium.
The local solution is obtained numerically using iterative finite element difference techniques, subdividing each structural element in an appropriate number of intervals, depending on the required precision.

The global balancing is simply obtained, in each step, by solving the linear problem, till the fixed convergence rate is achieved.
In this way the mesh refinement only influences the local solutions, so that the total solution time is practically independent on the element internal subdivision, while the element itself is modelled in the frame with an equivalent classical linear elastic element.

The results obtained on several benchmark examples, widely discussed in the paper, agree satisfactorily with those obtained using classical FEM codes, like COSMOS/M or ADINA, validating the procedure.

NONLINEAR FINITE ELEMENT ANALYSIS OF
DAMAGED CONCRETE STRUCTURES INCLUDING
THE EFFECTS OF CONCRETE REMOVAL AND CORROSION

G. Horrigmoe and A. Tørlen
NORUT Technology
P. O. Box 250, N-8501 Narvik, NORWAY

Phone: 76 96 53 50
Fax   :   76 96 53 51
E-mail: geir@tek.norut.no

During recent years, widespread problems of deterioration of reinforced concrete infrastructure have been experienced in many countries. The principal cause of this deterioration is corrosion of embedded reinforcement resulting from the diffusion of chlorides through the
concrete cover. Bridge structures are particularly affected due to the use of de-icing salts in the
winter season.

The majority of the repair work continues to be performed by removing the chloride contaminated concrete around the corroded reinforcement and replacing it by a suitable concrete or
mortar. For this method to be successful, all contaminated concrete must be removed. In many
cases, this may require the braking out of concrete over wide areas combined with loss of
composite action between steel reinforcement and concrete.

The load-carrying capacity of damaged concrete structures is a question of vital importance during the period when the repair is carried out as well as after the repair has been completed. In some cases the choice of a particular method of repair may depend on the residual strength of the structure. For instance, the use of hydro demolition to remove damaged concrete leads to reduced strength during repair, and the load-carrying capacity of the structure after repair has been completed needs to be predicted with a high degree of accuracy. All of this have exposed the need for more reliable methods for calculating the mechanical behaviour and ultimate strength of damaged and repaired concrete structures.

The present paper deals with the calculation of the load-carrying capacity of damaged concrete structures by means of nonlinear finite element analysis. The nonlinear crushing behaviour of concrete is modelled by the William-Warnke failure criterion. Cracking of concrete in tension is also accounted for using a fixed smeared crack approach. The material properties of the reinforcing steel bars are assumed to be that of an elastic-ideally plastic material. The present method also covers the effect of corrosion of embedded steel reinforcement. The cross sectional area of the reinforcing bars are reduced based on semi-empirical formula validated against field measurements.

Numerical simulations have been carried out for reinforced concrete beams and columns for which test data have been reported in the literature. The real load and deformation history of the structure is reproduced using the following sequence of loading, unloading and structural changes. First, the external load is gradually increased up to its full service value. At this level, corrosion of the reinforcing steel takes place. The variable component of the load is removed, damaged concrete is removed and replaced by a suitable mortar. Finally, the loading is increased until failure. The results obtained by the present approach compare well with the experimental data even for fairly complex loading situations. It is believed that nonlinear finite element simulations can be developed into an accurate and reliable tool for predicting the safety of damaged and repaired concrete structures.

A Methodology for Automatization of the Cable-Stayed -Design Process within the Idea of Multiobjective Programming

Dr F. A. N Neves
COPPE/UFRI, Brazil

The cable stayed bridge design comprises many stages and involves the knowledge of different subjects in a such way that the final structure conceived in the project has to satisfy a set of constraints and requisites, such as: security, economy, functionality, etc.

Traditionally, these tasks have been executed by means of the design spiral technique, in what the designs have been improved, step-by-step, on the experience-based insight of the designer.

This work presents a different methodology from that usually adopted in designs, taking into account a technique that searches a compromise solution between many objectives furnished by the designer and a set of constraints, based on a multiobjective programming philosophy.

Firstly, it is described what comprises the cable-stayed bridge design task and the technique usually adopted to perform an analysis of it.

Next, the inter-relation between the elements that compose a cable-stayed bridge - cable systems, towers, grids, etc. - and how they interfere in the final structural design is shown.

The elements required for the design, such as: height of the grid, space between the cables, height of the towers, etc., and the way they are incorporated in the proposed methodology are also shown.

Finally, the elements taken into consideration in the methodology are set within a wider concept, according to the philosophy of multiobjective programming called goal programming. A complete analysis of a cable-stayed bridge model taking into account this methodology is presented in this work.

SHAPE OPTIMIZATION in BIOLOGICAL STRUCTURES

Jim Wood

University of Paisley

Abstract

Several exemplars of shape optimization and weight efficiency in biological structures are examined. It is postulated that natural shape and form presents challenges to the present generation of shape optimization tools available to engineers and scientists and that the examples presented could form the basis of useful benchmarks. The structures and shapes examined include the spider's web, the comb cell of the honeybee and the Baud curve. The difficulty of establishing objective functions and constraints for possible optimization scenarios is discussed.

1. Introduction

Michael French[1], stated that *living organisms are examples of design strictly for function, the product of blind evolutionary forces rather than conscious thought, yet far excelling the products of engineering.* How therefore, can the structure of such organisms possibly represent optimum solutions, given the lack of conscious thought in their development ? One answer is of course provided by Darwin's[2] theory of natural selection. The cumulative effects of hereditary variation, coupled with a natural selection process, it is argued, inevitably leads to organisms and structures that are fitter for the purpose of survival. The process of cumulative selection is effectively explained by Dawkins in his popular classic The Blind Watchmaker[3]. Whether, in the scheme of things, biological structures are considered simply *adequate* or *optimum* is open to debate and depends to some extent on how the problem is posed. The evolution of a species by natural selection is inherently linked to the competition that it experiences. The extent of evolution (and hence development towards an optimum) is therefore inextricably linked to such pressures. It must also be borne in mind, that biological structures evolve ... the option of a revolutionary change in approach is generally

1299

not available. That being the case, then at most, natural structures probably represent *local optima* or *best* solutions. Optimization scenarios can certainly be postulated, but whether the goals and constraints proposed are the criteria that influence the development of the organism may be difficult to ascertain.

Although examples in nature are common where the *aesthetic* qualities of a particular organism clearly affects success (mainly with respect to attractiveness to members of the same species for mating purposes or to a different organism completely as in the case of pollination), it is probably true to say that the form of a great many natural structures arrives, more often than not, as a consequence of economy rather than aesthetics. As Newton noted in his immortal work Philosophiae Naturalis Principia Mathematica[4] ... *Nature does nothing in vain, and more is vain when less will serve.* There are clearly energy costs associated with growing bigger, stronger and with moving about. In the latter case particularly, it is obvious that mass will play an important role in the success of such organisms. It is therefore perhaps not surprising that a diverse array of fascinating examples of weight efficient structures exists in nature.

In many instances it is not at all obvious that an optimum or even a best solution has been achieved. However, the fact that any such structure or organism does not appear to be optimum, is probably due to our lack of understanding of the particular optimization objective function(s) and constraints. This same observation was originally made by D'Arcy Thompson in his treatise On Growth and Form[5] .... *We have dealt with problems of maxima and minima in many simple configurations, where form alone seemed to be in question; and when we meet with the same principle again wherever work has to be done and mechanism is at hand to do it. That this mechanism is the best possible under all the circumstances of the case, that its work is done with a maximum of efficiency and at a minimum of cost, may not always lie within our range of quantitative demonstration, but to believe it to be so is part of our common faith in the perfection of Nature's handiwork.*

Optimization of shape and form in nature embraces far more than the simple goal of minimum mass so often encountered in engineering structures. However, as has already been stated, for many biological structures it will be metabolically advantageous to reduce weight and all of the examples contained herein are presented in this context. The natural constraints arising from material and structural performance may also be different to those normally encountered by the engineer. The requirements for reproduction and growth impose significant constraints on the variables available for a natural solution. It may be argued that the ultimate unconscious goal for all biological structures is to **survive and reproduce** ... not an option available in the current generation of engineering optimization software! In biology, the concepts of survival and reproduction are referred to as *fitness* and Alexander[6] discussed the difficulty of using this as the objective function in a range of optimization studies in animals.

Although many biological structures have the ability to repair damage, it is also apparent that catastrophic failures do occur. The fact that trees blow down in the wind and animals break their bones, does not necessarily mean that such structures are not optimum. The problem lies in our understanding of the objective function and constraints. Obviously biological structures normally have reserves of strength against the various environmental loadings and failure mechanisms that they may be subjected to. However, such reserves of strength are not predetermined by some design code, as is generally the case in engineering, but are determined by natural selection for each individual case[7][8]. It is argued that nature in effect *chooses* not to design against such eventualities. Being *stronger and stiffer* will incur an energy or metabolic cost and if such cost impairs an organism's performance in other areas e.g. to gain food and to reproduce, then this may affect its overall chances of survival in competition with other species ... the solution is always a compromise and may not be obvious. Thus in nature's complex survival algorithm, the frequency of structural failure may be tolerated to a higher degree than that acceptable to the engineer. High incidences of failure in a particular biological structure may also be evidence of a change in environment and a species' inability to adapt quickly enough.

In its simplest form, a shape optimization problem will have the following characteristics :

> **A goal or objective ..... e.g. minimum mass.**
> **Geometry and material variables ..... with limits on range.**
> **Constraints ..... e.g. on stress, strain, deflection etc.**
> **Loads ..... e.g. self weight, hydrostatic pressure etc.**

Optimisation scenarios involving biological structures may often bring the additional complication of constraints and variables that may vary with time. In addition, the natural *forces* involved may vary in a random and non-linear manner.

It is a typical characteristic of the software tools available to the engineer, for this type of problem, that the solution is iterative and time-consuming ... invariably requiring intervention by the user and never as straightforward as the software vendor's demonstration! An increase in the number of variables and constraints often leads to increased difficulty in finding a solution. In addition, at the end of the day, the solution found may not in fact be the optimum one. However it is usually quite straightforward for the user to demonstrate quantifiably that the solution is a better one and many may be content at that.

The existing generation of shape optimization tools can be placed in one of two categories :
(a)   Those that do not use a *geometry master* and attempt to achieve the solution by direct manipulation of the finite element mesh. Examples of this type, are

the systems and procedures based on biological growth developed by Umetani & Hirai[9] and Mattheck[10].

(b)   Those that use a *geometry master* and attempt to obtain an optimum solution by manipulating the geometric entities. At each iteration, automatic meshing is used to re-mesh the current geometrical shape. An example of this type is Parametric Technology's *Mechanica*[11] system.

Apart from any disadvantages associated with particular systems, the former category suffers from the generic drawback of severing the link between the finite element model and any geometry model held in a CAD system. On the other hand, the latter approach may also suffer, to some extent, from constraints imposed by the maintenance of a link to a CAD system. For such a system to be truly effective, the user should have the ability of specifying, as an option, that certain geometrical entities be allowed to change their basic form. For example, it should be possible to specify that a conic in the CAD system be allowed to develop into a free-form curve with control over its shape. A unified representation of geometry, such as NURBS, would clearly be an advantage if a robust algorithm could be found to facilitate the basic manipulation of the selected geometric entities within the optimization process. Tools to allow the user to identify and constrain conic entities before optimisation and to identify and define, with a specified tolerance, conic entities after optimization would also prove useful. A system that recognised manufacturing constraints and the existence of *preferred sizes* would also find practical application. Another desirable facility in any system offering shape optimization would be the ability to *grow holes where none grew before*. It has long been recognized that an algorithm involving setting the density of each individual element as a design variable and then *killing* those elements with negligible mass during the optimization procedure, has the potential of creating holes and cavities. Alternatively the element moduli can be adjusted in a number of ways, as outlined by Mattheck[12] in his discussion of a *Soft Kill Option*. The *Optistruct* system from Altair[13] incorporates a development of the former approach using the concept of *homogenisation* - a microstructural modelling method that alters the size and orientation of voids, in an assumed porous media, to distribute the material density.

Some of the deficiencies in the current generation of commercial systems would be apparent in the study of biological structures.

2.   The Spider's Web

The spider's web shown in figure1 illustrates several different examples of optimization / minimisation. D'Arcy Thompson in his fascinating discourse on soap bubbles and raindrops[5] also notes that the beads of adhesive or dew on such webs are examples of minimum surfaces.

In the last century, mathematicians gave considerable attention to such surfaces and D'Arcy states as a fundamental *law of capillarity* that a liquid film in equilibrium assumes a form which gives it a minimal area under the conditions to which it is subjected. On this basis he describes how, under surface tension effects, the liquid coating on the web first of all changes to an unduloid and then finally to the string of spherical beads that are so obvious on dew covered orb webs. The orb web is essentially a planar tension structure which consists of structural elements (guy, frame, chords and radii), composed of dry thread drawn from the spider's ampullate glands.

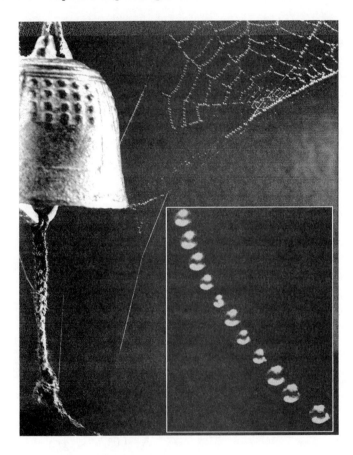

Figure 1 : Typical orb web of a spider

Each structural member is also comprised of differing numbers of strands, with the *guys being thickest.* The spiral thread is coated with adhesive material from the spider's aggregate glands. The actual construction of the web is quite a fascinating process and it has been found that it is built without either visual feedback or reference to gravity[14][15]. It is interesting to observe that our knowledge of the spider's complex behaviour and the sensory and motor apparatus

1303

which underlies the web building behaviour seems to exceed our knowledge of the somewhat similar activities of the honey bee in building the comb structure discussed in a later example. This is perhaps a reflection of the fact that the bee activity occurs in the darkness of the hive, within a cluster of active bees and is therefore more difficult to observe.

It is not untypical for a web to be constructed in under half an hour, use 20m of 1-3μm diameter thread and for an entire web to weigh between 0.1 and 0.5mg. The spider itself may weigh in excess of 500mg, although 100-150mg is typical for *Araneus*.

The structure of the orb web itself has also been examined in terms of its weight efficiency. Wainwright et al[7] present and discuss the force system in an abstract web, shown in figure2, when one of the guys is subjected to a tension of 2 units. The authors point out that this produces a remarkably uniformly stressed structure, given that the number of threads in each radius, frame and guy element is 2, 10 and 20 respectively. There is some natural variability in these figures[16], but the observation would appear to be generally valid nonetheless.

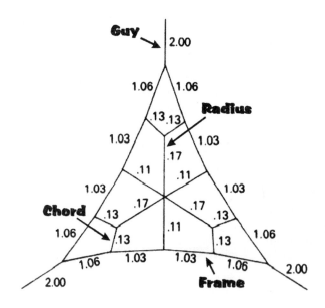

Figure 2 :   Schematic representation of web showing forces
(Reproduced from Wainwright et al [7])

Shown in figure 3 is a *Mechanica* finite element model of a symmetrical section of the web. The structure was assumed to be composed of pin-jointed

1304

tension bar elements. A linear elastic, small displacement, optimisation problem was set up in which the cross sectional area of each element was a variable. The goal was set as minimum mass, with the constraint that the tensile stress in each element should be the same.

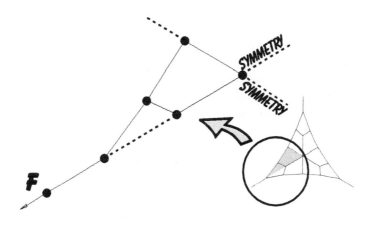

Figure 3 : Finite element model of web section

The results from the study are shown in figure 4 and it may be observed that there is quite good agreement with Wainwright et al ... given the accuracy of the input web data. Other interpretations of available data[8][16] show approximately a factor of 2 variation in stress throughout such webs .... with the radial elements showing the greatest variation. The question arises whether this

WEB SECTION	Nominal area from finite element study	Nominal area from Wainwright et al
Guy	20	20
Frame	10.8	10
Radius	1	2

Figure 4 : Comparison of web section cross-sectional areas

is due to natural and/or experimental variation, or whether it is due to evolved differences in factors of safety for the various web elements.

The silk used in the web construction is no doubt a *best* material, given the constraints on the spider's method of production (statement of faith). Its method and rate of production is also remarkable and has been widely studied[14][15]. Denny[16] reported that the framework threads of *Araneus seracatus* have on average, a true breaking strength of around 1GPa, a tangent modulus of 4GPa and a corresponding breaking strain of approximately 0.25. The sticky spiral threads, that have to deal with the struggling prey on the other hand, have the same breaking strength, a tangent modulus of 0.6GPa, and can withstand strains up to 2. The framework silk was found to be strain rate dependent, whereas the viscid spiral silk was found to have properties that were insensitive to strain rate. These significantly different properties are produced from materials that are chemically very similar.

When the engineering tensile strength of the framework silk is evaluated per unit density, it is found that the silk (σ_u=0.8GPa & SG=1.26) is approximately 12 times stronger than typical carbon steel (BS4360 Gd43 : σ_u=0.43GPa & SG=7.8) and 4 times stronger than aerospace standard high strength aluminium alloy (BS2L93 Gd 2014A : σ_u=0.41GPa & SG=2.7).

Also noteworthy is the fact that many such spiders tear down and rebuild their web daily .... and to reduce the metabolic cost, they eat the old web!

Whether this is an optimum structure or not depends on how the problem is posed. It is certainly approaching a least volume structure as defined by Maxwell[7], in that all members are in pure tension and are equally stressed near their breaking stress. However, there are spiders that manage to catch prey and survive as a species with a greater economy of silk ... in fact there are species of spider that do not produce a web at all! As mentioned earlier, identifying the optimum solution and indeed the optimization problem itself, is not always easy in nature.

3.   The Baud Curve

At the end of the last century and the beginning of this century, the curves and shapes that appear in nature held a particular fascination for mathematicians, biologists and natural historians of the time[5][17]. Treatises on logarithmic spirals and catenary curves abound! A particular curve that is extremely common in nature was reported by Baud in 1934[18] and was also discussed by Peterson[19] in relation to reducing stress concentration effects at changes of section. The curve is described as a form of fillet based upon the contour produced by an ideal frictionless fluid flowing by gravity from a circular opening in the bottom of a tank, as shown in figure 5.

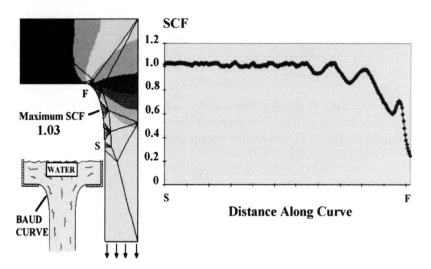

Figure 5 :   Stresses for a Baud fillet

To illustrate the weight efficiency of the Baud curve, a stepped shaft subjected to an axial tension was examined. The contours of maximum principal stress are shown in figure6 for the case of a reference circular fillet.

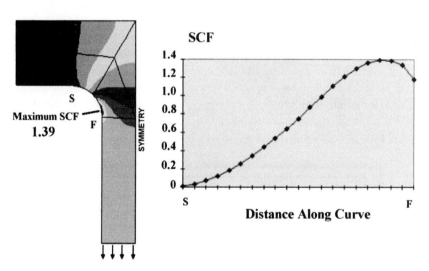

Figure 6 :   Stresses for a reference circular fillet

Also shown in this figure, is the variation of stress around the fillet. Similar results for a Baud fillet, with the **same swept volume**, are shown in figure5. It may be observed that in this case, use of the Baud curve has produced a *26% reduction* in the magnitude of the stress concentration factor.

1307

Circular fillets in nature are a rare occurrence and the reasons are perhaps apparent. Topping[20] used the growth reforming technique to examine such a profile in a tree trunk and Mattheck[21] has studied the shape extensively in relation to trees and bones.

Such a relatively simple problem provides a significant challenge to shape optimization software based on a geometry master and it is suggested that this could form the basis of a worthwhile benchmark.

4. The Comb of the Honey Bee

The comb of the honeybee has fascinated man for centuries and D'Arcy Thompson[5] cites early references by Virgil, Ausonius, Pliny and Pappus and devoted some 19 pages of his seminal work *On Growth and Form* to this particular topic. Charles Darwin[2] also studied the subject extensively and 9 pages of *The Origin of Species* provides a record of his observations. Darwin notes that *... He must be a dull man who can examine the exquisite structure of a comb, so beautifully adapted to its end, without enthusiastic admiration.*

The comb of the honeybee has probably received greater interest from scientists and mathematicians over the centuries than any other natural structure. The references are too numerous to mention and without doubt stems from man's historical associations with the bee as a source of honey. Of particular historical note is D'Arcy's reference to the studies of Pappus of Alexandria around AD3 .... his conclusions regarding the hexagonal shape of the cell arising from a consideration of economy of wax, D'Arcy notes, is probably the earliest record of such a minimisation principle and predates the principle of least action that guided Leibniz, Maupertuis and other 18th century physicists, mathematicians and philosophers such as Bernoulli, Euler, Lagrange and Koenig.

SHAPE	●	⬡	■	▲
Circumference	3.545 $\sqrt{A}$	3.795 $\sqrt{A}$	4.000 $\sqrt{A}$	4.559 $\sqrt{A}$
Circumference / Packing Efficiency	3.911 $\sqrt{A}$	3.795 $\sqrt{A}$	4.000 $\sqrt{A}$	4.559 $\sqrt{A}$

For the maximum storage area .... the shortest circumference means less wax, less work and less energy !

Figure 7 :   *The optimum cell shape for a honeycomb*

That the hexagonal shaped cell is the optimum in terms of honey storage for the least quantity of wax used, is simply illustrated in figure 7. For a given cross-sectional area A, the circumference of the hexagon is the smallest, after due allowance has been made for the packing efficiency of the circle.

The cell of the comb of the honeybee was also the basis of a celebrated optimization problem for 18th century mathematicians. The problem in this case concerned the shape of the bottom of the cell. The problem was effectively stated by the French naturalist Rene Antoine Ferchault R'eaumur and became known as *The Problem of the Bees* .... *A cell of regular hexagonal cross-section is closed by three equal and equally inclined rhombs : calculate the smaller angle of the rhombs when the total surface area of the cell is the least possible.* The first widely published value had previously been attributed to an astronomer working in Paris in the 1730's named Maraldi and the angle of 70 degrees 32 minutes became known as the *Maraldi Angle*. Maraldi also noted that this was *exactly* the angle built by the bees. Subsequent work has rightly noted that the use of the term *exact* was not appropriate, given the variability of the natural structure. Furthermore, the precise angle was later shown to be that of a rhombic dodecahedron and given by $\cos^{-1}(1/3)$. Koenig, the Swiss mathematician, solved the problem using calculus in 1739 and then asserted that *the bees had solved a problem beyond the reach of the old geometry and requiring the methods of Newton and Leibniz.* However Colin MacLaurin, the Professor of Mathematics at Edinburgh University, demonstrated[22] in 1743 that a geometry based solution was possible and concluded his presentation to the Royal Society with the observation that *what is most beautiful and regular, is also found to be most useful and excellent.*

These observations led to equally fascinating studies on insect intelligence and Bernard le Bovyer de Fontenelle, the French philosopher, is credited with the judgement in which the bees were *denied intelligence but were nevertheless found to be blindly using the highest mathematics by divine guidance and command !* The final compliment paid to the comb of the honeybee, is made by Charles Darwin when he wrote[2] ... *beyond this stage of perfection in architecture, natural selection could not lead; for the comb of the hive-bee, as far as we can see, is absolutely perfect in economizing labour and wax.*

D'Arcy Thompson stated in the introduction to his 1942 edition of **On Growth and Form** that ... *It is no wonder if new methods, new laws, new words, new modes of thought are needed when we make bold to contemplate a universe within which all Newton's is but a speck.* It is with some pride therefore that a solution to *the problem of the bees*, using one of the latest computer based shape optimization tools, is presented in figure 8 .... to D'Arcy Thompson's generation, a new method using new words and new modes of thought. A shape optimization benchmark with a finer pedigree surely could not be found !

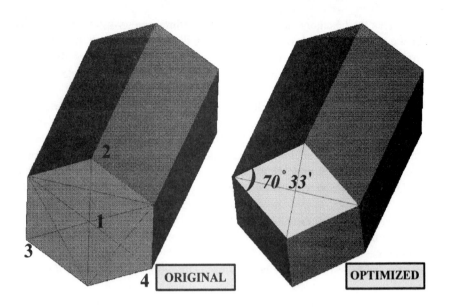

Figure 8 : Mechanica minimum mass solution for honeybee cell

The angle shown in figure 8 was obtained using Parametric Technology's *Mechanica* system with a goal of minimum mass and points 1 - 4 set as translation variables along the axis of the cell. It is also interesting to observe that this also represents a minimum stress solution for a cell subjected to hydrostatic pressure loading, as illustrated in figure 9.

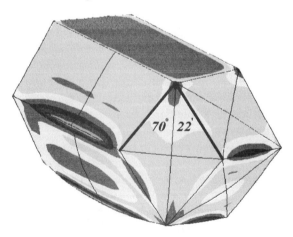

Figure 9 : Mechanica minimum stress solution for honeybee cell

It may be noted that any differences between the above angles are within the convergence tolerance set in the analysis system.

In 1781 Glaisher[23] discussed the work of L'Huillier, who had extended the problem to consider the *minimum minimorum* cell and examined the proportion of the depth of the cell to its width, for a given volume and minimum wax. Similar variations to the problem have also been reported more recently[24][25]. However, it is true to say that such variations, while they may be of interest from a mathematical viewpoint, neglect the fact that in a feral nest, the cells are also used to rear brood as well as to store honey and pollen. The cell width and depth is therefore related to the size of the bees themselves and whether the cell is to be used to rear workers or drones. The *problem of the bees* therefore starts from the premise that the cell width and depth are fixed and that the bottom of the cell is to be closed by three equal rhombs.

5.   The Pelvic Bone of a Sloth

The final example chosen to illustrate weight efficiency and shape optimization in nature is one which would pose a particular challenge to the present generation of software available to the engineer. The optimum shape of holes has been extensively studied in the past[26], but the remarkable structure shown in figure 10 would certainly provide a challenge to today's technology .... were we able to specifically define the problem !

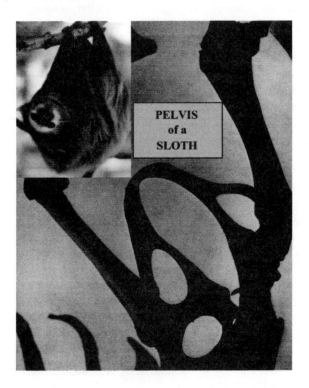

Figure 10 : Pelvic bone of a sloth (Modified from Feininger[27])

While this may indeed be beyond our range of quantitative demonstration, we can surely marvel at Nature's handiwork in the effective use of holes to produce a weight efficient structure.

The observant reader will already appreciate that the above examples represent but a small fraction of the fascinating array of *optimum* structures readily available in nature. While some optimization problems may be too difficult to formulate, others would no doubt make a worthwhile and interesting addition to traditional shape optimization benchmarks. In addition, it is also probable that most of the structural finite elements, from one dimensional through to three dimensional, could be accommodated.

It is the widespread occurrence of free-form curves in such natural structures that presents the difficulty to shape optimization systems developed for engineering purposes. However, in reality, representing the true behaviour of many of these structures and their constituent materials would also present severe challenges, in that non-linear behaviour is common. This need not present an insurmountable problem in the present context and it is likely that many interesting and varied problems could be posed through the simplifications of assuming small displacement behaviour and a linear elastic, homogeneous and non time-dependent material. The general optimization problems associated with biological systems, which could often theoretically be classed as *constrained, stochastic, non-linear, multivariable and dynamic*, are also invariably simplified and posed as special cases.

> The author would like to dedicate this paper to the Scottish scholar-naturalist D'Arcy Wentworth Thompson 1860-1948 .... *Hic erat vir* !

References

(1)   FRENCH, M. - Invention and Evolution, Design in Nature and Engineering; Cambridge University Press, 2nd Ed., 1994.

(2)   DARWIN, C. - The Origin of Species by Means of Natural Selection or the Preservation of Favoured Races in The Struggle for Life; Oldham Press, 6th Ed., 1872.

(3)   DAWKINS, R. - The Blind Watchmaker; Longman, 1986.

(4)   HUTCHINS, R.M. - Great Books of the Western World, Volume 34, Isaac Newton's Mathematical Principles of Natural Philosophy, p270, Encyclopaedia Britannica Inc., 1952.

(5)   THOMPSON, D.W. - On Growth and Form; Cambridge University Press, 2nd Ed., 1942.

(6)   ALEXANDER, R.Mc. - Optima for Animals; Edward Arnold, 1982.

(7)   WAINWRIGHT, S.A. et al - Mechanical Design in Organisms; Edward Arnold, 1st Ed., 1976.

(8) ALEXANDER, R.Mc. - Factors of Safety in the Structure of Animals; Sci. Prog., Oxf., 67, pp109-130, 1981.

(9) UMETANI, Y. et al - An Adaptive Shape Optimization Method for Structural Material Using the Growth Reforming Procedure; Proceedings of the Joint JSME-ASME Applied Mechanics Western Conference, pp359-365, 1975.

(10) MATTHECK, C. et al - A New Method of Structural Shape Optimization based on Biological Growth; Int J Fatigue, May 1990.

(11) PTC Corp. - Mechanica Applied Structure; Massachusetts, USA, 1997.

(12) MATTHECK, C. - Design in Nature, Learning from Trees; Springer-Verlag, 1998.

(13) ALTAIR COMPUTING Inc. - Optistruct; Michigan, USA, 1997.

(14) CHINERY, M. - Spiders; Whittet Books, 1st Ed., 1993.

(15) WITT, P.N. et al - A Spider's Web, Problems in Regulatory Biology; Springer-Verlag, 1956.

(16) DENNY, M. - The Physical Properties of Spider's Silk and Their Role in The Design of Orb Webs; J. Exp. Biol., 65, pp483-506, 1976.

(17) COOK, T.A. - The Curves of Life; Dover Publications, 1st Unabridged republication, 1979.

(18) BAUD, R.V. - Beitrage zur Kenntnis der Spannungsverteilung in Prismatischen und Keilformigen Konstruktionselementen mit Querschnittsubergangen; Eidgenoss, Materialpruf, Ber 83, Zurich, 1934.

(19) PETERSON, R.E. - Stress Concentration Design Factors; Wiley, 1953.

(20) TOPPING, B.H.V. - Fully Stressed Design of Natural and Engineering Structures; Natuerliche Konstructionen, Proc. Sonderforscungsbereiches 230, Stuttgart, 2, pp311-318, 1989.

(21) MATTHECK, C. - Engineering Components Grow Like Trees; Institut fur Material-und Festkorperforschung, Report KfK 4648, Kernforschungszentrum, Karlsruhe GMBH, Karlsruhe, November 1989.

(22) MacLAURIN, C. - Of the Bases of the Cells wherein the Bees Deposit their Honey; Phil Trans, p42, pp561-571, 1743.

(23) GLAISHER, J.W.L. - On the Form of the Cells of Bees; Phil Mag, v4, pp103-22, 1873.

(24) TOTH, L.F. - What The Bees Know and What They Do Not Know; Bull Amer Math Soc, 70 (1064), pp467-481, 1965.

(25) HILDEBRANDT, S. et al - The Parsimonious Universe; Springer-Verlag, 1996.

(26) HANG, E.J. et al - A Variational Method for Shape Optimum Design of Elastic Structures; Ch5 in New Directions in Optimum Structural Design by E. Atrek et al (Eds), John Wiley, 1984.

(27) FEININGER, A. - Anatomy of Nature; Thomas Yoseloff Ltd, 1956.

Hull Structural Analysis of *Titanic* -
Aft Expansion Joint Stress Analysis

David M. Wood, James L. Belshan, Christopher M. Potter
Gibbs & Cox Inc., USA

INTRODUCTION AND BACKGROUND

A significant amount of effort has been spent in developing structural failure scenarios for *Titanic* that have been based entirely on conflicting survivor testimony, limited visual evidence from the wreck site, and conjecture. What is clear beyond dispute is that the *Titanic* collided with ice, and obtained significant structural damage along the starboard side shell which was substantial enough to cause flooding throughout her watertight subdivisions. This flooding was so extensive that the ship was lost over the course of approximately 2-1/2 hours. With the discovery of the wreck, the myths of the famous 300-ft gash, and of the ship sinking intact were proven to be groundless. The *Titanic* rests at a depth of about 2-1/2 miles primarily in two sections (the bow section and the stern section) which are upright and face opposite directions.

Not until recently have investigations into the demise of the *Titanic* employed modern analytical methods to address many questions that still remain. These methods include metallurgical testing and structural investigations to determine whether the *Titanic* sank intact versus breaking apart at the surface. The initial modeling effort for the Discovery Channel documentary, *Titanic, Anatomy of a Disaster*, focused on determination of the location and magnitude of high stress areas that developed in the hull while she remained on the surface. Also, since much is known about how flooding progressed in the forward region of the ship [References (1) and (2)], stress changes from the undamaged condition to just prior to sinking could be visualized and quantified. Three distinct snapshots of the ship's condition were investigated:

(1) Undamaged Ship - Stresses developed within the hull are due to the interaction between the ship's weight and water buoyancy forces. Under normal conditions, stresses would be low and within the design stress tolerances.

(2) Intermediate Flooded Condition - A buildup of stress in the ship's midsection is clearly evident as flooding progresses in the forward subdivision of the ship. Loads correspond to Condition B3 flooding as described in Reference (2). Here, the forwardmost three subdivisions of the ship are fully flooded [see Figure 1].

(3) Prior to Sinking Condition - A significant increase in stress is seen as the forward part of the ship continues to take on water and the stern is lifted into the air. The stresses in the ship's midsection continue to worsen. Loads correspond to Condition B6 flooding as described in Reference (2). Here, the forwardmost six subdivisions of the ship are fully flooded [see Figure 2].

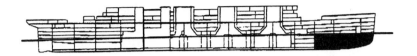

Figure 1. Three-Compartment Flooding Load Condition

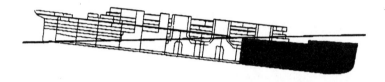

Figure 2. Six-Compartment Flooding Load Condition

The initial investigation determined that stress levels in the midsection of the ship were up to at least the yield strength or capacity of the steel just before she went down. When considered alone, these stresses do not indisputably imply catastrophic failure. Additional work, which focuses on areas where initial failures would have been likely to occur in the hull, was needed to address whether it was possible for the ship to suffer catastrophic failure at the surface and break in two versus sinking intact.

OBJECTIVE AND SCOPE OF INVESTIGATION

The purpose of these analyses is to provide greater insight and additional technical basis for the current theory that the *Titanic* broke apart at the surface prior to sinking. The intent is not to put forth a theory for the sinking or hull breakup, but to determine if loads developed in the hull girder during the Reference (2) flooding scenarios would be at a sufficient level to initiate critical modes of hull girder failure such as structural plasticity and cracking. Initial structural analyses were previously conducted to determine hull girder bending stresses for a full-ship representation of the *Titanic*. These intitial analyses determined that global hull girder stresses in the midships area were approximately at the yield strength of the material just prior to sinking. Additionally, stress concentrations in way of structural discontinuities were noted, yet could not be accurately characterized due to the coarse element discretization and scope of the initial finite element analysis. To more accurately revisit stresses in the midships region and investigate regions where failure was likely to occur, the initial finite element model (FEM) was refined in way of the aft expansion joint and further analyses were conducted.

Expansion joints [see Figure 3] are utilized in ship design to decouple the lighter deckhouse structures from the rest of the hull girder to ensure that this structure is not effective in longitudinal hull girder bending. The advantages of deckhouse expansion joints include:

- Lighter superstructure scantlings
- Greater flexibility in deckhouse arrangements – longer superstructures are possible (advantageous for passenger ship design)

However, expansion joint toe design details are critical to the overall success of the installation. A deficient design can quickly lead to fatigue cracks in way of the highly stressed toe region which can propagate down into the primary hull girder plate. With consideration given to the quality of steel, design details, and magnitude of the load, such a failure could initiate uncontrolled catastrophic crack propagation which could jeopardize the ship. Given such circumstances and the proximity of the aft expansion joint to the highly stressed midship region, it is evident why this design detail was investigated further.

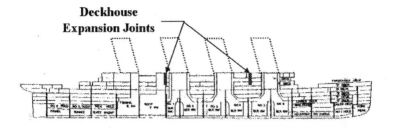

Deckhouse Expansion Joints

Figure 3. *RMS Titanic* – Location of Deckhouse Expansion Joints

THE FINITE ELEMENT MODEL

The initial full-ship FEM of the *Titanic* was developed in accordance with available structural data [References (1) - (4)] and drawings to characterize hull girder bending of *Titanic* during the sinking process. The plate and beam structure of the ship was modeled in the FEM exclusively with plate elements. The stiffness contribution from associated shell, deck, and bulkhead stiffeners was modeled through the use of an additional bending stiffness in the plate elements, and was determined by stiffener scantlings and nominal spacing. Mass distribution for non-structural and unmodeled structural weights was determined from the estimated distribution of weights and loads at the moment of rupture [Reference (5)]. This mass was distributed across each main transverse bulkhead, which is considered a reasonable assumption for global weight distribution over the hull girder model. The plate element discretization in the initial FEM yields typical deck plate elements of approximately 15 feet in length and 11 feet in width, and typical hull plate elements of 15 feet in length and four feet in height.

For this study, modifications to the initial FEM include refinement of the element mesh in way of the aft expansion joint to more accurately predict stresses in way of this structural discontinuity. Quadrilateral plate elements were used in way of the aft expansion joint wherever possible, with triangular element transition to the nominal mesh of the initial FEM to provide the most accurate stress results. Additionally, plating thicknesses above "C" Deck, deck heights in way of the forecastle and poop deck, and supporting structure in the deckhouse were modified from the structure in the initial analyses to more accurately represent the *Titanic* structure. A refined FEM that incorporated these changes was produced. Figure 5 shows the level of detail of the aft expansion joint, and its transition to the nominal mesh of the hull structure.

Mechanical properties for *Titanic* steel, as determined by Charpy V-notch and tensile tests conducted at the University of Missouri at Rolla, Missouri, Homer Laboratories, and the National Institute of Standards and Technology (NIST) in Gaithersburg, Maryland [Reference (6)], were used to analyze the FEM. The chemical analysis of the hull structural steel revealed that the hull steel is roughly equivalent to modern AISI 1018 mild steel [Reference (6)]. These tests determined the yield strength and ultimate tensile strength of *Titanic* steel to be approximately 38 ksi and 62.5 ksi, respectively. Additionally, the Charpy tests revealed that fracture of *Titanic* steel was almost entirely brittle, at ice-brine temperatures.

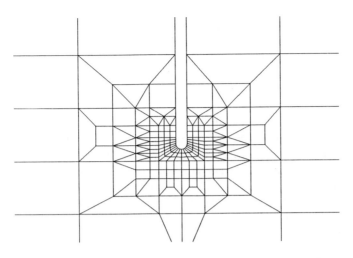

Figure 5. FEM Representation of *Titanic* Aft Expansion Joint

A primary assumption in development of the FEM relates to welded construction, as opposed to the actual riveted structures of *Titanic*. This simplification, while beneficial to the finite element analysis process, does not address the fact that rivet failures may have contributed to the sinking of the ship.

Additionally, the riveted construction details may have proved beneficial or deleterious to rapid crack propagation following the initial cracking in the aft expansion joint or other high stress areas. An optimistic approach would assume that cracks in the riveted plates propagate to the plate edges and stop. Another possibility would be where cracks propagated to rivet holes where they continued to follow the riveted seams until catastrophic hull collapse occurred. Cold punching of *Titanic* rivet holes [Reference (6)] resulted in microcracks in way of the holes, thereby exacerbating the likelihood of crack migration along riveted seams.

RESULTS

The *Titanic* FEM was analyzed for multiple loading conditions, as defined per Reference (2). Loading conditions were developed based on a still water condition, intermediate flooding condition and the flooded condition just prior to sinking. Corresponding forces were applied along the length of the ship, to simulate these loading conditions.

Results of the analysis for the case just prior to sinking show that the maximum Von Mises stress in the hull plating is approximately 61 ksi. This peak stress occurs at the aft expansion joint and is 61% above the minimum yield stress of *Titanic* steel as determined by the Reference (6) analysis. This peak stress is also 98% of the ultimate tensile strength of 62.5 ksi as determined by Reference (6). A stress contour plot of the aft expansion joint for the loading condition just prior to sinking is shown in Figure 6.

Figure 6 illustrates that the peak stress in way of the aft expansion joint for the loading condition just prior to sinking occurs where the joint terminates just above "B" deck. However, the stress contours also show stresses at or above the material yield stress in the regions immediately forward, aft and below the expansion joint, indicating that permanent set would be expected to occur in these regions.

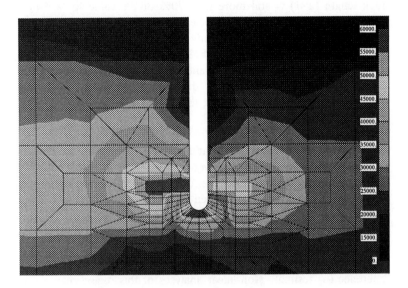

**Figure 6. Aft Expansion Joint Stress: Just-Prior To
Sinking Condition**

The loading condition just prior to sinking was also used to
predict crack propagation for a 4-inch initial crack in way of the aft
expansion joint using the principles of linear elastic fracture mechanics.
In the absence of a value for the plane strain fracture toughness (K_{Ic}) of
Titanic steel, the value for modern A36 steel was used for comparison.
The critical stress intensity factor (K_c) for modern steel is
approximately 101 ksi-in^(1/2), and is directly proportional to the
stress in way of the crack tip. This value represents the threshold of
stress in way of an existing crack where rapid propagation would be
expected to occur. For this initial 4-inch crack, subjected to the
loading condition just prior to sinking, the stress intensity factor was
determined by the finite element program to be approximately 198 ksi-
in^(1/2), well above the critical value for modern steels. In fact, using
the equation,

$$C\sigma\sqrt{\pi a} = K$$

1323

it can be seen that the stress intensity factor for a 2-inch instantaneous crack length (a=1) is still more than 70% of the value for a 4-inch initial crack (a=1/2). Consequently, even for a 2-inch initial crack, the loading condition just prior to sinking yields a stress intensity factor of approximately 140 ksi-in^(1/2), 138% of the critical value for modern A36 steel. It is anticipated that plane strain fracture toughness of the actual steel from *Titanic* will be determined during the additional materials testing planned following the 1998 expedition to the wreck site.

It is important to note that linear elastic fracture mechanics assumes that only small scale yielding occurs in way of the crack tip. If the plastic zone surrounding the crack tip grows too large, as compared to the crack length, then the basic assumptions of linear elastic fracture mechanics are violated and error is introduced. This violation of basic fracture mechanics theory may apply to this analysis of *Titanic* expansion joint cracking since the magnitude of free field stress is well above the material yield stress, as determined by the Reference (6) testing. Non-linear analyses of this region may reveal more accurate post-yield behavior for this particular loading which corresponds to the Condition B6 flooding [Reference (2)].

CONCLUSIONS

The aft expansion joint of the *Titanic*, under the loading conditions just prior to sinking, is a strong candidate for structural failure based both on the material strength and fracture mechanics investigations presented herein. Stress levels in way of the aft expansion joint are significantly higher than the yield strength of *Titanic* steel, and are very nearly equal to the ultimate tensile stress, when the ship is subjected to the loading scenario described in Reference (2).

It should be noted that the Reference (2) analysis predicts a stable ship for Condition B6 flooding (the ship sinks intact at Condition B8 flooding). Therefore, this loading condition, which corresponds to a stable ship, results in a large region of material overstress in way of the aft expansion joint. It can therefore be concluded that there would have existed a stage of flooding prior to Condition B6 where the ship would have been stable, and the stresses in way of a crack introduced in the aft expansion joint would just have exceeded the yield strength of the material. Under such conditions, the technical requirements for the fracture mechanics analysis would be satisfied, and accurate stress intensity factors could be determined in way of an assumed crack tip. Under such a scenario, it is expected that the calculated stress intensity factor would still exceed the critical stress intensity factor for *Titanic* steel, thereby predicting rapid crack propagation. It is recommended that this analysis be performed upon receipt of the toughness data from testing of *Titanic* steel.

Given the nature of the two *Titanic* steel specimens examined to date and the fact that notch sensitivity was not as well-understood as it is today, it is conceivable that there were pre-existing cracks in way of the expansion joint, and especially in way of cold-punched rivet holes. The combinations of high stress, stress concentrations, and low fraction toughness make the region in way of the termination point for the aft expansion joint highly susceptible to material yield and rapid crack propagation, and may have contributed significantly to catastrophic structural hull failure.

The forward expansion joint is intact on the bow wreck of the *Titanic*. It was observed during the sinking scenario described via Reference (2) that this joint was subjected to high stresses as well. Although, these stresses never reached a magnitude such that catastrophic failure was a serious concern. Further investigation of the forward expansion joint at the wreck site, however, may provide additional insight into these structural investigations. Of particular interest is the buckled region of shell plating directly below the expansion joint. Further work is in progress to determine whether it was probable for this damage to occur at the surface, or later as the bow wreck struck the sea floor.

REFERENCES:

(1)   Private correspondence between John Bedford of the Ulster Titanic Society and David K. Brown, William H. Garzke, Jr. and Arthur Sandiford [Mr. Bedford is a retired Chief Naval Architect of Harland and Wolff Shipyard].

(2)   C. Hackett and J.G. Bedford, "The Sinking of the Titanic, Investigated by Modern Techniques," The Northern Ireland Branch of the Institute of Marine Engineers and the Royal Institution of Naval Architects, 26 March 1996 and the Joint Meeting of the Royal Institution of Naval Architects and the Institution of Engineers and Shipbuilders in Scotland, 10 December 1996.

(3)   Garzke, William; Harris, Stewart; Yoerger, Dana; Dulin, Robert; and Brown, David; "Deep Ocean Exploration Vehicles, Their Past, Present, and Future," Centennial Transactions, The Society of Naval Architects and Marine Engineers, 1993, pp 485-536.

(4)   "The Titanic and Lusitania, A Final Forensic Analysis," Garzke, William; Brown, David; Sandiford, Arthur; Hsu, Peter; and Woodward, John; Marine Technology, The Society of Naval Architects and Marine Engineers, October 1996.

(5)   "RMS Titanic, Estimated Distribution of Weights and Loads at Moment of Rutpure," Sandiford, Arthur, 9 March 1997.

(6)   "Metallurgy of the RMS Titanic," Tim Foecke, National Institute of Standards and Technology, Gaithersburg, Maryland.

Recent Trends in FEM

J. Marczyk*

Abstract

The paper provides an overview and examples of management of uncertainty in numerical simulation. It is argued that the classical deterministic approaches in the Finite Element science have exhausted their potential and that it is necessary to resort to stochastic methodologies in order to boost a new more physics-based way of doing engineering design and analysis, namely Computer-Aided Simulation, or CAS. Stochastic techniques, apart from constituting a formidable platform of innovation in FEM and other related areas, allow engineering to migrate from a broad-scale analysis-based approach, to simulation-based broad-scope paradigms. The paper also illustrates that neglecting uncertainty leads to artificially smooth and simplistic problems that may often produce misleading results and induce exaggerated and unknown levels of optimism in engineering design.

1 Introduction

Most mechanical and structural systems operate in random environments. The uncertainties inherent in the loading and the properties of mechanical systems necessitate a probabilistic approach as a relistic and rational platform for both design and analysis. The deterministic and probabilistic approaches of Computational Mechanics differ profoundly in principles and philosophy. The deterministic approach completely discounts the possibility of failure. The system and its environment interact in a manner that is, supposedly, fully determined. The designer is trained to believe that via a proper choice of design variables and operating envelopes, the limits will never be exceeded. The system is therefore designed as immune to failure and with a capacity to survive indefinitely. Arbitrarily selected safety factors, an artifice that does, to a certain extent, recognize the existence of uncertainty, build overload capability into the design. However, since elements of uncertainty are inherent in almost all engineering problems, no matter how much is known about a phenomenon, the behaviour of a system will never be predictable with arbitrary precision. Therefore, absolutely safe systems do not exist. It is remarkable, however, that in modern computational mechanics, this fact is inexplicably ignored. It is a pity since the recognition of uncertainty not only leads to better design but, as a by-product, enables to

---

*CASA Space Division, Madrid, Spain, Senior Member AIAA

impulse innovation in Computer Aided Engineering on a very broad scale.

Computational mechanics, whether structural or fluid, has been sustained by multi-million dollar R&D initiatives and yet it has not reached the levels of maturity and advancement that one may have envisaged or expected a few decades ago. Computers and computer simulation play an increasing role in computational mechanics. Computers have developed at fantastic rates in the last few years and yet, due to an alarming stagnation in terms of innovative solvers, methodologies and paradigms, nothing spectacular has appeared on the scene that would prompt promising and attractive engineering solutions. It appears, paradoxically, that the progress in information technology has induced a devastating intellectual decadence in simulation science. Engineering simulation is currently going through an identity crisis, the reason being that it hinges more on the computer and numerical aspects rather than on the physics. In fact, as far as physical content is concerned, structural finite element models, for example, are subject to a constant erosion and devaluation given that most of the available computing horespower is being invested in simply increasing model size. Some think this race for huge models is synonymous to progress, instead, it is nothing but an intellectual *techo-pause*.

Whether computational science can stand beside theory and experiment as a third methodology for scientific progress is an arguable philosophical debate. Maybe yes, but in the current intellectually crippled scenario, the proposition is open to serious question. In fact, what new knowledge in structural mechanics has been generated since the dawn of the computer age? Have new discoveries come about because we can build huge numerical models or because parallel computing exists? Clearly the embarassing answer is no. Today, the simulation market has trapped itself in a mono-cultural deterministic dimension from which the only perspective of escape is via new methodologies and fresh ideas.

2   Uncertainty and Complexity Management

Complexity is a dominating factor of nearly every aspect and facet of modern life. From macro-economics to the delicate and astounding dynamics of the environmental equilibria, from politics and society to engineering. However, Nature, unlike humans, knows very well how to treat and resolve its own complexities. Man-made complexity is different. Today, things are getting complicated and complex and often run out of hand. Engineering in particular is suffering the effects of increasing complexity. The conception, design, analysis and manufacturing of new products are facing not only stringent performance and quality requirements, but also tremendous cost constraints. Products are becoming more complex, sophisticated and more alike, and at the very highest levels of performance it is practically impossible to distinguish competing products if not for minor often subjective differences. When complexity comes into play, new phenomena arise. A complex system requires special treatment and this is due to a newcomer to the field of engineering, namely *uncertainty*. Uncertainty, innocuous and inoffensive in simple systems, becomes a fundamental component of large and complex systems. The natural manifestation of uncertainty in engineering systems is parameter scatter. In practice this means that a system cannot be described "exactly" because the values of its parameters are only known to certain levels of precision (tolerance). In small doses this form of scatter may be neglected altogether and presents no particular problems. However, when the system becomes large and complex, even small quantities of scatter may create problems. The reason for this is quite simple. A system is said to be complex if it depends on a large number of variables (or design parameters). Uncertainty in these parameters means that the system can find itself functioning in a very large number of situations or modes (i.e. combinations of these parameters). In systems depending only on a few parameters, the number of these modes is of course reduced and therefore more controllable. It is also evident, at this stage, that if the system parameters have high scatter, then the number of these modes increases and their relative "distance" increases. The situation therefore becomes more difficult to handle in that the system may end up in an unlikely and critical mode without any early warning.

There are four main levels at which physical uncertainty, or scatter, becomes visible, namely

1. Loads (earthquakes, wind gusts, sea waves, blasts, shocks, impacts, etc.)

2. Boundary and initial conditions (stiffness of supports, impact velocities, etc.)

3. Material properties (yield stress, strain-rate parameters, density, etc.)

4. Geometry (shape, assembly tolerances, etc.)

On a higher level, there exist essentially two categories of scatter:

1. Scatter that can be reduced. Very often, high scatter may be attributed to a small number of statistical samples. Imagine we want to estimate how the Young's modulus of some material scatters around its mean value. Clearly, five, or even ten experiments, can not be expected to yield the correct value. Obviously, a larger amount of experiments is necessary in order to stablize the statistics.

2. Scatter that can not be reduced, i.e. the natural and intrinsic scatter that is due to physics.

Another equally important form of uncertainty is *numerical simulation uncertainty* which exists regardless of the physics involved. Five types of this kind of uncertainty exist and propagate thoughout a numerical simulation:

1. Conceptual modeling uncertainty (lack of data on the physical process involved, lack of system knowledge).

2. Mathematical modeling uncertainty (accuracy of the mathematical model).

3. Discretization error uncertainties (discretization of PDE's, BC's and IC's).

4. Programming errors in the code.

5. Numerical solution uncertainty (round-off, finite spatial and temporal convergence).

At present, there exists no known methodology or procedure for combining and integrating these individuals sources into a *global uncertainty* estimate.

The level of scatter, or nonrepeatability, is expressed via the coefficient of variation $\nu = \sigma/\mu$, where σ is the standard deviation and μ is the average. Some typical values of the coefficient of variation for aerospace-type materials and loads are shown in the table below (see [2] for details).

property	$\nu = \sigma/\mu$ (%)
Metallic materials; yield	15
Carbon fiber composites rupture	17
Metallic shells; buckling strength	14
Junction by screw, rivet welding	8
Bond insert; axial load	12
Honeycomb; tension	16
Honeycomb; shear, compression	10
Honeycomb; face wrinkling	8
Launch vehicle thrust	5
Transient loads	50
Thermal loads	7.5
Deployment shock	10
Acoustic loads	40
Vibration loads	20

However, no matter under which circumstances scatter appears, it normally assumes one of the following forms:

1. Random variables, eg. ultimate stress of a material sample.

2. Random processes, eg. acceleration at a given point during an earthquake.

3. Random fields, eg. the height of sea waves as function of position and time.

A fundamental reason why the inclusion of scatter should become a routine exercise in any engineering process is not just because it is an integrating part of physics (although this is already a good enough reason!). The unpleasant thing about scatter is that the most likely response of a system affected by uncertainty practically never coincides with the response one would obtain if the system were manufactured with only nominal values of all of its parameters. Examples in the following section shall clarify this and other concepts.

3 Numerical Examples

Example 1.

Consider a clamped-free rod of length L with a force F applied at the free end. Suppose also that the beam is made of a material with Young's modulus E and has a cross-section A. Imagine that the cross-section A is a random Gaussian variable with mean A_o and standard deviation σ_A. The problem is to determine the *most likely* displacement of the beam's free end. It is clear that the displacement corresponding to the nominal (mean) section is

$$x_o = \frac{FL}{EA_o}$$

However, the most likely displacement, $\hat{x}$, is given by

$$\hat{x} = \frac{FL}{EA_o}(1 + \frac{\sigma_A}{A_o^2})$$

which, of course, does not coincide with x_o. The reason is quite simple. The mechanical problem, that is the computation of the displacement of the beam's free extremity is, of course, a linear problem. In fact, the displacement x depends linearly on the force F, i.e. $x = F/k$ where k is the stiffness of the rod. However, the *statistical problem* is nonlinear in that the displacement depends on A as $1/A$. This nonlinear dependency is responsible for the fact that the probability distribution of x is not symmetrical, i.e. it is skewed, even though the distribution of A is not. In practice this means that the most likely displacement is not the average displacement, fact which would occur if the corresponding probability distribution were symmetrical. It is evident that the rod's cross-section which corresponds to the most probable displacement is $A' = A_o(1 + \frac{\sigma_A}{A_o^2})^{-1}$. Monte Carlo simulation can be used to obtain this value. The importance of this information is immense. In fact, since this cross-section corresponds to the most likely displacement, it may be used in a simulation model in an attempt to *estimate a-priori* this displacement. In effect, suppose we have to provide this estimate in prevision of an experiment with a rod. Then, the most reasonable value of cross-section to use in a numerical model is, in effect, A'. Of course, since the problem is stochastic (which means that the probability of manufacturing two *identical* rods is, for all practical purposes, null) even A' does not guarantee that we shall effectively predict the outcome of the experiment. However, this particular value *does* increase our chances, and certainly more than A_o does. The unexpected result of this trivial case rings an alarm bell. The fact that if even in a simple case one cannot trust intuition, what happens in the more realistic and complex structures (and structural models)? The sad truth is that in practically all cases, the relationships between the design parameters (such as thicknesses, moduli, densities, spring stiffnesses, etc.) and the response properties (such as frequencies, stresses, displacements, etc.) is nonlinear. This means that in reality, introducing into structural models the nominal values of design parameters, will almost certainly diminish our chances to estimate more closely the true (i.e. most likely) behaviour of the structure we're modelling.

What is happening to engineering today is truly surprising and uninspiring. It looks like we have lost the drammatic vigor and inventiveness of the previous generations of engineers. Simulation science is in a state of and excessive admiration of itself. We are consistently overlooking the existence of uncertainty while computers are reaching fantastic cost to performance ratios. Models are growing in size, but not in terms of physical content. The engineering jargon is intoxicated with cynical glossaries and semantical abuses. Simulation in engineering must not become a battlefield where the fight is between a big finite element mesh and a host of CPUs running smart numerical algorithms. Such simulations are risking to become expensive numerical games and the new generations of engineers just hords of mesh-men. Engineers are obsessed with reducing the discretization error, from maybe 5% to 3% by insistently and pedantically refining FE meshes, while overlooking uncertainty in, say, the yield stress which in many cases may scatter even more than 15% around some nominal value. Many people overlook the fact that models are already fruit of sometimes quite drastic simplifications of the actual physics. For example, the Euler-Bernoulli beam is result of, at least, the following assumptions: the material is continuum, the beam is slender, the constraints are perfect, the material is linear and elastic, the effects of shear are neglected, rotational inertia effects are neglected, the displacements are small. This list shows that models are, in effect, only models, and this fact should be kept in mind while preparing to simulate a structure, or any other system for that matter. The difficult thing is, as usual, to reach and maintain a compromise and a balance. If we don't have accurate data, then why build a mesh that is too detailed? Modern finite element science has reached levels of maturity in which the precision of numerical simulation codes has greatly surpassed that of the data we manage.

Mesh resolution, or bandwidth, is for many individuals a measure of technological progress. This is of course not the case. Clearly, mesh refinenemt, or the development of new and efficient numerical algorithms, are just two aspects of progress but things have certainly gone a bit too far. In order to quench the thirst of fast and powerful computers, models have been diluted in terms of physics. Physics is no longer a concern! Maybe the irremissible vogue to build huge models, and impress mute audiences at conferences, is just a wicked stratagem that helps to escape physics altogether! Who has the courage to controvert the results obtained with a huge and detailed model and presented by some prominent and distinguished member of todays Finite Element Establishment? It is very appropriate to cite here John Gustafson (see [3]) who made the following statement on the famous US ASCI Program: "The Accelerated Strategic Computing Initiative aims to replace physical nuclear-weapon testing with computer simulations. This focuses much-needed attention on an issue common to all Grand Challenge computation problems: How do you get confidence, as opposed to mere guidance or suggestion, out of a computer simulation? Making the simulation of some physical phenomena rigorous and immune to the usual discretization and rounding errors is computationally expensive and practised by a vanishing, small fraction of the HPCC community. Some of ASCI's proposed computing power should be placed not into brute-force increases in particle counts, mesh densities, or finer time steps, buto into methods that increase confidence in the answers it produces.

This will place the debate of ASCI's validity on scientific instead political grounds."

Every mechanical problem may be characterized by three fundamental dimensions, namely

1. Number of involved "layers" of physics (i.e. single or multi-physics).

2. Number of involved scales (i.e. single or multi-scale).

3. Level of uncertainty (i.e. deterministic or stochastic) .

It often happens that not all of these aspects of a problem come to play at the same time and with full intensity. For example, a physical problem might be characterized by various simultaneous interacting phenomena, such as aeroelasticity, where a fluid interacts with a solid, or may even involve different scales such as the meso and macro scales in a propagating crack. Finally, on top of all this, phenomena may be either deterministic or stochastic. However, it is true that if these dimensions are concurrent, then a numerical model that is supposed to mimic a certain phenomenon must reflect all of these components in a balanced manner. Deliberate elimination of one of these dimensions must be a very cautious exercise. [1] Otherwise, it will lead to a sophisticated and perverse numerical game. The most common such games are:

1. Parametric studies.

2. Optimization.

3. Sensitivity analysis.

In the presence of scatter, the above practices loose practical significance and can, under unfortunate circumstances, actually produce incorrect or misleading results. In any case, they most certainly lead to overdesign and frequently to unknown levels of conservatism and optimism. A few simple examples shall show how this happens.

Example 2.

---

[1]Eliminating one of these fundamental facets of a physical problem has far-reaching implications as far as its correct interpretation is concerned. It is in fact obvious that if, for example, we eliminate one dimension, say scatter, then we shall force ourselves to "squeeze" our understanding of the problem out of the remaining two dimensions. We can overload, for example, the multi-scale facet of a phenomenon, in order to compensate for the lack of uncertainty (since we have "cleaned-up" the problem and made it forcedly deterministic). By doing so, we are attributing the effects of missing dimensions to other dimensions. Luckily, in physics, an experiment performed correctly will eventually unmask this violation. In numerical analysis there in nothing to alert the analyst. If, for example, the natural frequency of a simulated plate does not match the value obtained experimentally, instead of including the effects of air, why not increase the thickness a bit? Or maybe the density? Why bother with the physics if we can quickly adjust some numbers in a computer file. These arbitrary numerical manipulations are of no engineering value at all. People call them *correlation* with experiments!

Consider a vector $x \in R^n$. Suppose that its components x_i are random Gaussian variables with mean 0 and standard deviation 1. Imagine also that for some reason, the objective is to reduce the norm of the vector. This is in practice a trivial n-parameter optimization problem. In fact, classical practice suggests to simply neglect the random nature of the problem and to establish the following *associated deterministic problem*:

$$\min_{x \in \mathcal{X}} |x| = \min_{x \in \mathcal{X}} (\sum_{i=1}^{n} x_i^2)^{1/2}$$

Clearly, the solution of the problem is 0 and, obviously, requires all the components of x to also be equal to 0.[2] From a classical standpoint the problem is readily solved. Now, admiting that each x_i is a random variable, changes the problem completely since the objective function is stochastic. This means that it is not differentiable and therefore traditional minimization methods can not be used. A completely different approach is mandatory. The problem requires a paradigm shift. In fact, the problem is no longer a minimization problem if it is viewed from its natural stochastic perspective. It becomes a simulation problem. Let us see why. Suppose that $n = 10$, i.e. x has 10 components. Let us generate randomly the ten components of the vector and compute its norm. Since each x_i is a random number, clearly we must repeat the process many times. The reason for this is that if we generate the x_i's only once, or even twice, we could be lucky, or unlucky, depending on what luck is in this particular case. Therefore, the process must be repeated many times until we have give each of the ten components a fair chance to express itself with respect to all the other components. Let us suppose that we generate a family of one thousand such vectors. This means that each of the x_i's shall be generated randomly from a Gaussian distribution one thousand times and one thousand values of the vector norm shall be obtained.[3] It is obvious that the norm shall also be a random variable. Let us examine its histogram, reported in the figure below. The conclusions are surprising.

First of all, one notices immediately that the minimum value the norm attains is around 1, while the deterministic approach yields 0. A striking disagreement! Secondly, examining the histogram reveals that the most frequent value of the norm is approximately 3. This second piece of information is impossible to obtain using the deterministic approach and is, incidentally, of paramount importance. In fact, in systems that are driven by uncertainty one should not speak of a minimum but rather concentrate on the most likely state. Logically, the most likely behaviour of a system is the one the system shall exhibit most frequently and therefore this particular situation *should* be the engineer's objective. In the light of this, one quickly realizes that the minimum norm of the vector in question is of little practical value. In fact, even though the vector *can* indeed reach the lucky value of 0, this circumstance is highly unlikely and therefore useless. The situation gets worse if the dimension of the problem increases. The table below portrays the situation.

---

[2]The objective function, i.e. $|x|$, is smooth and differentiable and any gradient-based minimization algorithm will deliver the solution in a very small number of steps.

[3]This process is known as Monte Carlo simulation and was discovered by Laplace while casting Buffon's needle problem in a new light.

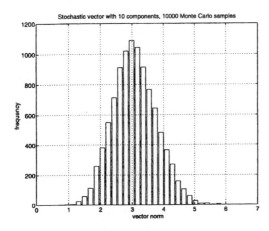

Figure 1: Plot of the stochastic vector norm of Example 3.

n	$|x|_{min}$	$\hat{x}$
10	1	3
50	5	7
100	7.5	10

where n is the dimension of the vector, $|x|_{min}$ the minimum norm and $\hat{x}$ the most likely norm. The results in the table have been obtained simulating the vectors 1000 times. The good news is that increasing the number of trials (i.e. Monte Carlo samples) does not alter the picture at all! The minima change slightly but the most likely norm does not. In fact, already with 100 samples, this quantity converges to the above values. It appears that the most likely norm is an intrinsic property of the vector. This is indeed the case.

The above example conveys an extraordinary message, namely that transforming a stochastic problem to assume a forcedly deterministic (and therefore easy and smooth) connotation prevents it from exhibiting its true nature. Consequently, basing any engineering decision on the stripped (deterministic) version of a problem is, evidently, a dangerous game to play. Moreover, the example shows that the minimum norm obtained by depriving the system of its intrinsic uncertainty, yields a value of 0, clearly an overly optimistic result. Handle with care!

A final conclusion stemming from the vector example hides, most probably, the essence and promise of what future simulation practice in mechanical engineering will look like. First of all, the example shows that in the case under examination, optimization, intended in its most classical form, does not make much sense. In fact, what should be minimized ? We have just seen that the vector in question has a most likely norm that is its invariant and intrinsic property. The problem is actually different. Clearly, the only way to change this property is to either changes the vector's lenght, or, alternatively, modify the probability density function of each component x_i. Assuming that the vector's length is a design constraint, all that can be done is to work at PDF level. Therefore, unless one is willing to make major design changes, such as topology for example, the only way to change the intrinsic properties of a mechanical system is to *shape* the PDF of its dominant parameters. Evidently, the most likely norm is a characteristic of the PDF of the vector problem, but it is not the only one. A PDF may in fact be described by other properties such as skewness, kurtosis and even higher order statistical moments. In general, therefore, the shaping of the PDF of the system's parameters shall achieve the shaping of the overall system PDF.

A final consideration, very often overlooked in engineering practice, is that regarding the *manufacturability of the solution*. The deterministic version of the vector problem states that the optimum is attained if all of the ten components of the vector are equal to zero. The problem is readily solved. All that remains to do is to manufacture ten perfectly null components and enjoy the best possible minimum! Reality, however, is less generous and forgiving. In fact, we have supposed that each component was a random variable and therefore we must acknowledge the existence of an unpleasant entity called *tolerance*. The practical implication of tolerances is that they rule out the existence of perfection. What this means in the case of our vector is that the probability of manufacturing a null component of our vector is 0! Clearly, to have all ten of them 0 at the same time is even more difficult. The seducing deterministic solution obviously gives not a single clue in this sense. What a pity! However, there is a solution if we are willing to sacrifice some perfection for a little bit of common sense. Let's in fact assume that we can manufacture components of the vector in the range -0.01 to 0.01 with a probability of, say, 90%. Then, according to basic probability, *all* the components will lie in this interval with a likelyhood of $0.9^{10} = .35$. In other word, 65 vectors out of every 100 we manufacture will have some components outside of the range (-0.01;0.01). If, for some reason, this results unacceptable, we will probably have to reject 65% of our production. The practical interpretation of this result is that the simultaneous manufacturing of ten components within the range (-0.01,0.01) is not that easy, only 35 out of a 100 will fulfill our requirement. The solution is, of course, to change the tolerances but this may require a major change in the way the components are manufactured. The situation gets worse in more complex engineering systems that depend on hundreds of design parameters. The table below illustrates the probabilities of manufacturing a certain number of components that fall simultaneously within a prescribed range of tolerance. Two cases have been chosen, i.e. where the probability of manufacturing a component with a compliant value is 80% and 90% respectively.

n	$p = 80\%$	$p = 90\%$
10	.11	.35
25	.004	.07
50	1.43×10^{-4}	.005
100	2.04×10^{-10}	2.66×10^{-5}

These values of course by no means represent true industrial standards.[4] However, the message conveyed by this simple example is that to obtain an optimal design is one thing, to manufacture it is another. Clearly, deterministic methods, which imprison and enclose a physical problem in an ideal and artifical numerical domain, are unable to furnish a single piece of evidence on whether that particular design is feasible or not. What we get with deterministic design is a result without pedigree.

Neglecting scatter where it really exists is a fundamental violation of physics because it leads to problems that are artificial, that do not exist in nature. Parametric studies, for instance, are another example of how an innocent, apparently sound practice, can actually transgress physics and almost surely produce overdesign (overkill). In parametric studies what one really does is to freeze all but one parameter (i.e. design variable) and to evaluate the response of a model (not the physical system!) while that parameter is changed in a specified range. The main flaw underlying this practice is that in reality, *all* the parameters change and *at the same time*. Let us see an enlightening example.

Example 3.

Consider a 5th order Wilkinson matrix.[5] Suppose that the diagonal terms $w_{1,1}, w_{2,2}, w_{4,4}$ and $w_{5,5}$ have an additive Gaussian term with 0 mean and standard deviation of 0.1. Consider these additive terms as design variables and imagine we want to examine the sensitivity of the third eigenvalue of the Wilkinson matrix with respect to the first parameter. The classical approach, adopted in parametric studies, is to freeze all of the parameters and to vary only the one under consideration. The correct way to approach the problem is to let *all* the parameters vary naturally, and at the same time. This approach respects of course the physics of the problem. In order to compare the physical and "artificial" parametric study-type approaches, the results of both approaches are superimposed on the plot below. The striking truth is that the parametric approach prevents the system from developing two bifurcations which lead to three eigenvalue clusters. In fact, in the parametric approach, only one such cluster exists at approximately 1.2. In the "natural" approach (which in practise is simply *simulation*) two additional clusters appear, namely at approximately 0.25 and 3.1. The impact of this result is obvious. Imagine that whatever this problem corresponds to from an engineering point of view, the system fails if the third eigenvalue falls above 2.0. Clearly, the deterministic approach will rule this case out and one would conclude that the system is safe! The stochastic approach, on the other hand, not only reveals that the probability of failure is far from negligible but also it exposes the true bifurcation-based nature of the problem.

A fundamental difference between the deterministic and stochastic approaches to struc-

---

[4]If, for example, we wish to manufacture a system that results compliant in 99 cases out of a 100 and its performace depends, simultaneously, on 100 design parameters, then each parameter must comply with its manufacturing tolerance in 9999 cases out of 10000.

[5]The Wilkinson matrix is J.H. Wilkinson's eigenvalue test matrix. It is symmetric, triadiagonal and has pairs of nearly equal eigenvalues.

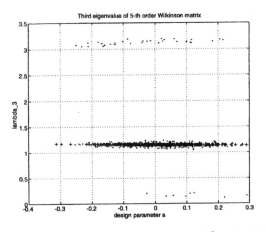

Figure 2: Plot of the third eigenvalue of the 5th order Wilkinson matrix.

tural design lies in the fact that deterministic design in based on the concept of *margin of safety*, while the more modern stochastic methods rely on the *probability of failure* or on *robustness*. The difference between the two approaches is drammatic and very profound. The margin of safety stands to the probability of failure as the determinant of a matrix stands to its condition number. Let us see why. The margin of safety of a structural system says nothing on how imminent failure is. It merely quantifies the *desired* distance that the structure's operational conditions have with respect to some failure state. There is no information in the margin of safety on how rapidly can this distance be covered. The analogy with a matrix determinant is obvious. The determinant expresses the distance of a matrix with respect to a singular one, however, it gives no clue on how easily can the matrix loose rank. The probability of failure of a structure is, clearly, something more profound than the margin of safety. It is also more expensive to compute but the investment is certainly a good one to make. The reasons are multiple. Knowing the probability of failure implies that one also knows the Probability Density Function (PDF) and, via integration, the Cumulative Density Function (CDF). Both provide an enormous quantity of information. The shape of either the PDF or CDF reflects, amongst others, the robustness of the design, something a margin of safety will never give. The analogy with a matrix condition number is now clear. Computing the condition number of a matrix, for example via the Singular Value Decomposition (SVD), is relatively expensive (twice as expensive as the QR decomposition) however it yields an enormous amount of information on the matrix. The condition number itself tells us how quickly can the matrix loose its rank, not merely its distance to a singular matrix. Knowing the rank structure of a matrix is fundamental towards understanding how the solution will behave, how robust it is, etcetera, etcetera.

4 What About Model Validation?

An embarassing issue in CAE is that of *model validation*. There are essentially four reasons for not dedicating time to the validation of models, namely

- The disjunctive thrust between experimentation and simulation.

- The arrogant belief that a fine mesh delivers perfect results.

- The false assumption that a test always delivers the "Gospel Truth".

- Lack of established model validation methodologies.

In reality, certain individuals do actually dedicate themselves to "validating" their models. The customary approach in the majority of the cases comes down to a one-to-one comparison of a simulation with the results of an experiment, and to the computation of the differences between the two. If the difference, usually expressed in terms of a percentage, is "small" enough then the numerical model is regarded as valid. This simplistic one-to-one comparison, often referred to as correlation, is yet another reflection of the deterioration and decline of CAE.[6] It is clear, however, that a numerical model is worth only as much as the *level of confidence* that the analyst is able to attribute to the results it produces. Meaningful validation of numerical models is expensive business (one good reason to say you don't need it!). In fact there exists an empirical relationship between the complexity of a numerical simulation and the complexity and cost of the associated validation process. Evidently, the possibility of applying increasingly complex numerical simulations to large industrial problems is related strongly to the development of new validation methodologies.

Validation of numerical models can only be performed, with our existing technology, via correlation (yes, correlation, not comparison!) with experiments. The experienced engineer, unlike the young mesh-man, is perfectly aware of the fact that both experiments and simulations produce uncertain results. Because of the numerical sources of scatter, simulations will yield non-repeatable results. Changing solver, computing platform, algorithm, even the engineer, will lead to a different answer. Similarly, due to measurement and filtering errors, sensor placement, analog to digital conversion and other data handling procedures, tests will also tend to furnish different results at each attempt. Of course, on top of both simulation and experiment we must not forget the existence of the natural (physical) uncertainty that is beyond human control and intervention. Therefore, it is evident that a sound and rigorous model validation procedure must take all these forms of uncertainty into account. Statistical methods, and Monte Carlo Simulation in particular, provide an ideal meeting point of simulation and experimentation and help reduce the danger of perfect, or lucky "correlation", that is hidden in classical one-to-one deterministic

---

[6]The computation of correlation between two variables requires multiple samples of each variable. It is therefore clear that it is incorrect to speak of correlation between two events if only one sample is available for each event.

approaches. A fundamental advantage of these methods lies in the fact that they overcome the major shortcoming of the conventional deterministic techniques, namely their inability to provide confidence measures on the results they produce. The soporific predilection of modern mechanical engineering for deterministic methods has eliminated from current practice such fundamental concepts as *confidence*, *reliability* and *robustness*. In fact, these concepts can not coexist with something deterministic and supposedly perfect. This quest for perfection, lost in the very beginning, is illustrated in figure 3 which portrays clearly the debility of the deterministic vision of life.

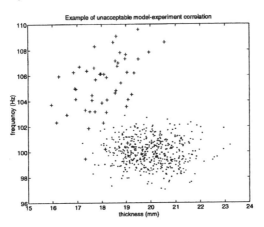

Figure 3: Multiple experiments versus Monte Carlo Simulation; a typical scenario. Only in a similar perspective is it possible to assess how the experimental results compare with those of a numerical simulation. The '+' corresponds to experiment and '.' to simulation.

One may observe in fact two distinct clouds of points, one resulting from a Monte Carlo Simulation and the other originating from a series of tests. The horizontal axis corresponds to a design parameter, say thickness, while the vertical axis represents some engineering quatity of interest, for example frequency of vibration. Clearly, the two clouds, do not stem from the same phenomenon although they overlap, at least in part. Without getting into details, intuition suggests that the *sine qua non* condition for two phenomena to be judged as similar, or stistically equivalent, is that their "clouds" have similar shapes and orientations. Clearly, if two such clouds come in contact only at their borders, or even if they overlap but have distinct shapes (topologies) there will be serious evidence of some major physical discrepancy or inconsistency between the phenomena they portray. From figure 3 another striking fact emerges, namely that there may exist situations in which a '+' (test) lies very close to a '.' (simulation) and both lie on the edges of two nearly tangent clouds. These cases, catalogued as *accidental or fortuitous correlation*, are extremely dangerous and misleading since a lucky combination of test circumstances and simulation parameters may prompt an impetuous and unexperienced mesh-man to conclude that his simultion has hit the nail on the head.[7] Obviously, a single test and a single simulation can not go further. It is impossible to squeeze out more information from where it does not exist. It is impossible to say anything on the validity of the numerical model in a single test-single simulation scenario. Not a single clue on robustness. Not one hint on confidence or reliability. All we can speak of in similar circumstances is simply the Euclidean distance between the model and the experiment. What complicates the situation is the fact that experiments are expensive because prototypes are expensive. A definitely better situation is portrayed in figure 4. Here the clouds overlap, have similar size and orientation. This is, of course, still no guarantee, but we are surely on the right track.

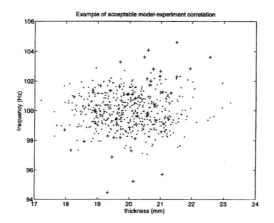

Figure 4: Example of statistically equivalent populations of tests and analyses. The '+' corresponds to experiment and '.' to simulation.

5 Conclusions

Statistics is, at least for engineers, part of the "lost mathematics", the mathematics now considered maybe too advanced for high school and too elementary, or useless, for college. Statistics has been kicked out of many university courses and school curricula. There is in fact a puzzling and obsessive adversion to statistics in engineering. One of the reasons is that statistics has been made repugnant and revolting to the student and engineer thanks to too much epsilonics and not enough examples of practical application, its usefullness and its tremendous power. Probability, on the other hand, is the mathematics of the 20th century. Its history goes back to the 16th century, but not until the present century did people fully realize that nature and the real world can be described exhaustively only

---

[7]In such cases we prefer to talk of unlucky, rather than lucky, circumstances.

by laws governing their randomness. Today, High Performance Computing technology, together with Monte Carlo Simulation techniques, offers a unique opportunity to push the FEM science, and CAE in general, into more physics-based domains and to abandon the idealistic and artificial vision of life upon which deterministic numerical analysis thrives. Monte Carlo Simulation is a monument to simplicity and constitutes a phenomenal vehicle for the incorporation of uncertainty and complexity management in engineering.

References

[1] Marczyk, J., editor, *Computational Stochastic Mechanics in a Meta-Computing Perspective*, International for Numerical Methods in Engineering (CIMNE), Barcelona, December, 1997.

[2] Klein, M., Schueller, G.I., et. al.,*Probabilistic Approach to Structural Factors of Safety in Aerospace*, Proceedings of the CNES Spacecraft Structures and Mechanical Testing Conference, Paris, June 1994, Cepadues Edition, Toulouse, 1994.

[3] Gustafson, J., *Computational Verfiability and Feasibility of the ASCI Program*, IEEE Computational Science & Engineering, Vol. 5, No. 1, January/March, 1998.